CONTINUED ON INSIDE BACK COVER

FORCE	LENGTH	MASS

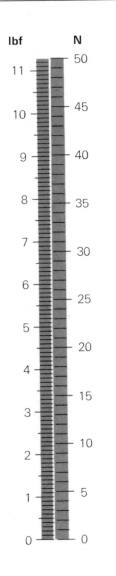

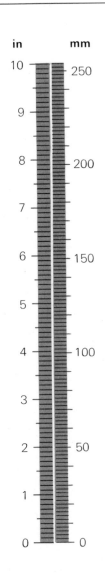

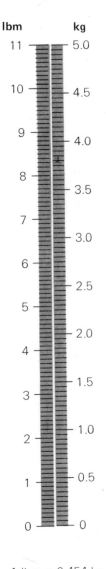

1 lbf = 4.45 N

1 in = 25.4 mm

1 lbm = 0.454 kg

0.225 lbf = 1 N

0.039 in = 1 mm

2.20 lbm = 1 kg

THE TESTING OF ENGINEERING MATERIALS

THE TESTING OF ENGINEERING MATERIALS

Fourth Edition

Harmer E. Davis

Professor of Civil Engineering, Emeritus
University of California, Berkeley

George Earl Troxell

Professor of Civil Engineering, Emeritus
University of California, Berkeley

George F. W. Hauck

Associate Professor of Civil Engineering
University of Missouri—Columbia

McGraw-Hill, Inc.
New York St. Louis San Francisco Auckland Bogotá
Caracas Lisbon London Madrid Mexico City Milan
Montreal New Delhi San Juan Singapore
Sydney Tokyo Toronto

This book was set in Times Roman by The Total Book (GTI).
The editor was Julienne V. Brown;
the production supervisor was Leroy A. Young.
The cover was designed by Hermann Strohbach.

THE TESTING OF ENGINEERING MATERIALS

Printed and bound by Book-mart Press, Inc.

13 14 15 BKM BKM 9 9 8 7 6 5

ISBN 0-07-015656-5

See Acknowledgments on pages xv-xvi.
Copyrights included on this page by reference.

Library of Congress Cataloging in Publication Data

Davis, Harmer Elmer, date
     The testing of engineering materials.

     Third ed. published under title: The testing and
inspection of engineering materials: New York :
McGraw-Hill, 1964.
     1. Materials—Testing.   I. Troxell, George Earl,
date   . II. Hauck, George F. W.   III. Title.
TA410.D3   1982   620.11'0287   81-8399
ISBN 0-07-015656-5   AACR2

CONTENTS

Preface   xiii

Acknowledgments   xv

Part 1   Principles of Testing

1   Control of Material Properties   3

1.1   The Role of Testing   3

1.2   Engineering Materials   4
     Properties • Selection

1.3   Testing of Materials   6

1.4   Inspection of Materials   8

1.5   Significance of Tests   9

1.6   Specification of Materials   10
     Specification Types • Standard Specifications

1.7   Standardizing Agencies   12
     Problem 1 Inspection of the Testing Laboratory   15

2   Mechanical Behavior   17

2.1   Fundamental Definitions   17
     Stress and Strain • Stiffness

2.2   Stress-Strain Diagrams   20
     Test Control • True Stress—Natural Strain Relations •
     Nondimentionalization

2.3   Elasticity   24
     Elastic Action • Yielding

2.4   Plasticity   27
     Plastic Action • Maximum Strength

2.5   Failure   30

2.6   Energy Capacity   32
     Problem 2 Compression Tests of Small Wooden Columns   34

v

3 Analysis of Data 37

3.1 The Management of Data 37
3.2 Variations in Data 38
3.3 Grouped Frequency 39
  Grouping of Data • Frequency • Diagrams
3.4 Distribution Characteristics 44
  Central Tendency • Dispersion • Skewness
3.5 Sampling and Errors 48
  Samples • Errors
3.6 Correlation 50
3.7 Control Charts 51
3.8 Statistical Summaries 53
  Problem 3 Sieve Analyses of Aggregates 54

4 Presentation of Results 56

4.1 Arrangement of Reports 56
4.2 Narration 57
4.3 Mathematics 58
4.4 Tables 59
4.5 Figures 60
4.6 Mechanics of Report Preparation 63
  Odds and Ends • Assembly and Duplication •
  Covering Correspondence
4.7 Oral Presentations 66
  Problem 4 Report Preparation 66

5 Testing Procedures 68

5.1 Purposes of Mechanical Tests 68
5.2 Types of Mechanical Tests 69
  Loading Methods • Test Conditions
5.3 Design of Tests 71
5.4 Test Specimens 72
  Selection • Preparation
5.5 Testing Apparatus 74
5.6 Measurements 74
5.7 Conducting Tests 76
  Testing and Common Sense
  Problem 5 Withdrawal Test of Nails 78

6 Testing Machines 80

6.1 Types of Testing Machines 80
6.2 Determination of Load 82
  Weights • Hydraulic Devices • Mechanical Devices •
  Electrical Devices
6.3 Static Testing Machines 85
  Screw-Gear Machines • Hydraulic Machines •
  Readout Devices • Speed Adjustment

6.4 Cyclic Loading Machines 90
Hydraulic Machines • Control Systems
6.5 Accessories 91
6.6 Calibration 93
Problem 6 Study and Calibration of a Testing Machine 97

7 Instruments 99
7.1 Linear Measurements 99
Micrometers • Interferometer
7.2 Inspection Gages 102
7.3 Determination of Displacement 103
Mechanical Devices • Optical Devices • Electrical Transducers
7.4 Electric Strain Gages 107
Strain and Resistance • Bonded Resistance Strain Gages •
Attachment of Gages • Stress Determination •
Special Arrangements
7.5 Strain Gage Instrumentation 111
Wheatstone-Bridge Hookup • The Strain Indicator •
Dynamic Testing • The Oscilloscope
7.6 Brittle Coatings 116
Factors Affecting Cracking • Test Method
Problem 7 Beam Deflection and Strain Measurements 120

Part 2 Characteristics of Tests

8 Static Tension 125
8.1 Tension vs. Compression 125
8.2 Specimens 127
8.3 Apparatus 131
8.4 Procedure 135
8.5 Observations 137
8.6 Effects of Specimen Variables 140
8.7 Effects of Test Variables 143
Eccentricity • Test Speed • Fracture
8.8 Low- and High-Temperature Tests 147
Tests at Low Temperatures • Tests at High Temperatures
Problem 8 Tension Test of Steel 149

9 Static Compression 151
9.1 Compression 151
9.2 Specimens 152
9.3 Apparatus 154
9.4 Procedure 156
9.5 Observations 157
9.6 Effects of Variables 159
Size and Shape • End Conditions • Test Speed
Problem 9 Compression Test of Concrete 162

10 Static Shear   164

10.1 Shear   164
    Direct Shear • Torsion • Pure Shear Stress • Shearing Strain
10.2 Scope and Applicability   166
10.3 The Direct Shear Test   168
10.4 Torsion Specimens   170
10.5 Torsion Apparatus   171
10.6 Torsion Procedure   172
10.7 Observations   174
    Failure Mode • Modulus of Rupture • Ductility
10.8 Effects of Variables   175
    Problem 10 Torsion Test of Steel   176

11 Static Bending   179

11.1 Bending   179
11.2 Scope and Applicability   182
11.3 Specimens   183
11.4 Apparatus   184
11.5 Procedure   186
11.6 Observations   187
    Rupture • Yielding
11.7 Effects of Variables   189
    Loading Type • Specimen Dimensions • Stability • Test Speed
11.8 "Bend" Tests   191
    Problem 11 Flexure Test of Wood   193

12 Hardness   195

12.1 Hardness   195
12.2 Scope and Applicability   196
12.3 Standard Tests   199
    Static Indentation Hardness Tests • Dynamic Hardness Tests • Scratch- and Wear-Hardness Tests
12.4 Brinell Tests   204
    Specimens • Apparatus • Procedure • Observations • Effects of Variables
12.5 Rockwell Tests   210
    Specimens • Apparatus • Procedure • Observations • Effects of Variables
12.6 Scleroscope Tests   215
    Specimens • Apparatus • Procedure • Observations
12.7 Correlations   218
    Problem 12 Hardness Tests of Steel   218

13 Impact   221

13.1 Impact   221
13.2 Scope and Applicability   223
13.3 Test Principles   226

13.4	Apparatus	227
13.5	Specimens	231
13.6	Procedure	233
13.7	Observations	235
13.8	Effects of Variables	237
	Velocity • Specimens • Temperature • Failure	
	Problem 13 Impact Flexure (Charpy) Test of Steel	242

14 Fatigue   244

14.1	Fatigue	244
14.2	Scope and Applicability	247
14.3	Specimens	249
14.4	Apparatus	251
14.5	Procedure and Observations	253
14.6	Effects of Specimen Variables	255
14.7	Effects of Test Variables	257
14.8	Correlation	259
	Problem 14 Axial-Fatigue Test of Steel	259

15 Creep   261

15.1	Creep	261
15.2	Scope and Applicability	264
15.3	Apparatus	266
15.4	Procedure and Observations	268
15.5	Effects of Variables	271
	Problem 15 Creep Test of Lead	272

16 Nondestructive Tests   274

16.1	Nondestructive Tests	274
16.2	Radiographic Methods	275
	Equipment • The Radiograph	
16.3	Electromagnetic Methods	279
16.4	The Magnetic-Particle Method	282
16.5	Electrical Methods	285
	Flaws in Rails • Flaws in Tubes • Surface Roughness • Moisture in Wood	
16.6	Sonic Methods	288
	Defects • Thickness • Stiffness	
16.7	Mechanical Methods	291
	Concrete Test Hammer • Hardness Tests • Proof Tests	
16.8	Visual Methods	293
	Surface Roughness • Penetrant Tests	
16.9	Photoelasticity	295
	Plane Polarization • Circular Polarization • Models • PhotoStress	
	Problem 16 Photoelastic Analysis	300

Part 3 Properties of Materials

17 Metal   305

17.1   Composition and Characteristics   305
Ferrous Metals • Nonferrous Metals

17.2   Production Methods   307
Iron and Steel Production • Aluminum Production

17.3   Properties of Cast Iron   311

17.4   Properties of Steel   313
Carbon Content • Alloying • Processing • Standards

17.5   Properties of Nonferrous Alloys   318
Copper • Aluminum • Other Alloys

17.6   Corrosion   319
*Direct Chemical Reaction • Electrolytic Corrosion •
Corrosion Protection*

17.7   Wire Rope   323
*Rope Construction • Manufacture • Mechanical Properties •
Durability • Specifications*

Problem 17   Heat-Treatment and Tests of Steel   328

18 Concrete   330

18.1   Composition   330

18.2   Materials for Concrete Making   331
Cement • Water • Admixtures • Aggregates

18.3   Manufacture of Concrete   336
Design of Concrete Mix • Mixing, Placing, and Curing

18.4   Properties of Fresh Concrete   340
Consistency • Bleeding • Air Entertainment

18.5   Strength Characteristics   342
Strength • Stiffness • Creep

18.6   Other Properties   346
Density • Durability and Impermeability • Volume Changes

Problem 18   Concrete Mix Preparation and Tests   349

19 Asphalt   351

19.1   Nature and Uses of Bitumens   351

19.2   Manufacture and Types of Asphalts   353

19.3   Asphalt Properties and Tests   356
*Physical Characteristics • Composition • Consistency •
Stability • Ductility*

19.4   Characteristics of Commonly Used Paving Asphalts   363

19.5   Mineral Aggregates   370
*Density, Voids, Porosity, Absorption • Size and Gradation of
Particles • Particle Shape, Surface Texture, and Surface Area •
Toughness and Abrasion Resistance • Soundness • Cleanness •
Mineral Composition and Surface Chemistry*

19.6	Asphalt Mixtures	378
	Mix Design • Selection of Type of Mixture • Selection of Aggregate • Selection of Asphalt • Selection of Trial Mix Proportions • Manufacture and Testing of Trial Mixes • Selection of Optional Mix	
	Problem 19 Marshall Test of an Asphaltic Mixture	389

20 Wood — 391

20.1	Types and Production	391
20.2	Structure	392
20.3	Physical Properties	395
	Hygroscopicity • Shrinkage • Density • Other Properties	
20.4	Mechanical Properties	397
	Effects of Density • Effects of Growth • Effects of Moisture • Effects of Orientation • Time Effects • Summary	
20.5	Durability	402
20.6	Plywood	405
	Adhesives • Wood Plies • Fabrication • Specifications • Physical Properties	
	Problem 20 Compression Tests of Wood	410

21 Plastic — 412

21.1	Types and Characteristics	412
21.2	Production Methods	413
	Raw Materials • Polymerization • Additives	
21.3	Processing Methods	416
	Extruding • Compression Molding • Injection Molding • Casting • Foaming • Other Forming Methods • Fabrication	
21.4	Engineering Plastics	419
	Thermoplastics • Thermosetting Plastics • Adhesives	
21.5	Physical Properties	422
	Problem 21 Specific Strength of Plastic	425

Appendixes

A	The International System of Units (SI)	429
A.1	General Features	429
	Units • Prefixes • References	
A.2	Special Notes	431
	Mass • Density • Force and Weight • Moment • Plane Angle • Energy and Power • Pressure or Stress • Time • Hardness • Toughness • Temperature • Volume • Cross-Sectional Properties • Viscosity	
A.3	Style and Usage	433
A.4	Computations	434

B   Useful Facts   436

B.1   Conversion Factors   436
B.2   Selected Chemical Elements   438
B.3   Standard Sieve Sizes   439
B.4   Formulas from Mechanics   440
        Simple Stress Conditions • Cross Sections • Beams •
        Columns • Mohr's Circle

C   Sources of Information   443

C.1   Addresses of Organizations   443
C.2   Selected ASTM Standards   444
C.3   Text References   448

Index   455

PREFACE

Experimentation and testing play an important role in developing new materials and devices and in controlling the quality of materials for use in design and construction. In the educational process they also serve, through direct involvement of the student, to give substance to theoretical concepts and to provide a means of augmenting insights gained from analytical studies. For these reasons, students of engineering, architecture, and other fields associated with design activities or resource utilization can benefit from experiences in obtaining first-hand information about materials and processes.

This text is intended for a course that includes laboratory testing of constructional materials. Such a course may include some or all of the following objectives: familiarity with the mechanical properties of engineering materials, skill in methods of observation, knowledge of accepted methods of testing, appreciation of the significance of data, strengthened grasp of the principles of mechanics of materials, acquaintance with standards and the technical literature, and ability to report the results of investigations. The prerequisites generally include the courses normally taken by engineering students during their first two years, but by adjusting the requirements or by adding explanations instructors can accommodate students with somewhat different backgrounds. As the text does not emphasize microproperties of materials, the course may well be preceded or succeeded by one in material science.

The fourth edition has been reorganized and is divided into three parts. Part One deals with general testing principles, procedures, and apparatus; it includes separate chapters on statistical data evaluation and on report preparation. Part Two discusses specific types of tests, namely tension, compression, shear, bending, hardness, impact, fatigue, and creep tests, and concludes with a chapter on nondestructive test methods. Part Three provides a brief overview of the properties of the common constructional materials—metals, woods, plastics, portland cement concretes, and asphaltic materials. It can serve as an introduction to the nature of these materials for those students who have not had previous instruction concerning them or as a convenient review for those who have had such instruction. In any case, the chapters on materials are provided to serve as reference material to support laboratory studies and as an aid in interpreting the results of laboratory experiments.

A new chapter on asphaltic materials and mixtures has been added in this edition. One of the purposes for its inclusion is to illustrate approaches to dealing with materials that at some stage may be solid, plastic, or liquid. This chapter, together with the one on portland cement concrete, serves also to illustrate materials whose composition must be designed, in the technical sense, for each individual condition of use and whose manufacture and quality control must usually be carried out under field conditions.

Each chapter is followed by a pertinent laboratory problem and by discussion questions for outside study. The individual chapters are largely self-contained, so that the order and extent of coverage can be adjusted to suit the needs of a particular course. Laboratory problems may be selectively used or modified to fit the laboratory resources available and to meet the purposes of the instruction.

Suggestions for further study in the subject matter of most chapters are given at the end of such chapters. References that provide substantiation for, or more detailed explanation of, particular points throughout the text are listed in Appendix C.

The book is being published at a time when much of U.S. industry is changing over to metric (SI) measurements or has already done so, and when students have become familiar with SI units during their earlier studies. Although the present edition still includes some inch-pound units where necessary, the general treatment is based on the assumption that the primary technical measurement language will be SI. Many features are included that will help those readers not yet conversant with SI units become fluent in this language. In using the book, students are encouraged to collect data in metric units whenever possible, and to convert other data before using them when not already metric.

The authors are indebted to many organizations and individuals for their permission to use published material and for their courtesy in supplying information on methods and materials and photographs of testing apparatus. A list of acknowledgments follows this preface. Throughout the text an attempt has been made to credit the sources of all material quoted or reproduced, but it is obviously impossible to identify all those whose spoken or written words have directly or indirectly contributed to this latest book on a large subject. Their labors are deeply appreciated.

Harmer E. Davis
George Earl Troxell
George F. W. Hauck

ACKNOWLEDGMENTS

Special thanks are due persons assisting in the preparation of this or earlier editions:

Professor H. D. Eberhart
Corazon L. Ferrer
Susan J. Fershee
Professor J. W. Kelly
Professor E. A. Mechtly
Professor Carl L. Monismith
The late Professor Clement T. Wiskocil

The contributions of the following organizations are also gratefully acknowledged:

American Concrete Institute
American Instrument Co.
American Society for Metals
American Society for Testing and Materials
The Asphalt Institute
Baldwin-Lima-Hamilton Corp.
Effects Technology, Inc.
Federal Products Corp.
Forest Products Laboratory, U.S. Forest Service
Forney's, Inc.
General Electric X-Ray Corp.
The Institute of the Plastics Industry
Magnaflux Corp.
Morehouse Machine Co.
MTS Systems Corp.
National Bureau of Standards
Riehle Testing Machine Div., American Machine and Metals, Inc.

Shore Instrument and Manufacturing Co.
Sperry Products, Inc.
The Starrett Co.
TecQuipment Ltd.
Tektronix, Inc.
Tinius Olsen Testing Machine Co.
Vertek Assoc., Inc.
Wiedemann Machine Co.
Wilson Mechanical Instrument Co.
U.S. Navy
U.S. Army Ordnance Dept.

PRINCIPLES OF TESTING

ONE

CONTROL OF MATERIAL PROPERTIES

1.1 THE ROLE OF TESTING

The extensive use of experimental studies preliminary to the design and construction of new mechanical or structural elements and the use of testing procedures for control of established processes of manufacture and construction are significant and well-recognized features of our technical development. Practically all branches of engineering, especially those dealing with structures and machines, are intimately concerned with materials, the properties of which must be determined by tests. Successful mass production depends on inspection and control of the quality of manufactured products, which implies a system of sampling and testing. The preparation of adequate specifications and the acceptance of material purchased under specifications involve an understanding of methods of testing and of inspection. Settlement of disputes regarding failures and substandard quality almost invariably calls for investigations involving physical tests. Engineering research and development function in large measure on an experimental basis and call for carefully planned, well-devised tests.

For the intelligent appraisal and use of test results, it is important for engineers, even those not engaged in actual testing work, to have a general understanding of the common methods of testing for the properties of materials, and of what constitutes a valid test. Further, in dealing with the specification and acceptance of materials, an understanding of the limitations imposed by methods of testing and inspection is important.

The following subjects are basic in the study of materials testing:

1. *Technique of testing.* How do commonly used types of equipment operate? Is the apparatus in widespread use? What are common variants of ordinary apparatus? What limitations are imposed by the apparatus on the accuracy that can be obtained? Lacking first-class equipment, how can one get rough but significant results from crude tests in the field? Has the theory of models any application in a given test?

2. *Physical and mechanical principles involved in the testing apparatus and procedure.* Are the assumed conditions satisfied? What is most likely to go wrong or produce incorrect results? What apparent crudities can be overlooked? What refinements can be made to obtain greater accuracy?
3. *Theory of measurements.* What is the precision of the results? Which of the measurements involved control the precision of the final results? Are time and effort wasted securing needless precision in some of the measurements?
4. *Variability of materials.* How many tests are necessary to give a significant average? What variation from the average is cause for rejection of individual values? What range in strength (or other property) may be expected from a given material as it is used under job conditions?
5. *Interpretation of results.* What is the significance of the test results? Can the numerical results be applied directly to design and similar uses, or are they of value only for comparison with other results? How can the results of arbitrary tests be interpreted? Do the results of arbitrary tests have meaning if the test conditions are outside the range of those for which correlations have been set up? Considering the methods of testing and the kind of material, what are the limitations of the test results, or how reliable are the test data? How are the limitations of the testing method and the variability of materials reflected in the specifications? How should a satisfactory specification for a given material be written?

With advances in our technological development have come notable improvements in the older type of materials, many discoveries of new materials, and a variety of new uses for all materials. These have greatly extended the scope of materials testing and have complicated its practice. However, the fundamental principles involved in conducting valid and reliable tests are common to all materials testing. In this book we will address these principles by discussing the ordinary methods of testing the common materials of construction.

1.2 ENGINEERING MATERIALS

The principal materials used in building structures and machines are metals, woods, portland cement concretes, bituminous mixtures, clay products, masonry materials, and plastics. The principal function of constructional materials is to develop strength, rigidity, and durability adequate to the service for which they are intended. These requirements largely define the properties that the materials should possess and hence broadly determine the nature of the tests made on the materials. A full appreciation of the significance of tests of constructional materials therefore requires some knowledge of both materials science and structural mechanics. A very brief description of the behavior of materials is provided in Chap. 2, but the many available treatises on materials and mechanics should be consulted (such as those listed at the end of Chap. 2) for a detailed review of these matters. The principal *macroscopic* properties of some materials are discussed in Part Three.

A distinguishing feature of auxiliary constructional materials, such as sealants, adhesives, and preservative coatings, and perhaps of the majority of specific-use materials, is the requirement for some particular qualities or properties, the determination of which is often made by special and sometimes arbitrary tests. The particular problems involved in specialized fields are not within the scope of this book. It is worthy of emphasis, however, that the basic considerations given in Art. 1.1 apply with as much force to specialized tests as to the common ones.

In a general sense fluids and soils, in addition to solids, are engineering materials whose properties must be found by testing, and the performance of machines and structures, as well as the characteristics of materials, are subjects of engineering testing.

Properties A partial classification of the properties of engineering materials is given in Table 1.1. In general the determination of any or all of these properties may be the subject of engineering testing. However, the major work of the ordinary materials-testing laboratory has to do with mechanical properties. This work is often referred to as "mechanical testing." Because the major factor in the life and performance of structures and machines is applied force, strength is of utmost importance; a first requirement of any engineering material is adequate strength. In its most general sense the term *strength* may be taken to refer to the resistance to failure of an entire piece of a material, a small part of it, or even its surface. The criterion of failure may be either rupture or excessive deformation. From a historical point of view, the earliest tests were concerned with the strength of materials; therefore today the term *testing machine*, used without qualification, refers to a machine for applying known loads.

Table 1.1 Properties of engineering materials

Class	Property	Class	Property
General	Density or relative density Porosity Moisture content Macrostructure Microstructure	Mechanical	Strength: Tension, compression, shear, and flexture Static, impact, and endurance Stiffness Elasticity, plasticity Ductility, brittleness Hardness, wear resistance
Chemical	Oxide or compound com- position Acidity or alkalinity Resistance to corrosion or weathering, etc.	Thermal	Specific heat Expansion Conductivity
		Electrical and magnetic	Conductivity Magnetic permeability Galvanic action
Physico- chemical	Water-absorptive or water- repellent action Shrinkage and swell due to moisture changes	Acoustical	Sound transmission Sound reflection
		Optical	Color Light transmission Light reflection

Complete knowledge of the behavior of a given material would involve study of all its properties under a very wide range of conditions, but conducting the exhaustive tests required to obtain complete information usually would not be necessary or economically feasible. The problem, then, is to secure data on the properties that have a bearing on the economic value and serviceability of a material, or a product made from the given material, for a given purpose. The relative efficiency of a material for a specific use depends on the extent to which pertinent properties are present. For some uses, a property may be highly desirable, whereas for other uses, it may be undesirable or even dangerous.

Selection Serviceability, in a broad sense, is the ultimate criterion in a choice of materials. One important object of materials testing is to aid in predicting or ensuring the desired performance of materials under service conditions. However, in the selection of materials for building of structures and machines, the problems of quality of material, of design, and of use are interwoven. It should be noted in passing that sound material and correct design can do no more than give assurance that a particular construction will be satisfactory within the limits of its intended use, although the material that can withstand the greatest abuse certainly has one advantage over its competitors.

In the selection of materials, designers have two sources from which they can obtain information: (1) knowledge or records of performance of materials in actual service and (2) the results of tests made to supply data on performance. On the basis of such information, a specification is prepared.

To transform a design into actual construction, it is necessary for the constructor (or manufacturer) to select, from a variety of available grades of material, the one that the designer had in mind and tried to specify. Tests are then required to identify the desired material.

The considerations involved in the selection of materials as regards the problems of design and manufacture include

1. Kinds of materials available
2. Properties of various materials
3. Service requirements for materials
4. Relative economy of various materials and of various forms of a particular material
5. Methods of preparation or manufacture of various materials or products and the influence of processing on their properties.
6. Methods of specification and their relation to uniformity and dependability of product secured
7. Methods of testing and inspection and their significance with respect to measures of desired properties

1.3 TESTING OF MATERIALS

The testing of materials may be performed with one of three objects in view: (1) to supply routine information on the quality of a product—commercial or control testing,

(2) to develop new or better information on known materials or to develop new materials—materials research and developmental work, or (3) to obtain accurate measures of fundamental properties or physical constants—scientific measurement. These objectives should be clearly discerned at the outset, since they generally affect the type of testing and measuring equipment to be used, the desired precision of the work, the qualifications of the personnel to be employed, and the costs involved.

Commercial testing is concerned principally either with checking the acceptability of materials under purchase specifications or with the control of production or manufacture. Generally, the type of test has been specified, although as a guide to measuring quality it may be entirely arbitrary; standard procedures are used, and the object is simply to determine whether the properties of a material or of a part fall within required limits. A high degree of refinement is not required, although limits of accuracy are often specified.

Common purposes of *materials research* are (1) to obtain new understanding of known materials, (2) to discover the properties of new materials, and (3) to develop meaningful standards of quality or test procedures. In addition, there may be the specific objective of choosing a material for a particular use, of determining principles to improve design with a chosen material, or of studying the behavior of the part or structure after is has been made. Although many investigations are more or less routine in nature, there are also many that call for a very wide variety of tests and measurements, require an appreciation of all phases of the general problem, and make extreme demands on the skill, ingenuity, and resourcefulness of the experimenter if success is to be attained.

The aim of what is here is called *scientific testing* is the accumulation of an orderly and reliable store of information on the fundamental and useful properties of materials, with an ultimate view to supplying data for accurate analyses of structural behavior and for efficient design. Work of this kind, above all, calls for care, patience, and precision. In a student laboratory, experiments may serve to give insight into principles developed in texts on the mechanics of materials.

Closely related to materials testing is "experimental stress analysis." This differs, however, in purpose: the effects of the shape of a stressed body and of the manner of loading, rather than of the material, are of primary interest.

We may distinguish between *experiments* and *tests*, although the two terms do overlap. Experimentation means that the outcome is uncertain, that new insights are to be gained. Testing indicates a more or less established procedure, with the limits of the results generally defined. Experimentation, especially on a planned or large scale, ordinarily involves many routine tests. Many of the large materials-testing laboratories serve the dual purpose of being experimental research bodies and routine-control testing agencies. Although the purpose, the point of view, and the method of attack may differ widely between research and routine testing, many of the detailed procedures are exactly the same in both kinds of testing. For example, in research on heat-treatment of steel plates, the tension-test specimens and the test method would very likely be the same as those used for acceptance tests of boiler plates.

For convenience, we may differentiate between *field tests* and *laboratory tests*. Because of difficult or hazardous working conditions, interference, time limitations, and variable weather conditions, tests done in the field usually lack the precision of similar tests conducted in the laboratory; however, working in a laboratory does not necessarily ensure precision or accuracy. Certain types of tests, as, for example, a sieve analysis of gravel, may be made just as accurately by an inspector on the job as by a technician in the laboratory. Some tests cannot be made in the laboratory and others cannot be made in the field, so that the question of field vs. laboratory is not always pertinent.

With respect to the general method of attack and to the interpretation of results, it is desirable to distinguish among

1. Tests on full-size structures, members, or parts
2. Tests on models of structures, members, or parts
3. Tests on specimens cut from finished parts
4. Tests on specimens of raw or processed materials

Testing on models demands knowledge of the similitude of the model to the prototype and the satisfaction of particularly exacting requirements.

As regards the usability of a material or a part after a test, tests may be classified as *destructive* or *nondestructive*. Tests that determine ultimate strength naturally mean destruction of the sample. Since an entire lot cannot be tested in this way, the problem enters of how to obtain a reliable indication of the strength of the lot using a sufficient number of specimens and at the same time keeping the expense of the sample material within reasonable bounds. For finished products it is desirable to use nondestructive tests if possible. Some hardness tests are of this type; for example, the scleroscope tests may be used to determine the surface hardness of ground-surfaced heat-treated steel shafting. Proof tests, applied to fabricated parts or to structural elements, are of the nondestructive type. For example, a proof test of a crane hook involves the application of a load somewhat in excess of the working load but less than a damaging load in order to give assurance that no harmful defects, which might cause failure in service, are present. Nondestructive tests are of particular interest to an inspector on the job. Several methods of nondestructive testing are introduced in Chap. 16.

1.4 INSPECTION OF MATERIALS

Despite certain common features of *testing* as such and *inspection*, it is desirable to distinguish between them. Specifically, testing refers to the physical performance of operations (tests) to determine quantitative measures of certain properties. Inspection has to do with the observation of the processes and products of manufacture or construction for the purpose of ensuring the presence of desired qualities. In many instances inspection may be entirely qualitative and involve only visual observation of correctness of operations or dimensions, examination for surface defects, or possibly the indication of the presence or absence of undesirable conditions such as excessive moisture or temperature. On the other hand, inspection may involve the performance of complicated tests to determine whether specification requirements are satisfied.

Often such tests are nondestructive, but not always. Sample concrete specimens, for example, are tested to failure. Inspection aims at the *control* of quality through the application of established criteria and involves the rejection of substandard material. In testing, the aim is to *determine* quality, that is, to discover facts regardless of the implications of the results.

In some organizations, the inspection forces make only simple examinations and send selected samples to the testing department. In other organizations, the testing engineer is also the chief of inspectors in addition to being charged with routine testing and with research and development work. Inspection involves human relationships as well as technical duties. Not all people can be good inspectors.

1.5 SIGNIFICANCE OF TESTS

Our concepts of the properties of materials are usually idealized and oversimplified. Actually, we do not *determine* properties, in the sense that some unchanging values are obtained that once and for all describe the behavior of some material. Rather, we obtain only *measures* or *indications* or *manifestations* of properties found from samples of materials tested under certain sets of circumstances. For example, in a tension test of steel, the percentage elongation in a given gage length is used as a measure of ductility. At the same time, to the extent possible, we attempt to avoid test conditions that arbitrarily restrict the general significance of the quantitative measure obtained from a test; for example, we believe that the results of the tension test give a reliable indication of the macroscopic property of tensile strength in uniaxial tension. The measures we obtain depend on the test conditions, which include the way the sample is taken and prepared, as well as on the particular procedures involved in making the test. The significance of a test is also affected by its *reliability* in yielding a measure of the property it is supposed to determine. For example, the use of the proportional limit as a practical measure of the elastic strength of a material has been questioned because the test results are affected by a number of factors that cannot be properly controlled in ordinary testing and that make the determination of the true magnitude of the proportional limit uncertain.

The real significance of any test lies in the extent to which it enables us to *predict the performance* of a material in service. A test may have significance in one of two ways: (1) it measures adequately a property that is sufficiently basic and representative that the test results can be used directly in design, or (2) the test, even though highly arbitrary, serves to identify materials that have been proved by experience to give satisfactory performance. For example, in connection with the design of a tension member for a bridge structure, a tension test on a properly selected sample of the steel will give a value that, when modified by a suitable factor of safety, can be taken as the allowable working stress. On the other hand, for example, the Charpy impact test of metals has significance in relation to the use of a material only when the test results are correlated with the performance of the material in service. This test gives values that are stated in terms of energy absorbed during the rupture of small, standardized, arbitrarily shaped specimens of the metal. The results cannot be utilized directly in any design, as can strength values. Yet it has been found that, for certain types of service, failures may be expected if the Charpy values fall much below a given value. This test, then, has significance in that it aids in the elimination of steels unsuitable for a particular use. The test that can be made to give a direct indication of expected performance depends to a large extent on the state of development of the arts of testing and stress analysis.

One striking fact to be noted in a study of detailed test data, and in the results of investigations in general, is the variation in the quantitative measures of given properties. This may be due in part to the lack of absolute precision in the testing operations, but it also may be due to actual variation in a given property within samples. Our materials are not homogeneous; within limits, their composition may be governed entirely by chance, so that a description of their behavior may rest to a great extent on a statistical basis. For an intelligent interpretation of results, both with respect to the meaning of variation between samples or parts of a sample and with respect to the relation of samples to an entire lot or the full-sized piece of material, due regard should be given to the requirements of sampling theory and to the statistical nature of the data (see Chap. 3). Further, in the interest of efficient testing and reliable results, the test should be designed so that the precision of the various measurements or operations involved is consistent throughout (see Chap. 5, especially Art. 5.6).

We have used the words "precision" and "accuracy" several times. They are not synonyms, and this may be a good time to note the difference. Precision denotes the repeatability of a measurement, accuracy its closeness to the true value. Measurements may be precise and inaccurate or imprecise and accurate. If an instrument consistently gives nearly identical but wrong readings, it is precise but inaccurate. If the readings vary considerably but do center about the true value, they are accurate but imprecise. Ideally, of course, measurements should be precise *and* accurate.

1.6 SPECIFICATION OF MATERIALS

A *specification* is the attempt on the part of the consumer to tell the producer what is wanted. Obviously, the skill and accuracy with which a thing *can* be specified depends on the state of knowledge concerning it and on the precision with which its qualities can be determined. As the art and science of testing are advanced, so is the basis for preparation of adequate specifications improved. However, the effectiveness with which a thing *is* specified depends also on how well the specification is written and how enforceable the provisions are. This problem, although it often concerns the testing engineer in particular, may confront any engineer.

Specification types At one time it was customary to specify merely a given brand, or equal; past performance and the integrity of the producer were the only guaranties of potential quality. Early specifications were often crude because consumers knew little about the material they tried to specify; some present-day specifications are just as crude, and for the same reason. With the increased complexity of our industrial system, more adequate specifications have become necessary, and with advances in our scientific knowledge of materials, more adequate specifications have become possible.

A specification is intended to be a statement of a standard of quality. The ideal material specification would uniquely define the qualities necessary to serve most efficiently for a given use, and it can be approached if truly significant tests can be made to determine the presence of the required qualities. A specification often falls short of the ideal for a number of reasons, some of which are the following: (1) it may be so loose as to admit material of inferior quality, (2) it may be overly restrictive and so exclude an equally or more efficient material, (3) it may be based on inadequate or improper criteria with respect to the type of service required, and (4) it may make no provision or inadequate provision for proper enforcement. Defects such as these lead not only to the procurement of unsatisfactory materials but often to disproportionate costs and endless disputes. It is important to note also that a specification may admittedly and necessarily be imperfect because it would be impracticable to produce the ideal material. All things considered, it may be just as inefficient to require too high a quality as to accept too low a quality. Practically, specifications are drawn up not for an ideal material but for a material that it is possible to obtain at reasonable cost under existing conditions of manufacture.

Several considerations fix the limits within which a specified property may be allowed to vary. The maximum and minimum to be set may be based on experiment but should take into account the limitations of the manufacturing process. These limits correspond to the size limits allowed in making machine parts, where such variation in the size of each part is allowed as leads to economy in manufacturing the parts without unduly impairing the efficiency of the assembled machine. In fixing these limits of tolerance for a material, care must be exercised to avoid ranges that are too narrow or too wide.

Specifications for materials of construction may define the requirements for acceptability of the material in one or all of the following ways:

1. By specifying the method of manufacture
2. By specifying form, dimensions, and finish
3. By specifying desired chemical, physical, or mechanical properties

Another type of requirement, although rarely used in the materials field, is that a product shall not exhibit stated defects within a certain period after purchase. Performance specifications are commonly used for machines. Often included in materials specifications are requirements with regard to methods of sampling, testing, and inspection.

Standard specifications A notable development of the twentieth century has been the preparation and use of standard specifications. A standard specification for a material is usually the result of agreement among those concerned in a particular field and involves acceptance for use by participating agencies. It does not necessarily imply, however, the degree of permanence usually accorded to dimensional standards, because technical advance in a given field usually calls for periodic revision of the requirements. Some of the various types of standardizing agencies are independent companies, trade associations, technical and professional societies, and bureaus and departments of municipal, state, and national governments. The breadth of acceptance depends to an extent on the scope of influence and authority of the standardizing agency. Under the standardizing procedure followed by important agencies, a period of negotiation, formulation, and trial usually precedes the use of a specification as a standard, so that it has the assurance of being workable.

A standard specification implies standard methods of testing and sometimes also standard definitions. In some instances, the methods of testing are incorporated within a materials specification. On the other hand, some standardizing agencies set up standard test methods separately from the materials specifications and make mandatory reference to the test methods.

Properly devised and enforceable standard specifications can be of immense value to industry. Some of the advantages that may be cited for standard specifications for materials are

1. They usually represent the combined knowledge of the producer and consumer and reduce the possibility of misunderstanding to a minimum.
2. They give the manufacturer a standard of production and so tend to result in a more uniform product and to reduce the number of required varieties of stock, lowering the attendant waste and therefore lowering the cost.
3. They lower unit costs by making possible the mass production of standardized commodities.
4. They permit the consumer to use a specification that has been tried and is enforceable.
5. They permit the designer to select a material that there is reasonable assurance of getting.
6. They simplify the preparation of special-use specifications because published standard specifications can be incorporated by reference.
7. They aid the purchasing agent in securing truly competitive bids and in comparing bids.
8. They set standards of testing procedure in commercial testing and hence permit comparison of test results obtained from different laboratories.

In the initial development of a standard test procedure, considerable research is often conducted by cooperating organizations to develop a procedure that will yield reproducible and meaningful test results.

The disadvantage of standard specifications is that they tend to "freeze" practices that may be in only the developmental stage and thus hinder progress where it is most needed. For this reason, standard specifications should be under the jurisdiction of a well-informed and thoroughly open-minded agency.

Specifications for both materials and methods of testing should be subject to continuous review to determine their suitability under changing conditions. Also, various codes based on these standards should be reviewed frequently.

1.7 STANDARDIZING AGENCIES

Because standardization has such an important influence on ordinary methods of testing, it is desirable for engineers to have some familiarity with the nature and the publications of the agencies that have promulgated some of the widely used materials specifications and methods of testing.

Standards always represent an effort by some organized group of people to agree on a common definition or procedure. Any such organization, be it private or public, then becomes a standardizing agency. Various levels of agency exist, ranging from a single firm or city government to national groups such as industry associations to international organizations. There are easily thousands of such agencies around the world.†

The desirable tendency is for a given standard to become more uniformly accepted. One method of increasing standardization is for a large agency to adopt a standard developed by a smaller one. In the United States, for example, thousands of standards developed by private and public groups are recognized by a national, yet private, coordinating organization, the American National Standards Institute (ANSI). Another method consists of the development of a standard by a large standardizing agency. Smaller agencies are then encouraged to adopt the standard by reference or duplication.

Private (professional and industrial) organizations in the United States with an interest in the development of engineering materials standards include the Society of Automotive Engineers (SAE), the American Petroleum Institute (API), the American Concrete Institute (ACI), the American Iron and Steel Institute (AISI), the American Society of Civil Engineers (ASCE), the American Association of State Highway and Transportation Officials (AASHTO), the American Railway Engineering Association (AREA), and various others. Many U.S. government agencies, too, have written standards specifications for their own use. These agencies include the General Services Administration (GSA), the Department of Transportation (DOT), and the Department of Defense (DOD) and its various agencies, such as the Army Corps of Engineers and the Naval Facilities Engineering Command.

The standards promulgated by the American Society for Testing and Materials (ASTM) are of particular interest and importance to those concerned with materials testing and inspection. This national technical society, formed in 1898, has a membership that may be roughly divided into three groups: consumers, producers, and a general-interest group comprising engineers, scientists, educators, testing experts, and the like. Organizations as well as individuals are members.

ASTM performs the dual function of (1) standardizing materials specifications and methods of testing, a process carried out by standing committees, each of which has under its jurisdiction engineering materials in a definitely prescribed field or some specific phase of materials testing; and (2) improving engineering materials, through investigations and research by committees and individual members, the results of which are made public through the society's publications. The specifications are published in a separate series of almost 50 volumes called the *Annual Book of ASTM Standards*.

ASTM's standardization work comprises in general (1) the development of methods of testing for materials, (2) the setting up of standard definitions, (3) the formulation of materials specifications, and (4) the formulation of recommended practices

---

†For a partial list of such agencies and their addresses see App. C.

having bearing on various processes in the utilization of materials. Committees concerned with the development of specifications first study the materials in their respective fields and foster the necessary research on which the standardization work must be based. On committees dealing with materials having a commercial bearing, the policy is generally to maintain a balance between representatives of producer and consumer interests.

After completion of studies involving methods of testing, nomenclature, material requirements, and the like, a proposed standard is evolved. If it is approved by the committee responsible for the material or field and adopted by the Society in accordance with the *Regulations Governing ASTM Technical Committees*, it is published for voluntary use by anyone. An ASTM standard, which may be in the form of a specification, test method, definition, classification, or practice, may be revised at any time but must be reviewed at least every five years; unless it is revised or reapproved it must be withdrawn. Thus the standards constantly change, and some of the many cited in this book will have been revised or discontinued by the time it is read.

All ASTM standards carry an identifying designation, for example, C 150-78a. The initial letter indicates the general group to which the specification belongs:

A: Ferrous metals
B: Nonferrous metals
C: Cementitious, ceramic, and masonry materials
D: Miscellaneous materials
E: Miscellaneous subjects
F: Materials for specific applications
G: Corrosion, deterioration, and degradation of materials
ES: Emergency standards

The numerals immediately following the letter are serial numbers that show the order in which the standard was adopted in a given letter group. The numerals following the dash indicate the year of adoption or of the latest revision. If a lowercase letter follows, it denotes a revision during the same year. The letter M after the numerals indicates a hard metric equivalent of a nonmetric standard. If the letter T follows the numerals, it indicates that the specification is still in the tentative stage. For many purposes, only the first part of the designation, e.g., C 150, is necessary to identify the specification: it is then assumed that the latest revision is meant. A brief list of ASTM standards is given in App. C.

The American National Standards Institute (ANSI), formerly known as the American Standards Association, was organized in 1918 by ASCE, ASTM, the American Society of Mechanical Engineers (ASME), the American Institute of Electrical Engineers [now the Institute of Electrical and Electronic Engineers (IEEE)], and the American Institute of Mining, Metallurgical and Petroleum Engineers (AIME) to provide a means for industry, technical organizations, and governmental departments to work together in developing national industrial standards acceptable to all groups and to provide a means whereby standardizing organizations might coordinate their work and prevent duplication of effort. Using one method, ANSI standards are developed and approved in a manner very similar to that of ASTM. Using another method, ANSI may approve existing proprietary standards as American National Standards; many of the ASTM standards have been accepted by ANSI in this manner. See App. C.

The Department of Commerce (DOC) plays a particular role in promulgating standards in that it is charged with the administration of standards that are statutory. The agency within DOC that deals with these matters is the National Bureau of Standards (NBS). So, for example, the Metric Conversion Act of 1975 charges the Secretary of Commerce with the interpretation of SI for the United States, a function performed by NBS.

At the international level, the organization of primary interest in the present context is the International Organization for Standardization (ISO), which is responsible for standards in all fields except electricity and electronics. Standards in the latter areas are administered by the International Electrotechnical Commission (IEC). All developed nations are represented in these organizations, usually by governmental but sometimes by private agencies. The United States, for example, is represented in ISO not by NBS but by ANSI. The technical work of ISO is performed by over a hundred technical committees (TCs) relating to various industries. The secretariat of each technical committee is held by a member body; for example, the secretariat for plastics (ISO/TC.61) is held by the United States. Each member country is given the option of joining any technical committee as a participating member with voting rights at committee meetings, as an observer member or not at all.

The addresses of the headquarters offices of several of the standardizing agencies interested in materials specifications are listed in App. C.

No matter who develops a standard, the words written by C. L. Warwick about ASTM over half a century ago are still true. In the development of specifications,

> full account must be taken of the influence of manufacturing processes, the nature of the stresses and other conditions to which the materials will be subjected in service, and the particular properties of the material that enable it to give satisfactory service. Painstaking investigation and study of experience over years of service are often required before an adequate specification can be prepared. The committee must come to an agreement upon the properties of the material to be specified, the methods of test, such details of manufacture as may be necessary, methods of inspection and of marking, etc. In all these things it seeks to follow the best commercial practice that has been developed in supplying the particular material or commodity to the trade. Specifications for materials upon whose strength and reliability the safety of human life may depend must be especially carefully drawn and provided with adequate safeguards in testing and inspection. At times, a compromise between the somewhat extreme views that may be held by the producer and consumer is necessary in reaching at least a tentative agreement on certain details, although the more clearly the problems involved are understood and the more complete are the technical data that can be presented on the subject, the more easily can a logical rational agreement be reached [99].[†]

PROBLEM 1 Inspection of the Testing Laboratory

Object: To become acquainted with the materials testing laboratory, the equipment available, and the requirements of the course.
Preparatory reading: Apps. A, B, and C, ASTM E 380.

[†]Numbers in brackets refer to the list of references in App. C.

Assignment:
1. Under the guidance of a staff member, visit the laboratory and notice where general equipment is located.
2. Ask to be instructed in the operation of a universal testing machine.
3. Make a list of the major types of equipment available. Note the units of calibration and the dial divisions.
Report: Write an informal report that includes
1. A guide to the laboratory, with the major features indicated on a sketch.
2. A brief description of each major testing machine. This should include, where appropriate, the factor necessary to convert from the calibration units to the correct SI units.
3. An assessment of the role of the course in your education.

DISCUSSION

1. What are the principal objectives in the testing of engineering materials?
2. Discuss the relative merits of various types of testing machines. What is the most common type of machine used in the testing of engineering materials? Why is it used so commonly?
3. What item other than load is often progressively observed for a specimen under test? What equipment is used to aid these observations?
4. Why is it essential that standard methods of testing be used?
5. What considerations are involved in the selection of materials for machines and structures?
6. Discuss the function of inspection.
7. What is meant by the "significance" of a test?
8. Using an everyday example, explain the difference between "precision" and "accuracy."
9. Discuss the problem of specification of materials.
10. What are the functions and objectives of ASTM, ANSI, NBS, and ISO?

READING

At the end of most chapters you will find a few suggestions for additional reading. Consult the library for other books (especially those published after this one) and check recent periodicals for current thinking about the subjects discussed in this book.

Cordon, W. A.: *Properties, Evaluation, Control of Engineering Materials,* McGraw-Hill, New York, 1979.

Gilmore, H. L., and H. C. Schwartz: *Integrated Product Testing and Evaluation,* Interscience-Wiley, Somerset, N. J., 1969.

Jones, S. W.: *Materials Science–Selection of Materials,* Butterworths, London, 1970.

Parkinson, A. C.: *Engineering Inspection,* 4th ed., Pitman, London, 1967.

Rosen, H. J.: *Construction Materials Evaluation and Selection*, Interscience-Wiley, Somerset, N.J., 1979.

Sharp, H. J.: *Engineering Materials, Selection and Value Analysis,* American Elsevier, New York, 1966.

TWO

MECHANICAL BEHAVIOR

2.1 FUNDAMENTAL DEFINITIONS

In mechanical testing, as in the field of mechanics in general, we attach rather specific meanings to certain terms, some of which are interpreted more loosely in everyday life. For example, *mass* is a property of a body, expressed in kilograms. It is not to be confused with *weight*, which is the gravitational force exerted on a body by the earth, moon, or another celestial body. Weight thus varies with location and is not a property; it is expressed in newtons.

Density is mass per unit volume, expressed in kg/m^3 or Mg/m^3. Although it varies with temperature, pressure, and other conditions, it is considered a property of a material. The "unit weight," given in N/m^3, may be useful in computing load, but it is not a material property. The ratio of the density of a material to that of water in its densest state is called the *relative density* (formerly specific gravity) and is, of course, dimensionless. Since water has a density of just about $1.0 \ Mg/M^3$,[†] a material's density, when expressed in Mg/m^3, is numerically equal to its relative density (see also ASTM E 12).

Stress and strain In testing materials, *loads* are applied and measured by means of testing machines such as those described in Chap. 6. Loads are usually expressed in force units, for example, as kilonewtons, although for certain tests, such as torsion tests, the load may be expressed in units of moment, for example, as kilonewton-meters per radian.

Stress is defined as the intensity of the internally distributed forces or components of forces that resist a change in the form of a body. Stress is measured as force per unit area. For structural materials the unit commonly used is the meganewton per square meter, or megapascal (MPa), which is numerically equal to a newton per square millimeter. There are two basic kinds of stress: direct (tension and compression) and shear.

[†]More precisely, the density of water in its densest state, at 4°C, is $0.999 \ 973 \ Mg/m^3$.

It is customary to compute stresses on the basis of the cross-sectional dimensions of a piece before loading, which are usually called the original dimensions. In simple tension and compression tests, where the specimen is subjected to uniformly distributed stress, the stress is computed by dividing the (known) load by the minimum original cross-sectional area; if the dimensions vary slightly, the value for the area may be based on the average of critical dimensions.

In cases where the stress distribution is not uniform, the stress at specified locations may be determined by indirect methods. In flexure and torsion tests, stresses may be computed by means of theoretical relations. Within the elastic range, stresses may be evaluated from measured strains through the use of the modulus of elasticity.

The term *deformation* is used as a general term to indicate the change in form of a body; the alteration may be due to stress, to thermal change, to change in moisture, or to other causes. In conjunction with direct stress, deformation is usually taken to be a linear change and is measured in length units. In torsion tests, it is customary to measure the deformation as an angle of twist (sometimes called a *detrusion*) between two specified sections; from a consideration of the dimensions of the piece, the angle of twist in a cylindrical piece may be translated into terms of shearing strain. In flexure tests, the deformation may be expressed in terms of the deflection of some specified point of a beam from its original position.

Strain is defined as the change per unit length in a linear dimension of a body, the change accompanying a change in stress. Strain is a unit deformation due to stress. It is a ratio, or dimensionless number, and usually quite small. A convenient way of expressing (unit) strain is in terms of mm/m or μm/m. Under direct stress, unless otherwise specified, strain is measured parallel to the direction of the force and parallel to the dimension to which it is referred. Shearing deformation is measured parallel to the direction of the shearing force, but shearing strain is computed with respect to the dimension perpendicular to the direction of the force; shearing strain is therefore an angle, expressed in radians. Figure 2.1 illustrates these definitions of strain. Methods of measuring deformations and strain are described in Chap. 7.

Set, or *permanent set*, is the deformation or strain remaining in a previously stressed body after release of load.

Stresses and strains based on the original dimensions are sometimes called the *nominal* stresses and the *nominal* strains, or the *engineering* or *conventional* stresses and strains, to distinguish them when necessary from the "true stresses" and "natural strains," which are calculated on the basis of the instantaneous dimensions under given loads. This will be discussed more fully in Art. 2.2.

If a body is subjected to tensile or compressive stress in a given direction, strain takes place not only in that direction (axial strain) but also in directions perpendicular to the direction of stress (lateral strain). Within the range of elastic action the ratio of lateral to axial strain under conditions of uniaxial loading is called *Poisson's ratio*. Axial extension causes lateral contraction, and vice versa. For most structural materials, Poisson's ratio has values that lie between one-third and one-sixth; hence, with ordinary measuring devices, the precision of lateral-strain measurements is not as high as that of corresponding axial-strain measurements.

Occasionally, volumetric deformations are determined. For solid bodies, the

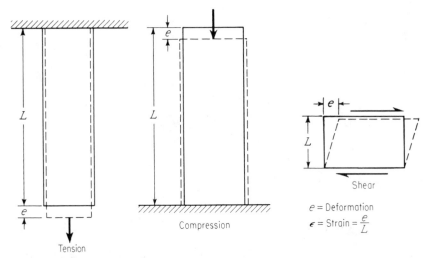

Figure 2.1 Relation between strain and deformation.

volumetric strain (sometimes called the cubical dilation, or the dilatation) is usually computed from measurements of linear strains. For porous or highly deformable bodies, the volume change is often measured by the displacement of a fluid.

Stiffness   Stiffness has to do with the relative deformability of a material under load. It is measured by the rate of stress with respect to strain. The greater the stress required to produce a given strain, the stiffer the material is said to be.

When a material is under simple stress within the proportional limit, the ratio of stress to corresponding strain is called the *modulus of elasticity* (*E*). This term is somewhat of a misnomer, since it refers to stiffness in the elastic range rather than to elasticity. Corresponding to the three fundamental types of stress, there are three moduli of elasticity: the modulus in tension, in compression, and in shear. When a material is under tensile stress, the measure of stiffness is sometimes called *Young's modulus*, after the English physicist who first defined it. The stiffness of a material under simple shear is sometimes called the *modulus of rigidity*. In terms of the *stress-strain diagram* shown in Fig. 2.2*a*, the modulus of elasticity is the slope of the stress-strain diagram in the range of linear proportionality of stress to strain.

Because many materials are imperfectly elastic, special definitions of the modulus of elasticity may be necessary. Figure 2.2*b* shows a stress-strain diagram that is continuously curved, the type of diagram that is obtained for such materials as concrete or cast iron. The slope of a line (*OA*) drawn tangent to the curve at the origin is the *initial tangent modulus*. The slope of the curve at, say, point *B*, is the *tangent modulus* at a stress *b*. The ratio of any given stress to the corresponding strain, which is equivalent to the slope of the line *OC*, is the *secant modulus* of elasticity at stress *c*.

The modulus of elasticity of materials is ordinarily determined directly by tests involving stress-strain measurement of specimens subjected to simple stresses. But modulus determinations are sometimes made through the use of special properties and

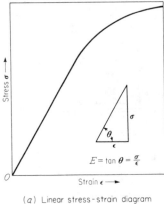

$$E = \tan \theta = \frac{\sigma}{\epsilon}$$

(a) Linear stress-strain diagram

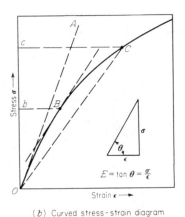

$$E = \tan \theta = \frac{\sigma}{\epsilon}$$

(b) Curved stress-strain diagram

Figure 2.2 Moduli of elasticity.

theoretical relations. For example, the modulus is computed from observations of the deflection of beams, the period of vibration of wires and rods, and the sound emitted by vibrating bars. The modulus of elasticity is expressed in units of force per unit area, usually in gigapascals (GPa). The modulus in shear for metals is about two-fifths of the modulus under axial stress.

There is no established measure of stiffness in the plastic range.

The term *flexibility* is sometimes used as the opposite of stiffness, or rigidity. However, flexibility usually has to do with flexure or bending; also, it may connote ease of bending in the plastic range. The effective or overall stiffness or flexibility of a body or structural member is obviously a function of the dimensions and shape of the body as well as of the characteristics of the material.

The measure of resistance to change in volume is called the *coefficient of compressibility* or the *bulk modulus* and is taken as the ratio of (hydrostatic) stress to the corresponding unit change in volume.

2.2 STRESS-STRAIN DIAGRAMS

Normally, a *stress-strain diagram* is a graph plotted with values of stress as ordinates and values of strain as abscissas. However, the use of the term stress-strain diagram is often extended to cover diagrams in which the ordinates are values of applied load or applied moment and the abscissas are values of extension, compression, deflection, or twist.

The basic procedure in obtaining a stress-strain diagram is to plot data from a series of load readings against corresponding data from the readings of a strainometer. Stress-strain diagrams may be obtained directly by attaching an autographic instrument to the testing machine.

Test control In planning a test requiring stress-strain data, it is necessary to select the increment of load or the increment of strainometer reading to be used between succes-

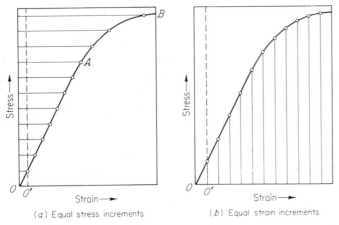

(a) Equal stress increments   (b) Equal strain increments

Figure 2.3 Determination of stress-strain relations.

sive readings. In Fig. 2.3, which shows an idealized stress-strain diagram, two methods of scheduling stress-strain readings are illustrated. Figure 2.3a shows the method employing equal increments of load. It can be seen from the figure that if this method is used, sufficient data for adequately locating the curve between points A and B may be lacking. That particular section of the curve between A and B (the "knee" of the curve) is a part for which it is often especially desirable to locate several points on the diagram.

Sometimes smaller load increments are used in the stage of loading that corresponds to the knee of the stress-strain diagram than in the early stage of load application. This procedure is not always satisfactory, however, since it may involve considerably more readings than are necessary, and in a test of an unknown material there is the possibility that the stage of small-increment loading will not have been started soon enough.

Figure 2.3b illustrates the method employing equal increments of strain. Using this method, several points are located near the knee of the curve, and thus the shape of the diagram in this region is more definitely determined than it is by the method of equal load increments.

The use of increments of load rather than increments of strain has been common practice because, in general, it is simpler to schedule load increments than increments of strainometer reading. The load necessary to stress the test specimen to a value corresponding to the knee of the curve is estimated, and a fraction (frequently one-tenth) of this value is taken as the increment to be employed. The determination of a suitable increment of strainometer reading is somewhat more complicated but not difficult. The following procedure is suggested: the load corresponding to the knee of the stress-strain diagram is estimated, and one-tenth of this load is applied to the test specimen. The change in extensometer reading is noted, and an increment of strainometer reading is then chosen, for the remainder of the test, that is equal to some convenient interval on the scale of the strainometer and is approximately equal to this initial strain increment. In routine tests of a given material, the appropriate strain increment is known from experience. The use of equal increments of strain is considered preferable.

In connection with the practical determination of stress-strain relations, it should be noted that the initial or datum strainometer reading is often taken after some small initial load has been applied. Such a procedure is desirable in order that firm gripping of or bearing on the test specimen can be attained and so that firm seating of the strainometer can take place. Data taken during this procedure, when plotted, give a stress-strain diagram that shows a finite value of stress at zero strain; this condition is shown in Fig. 2.3, provided the stress axis has its origin at O'. The effective origin of the stress-strain diagram may be obtained by shifting the stress axis to the left so that it passes through O, the intersection of the projection of the stress-strain diagram with the strain axis. For convenience in further discussion, stress-strain diagrams will be drawn from the effective origin O.

True stress–natural strain relations When a ductile material is loaded beyond the yield strength, through the plastic range, the dimensions change appreciably; as the fracture load is approached, particularly after necking starts in the tensile test, the stress at the critical cross section departs more and more from the nominal stress calculated on the basis of the original cross section. Thus in studies of the behavior of materials subject to large deformations, such as metals in the plastic range, it has been found desirable to calculate stress under a given load on the basis of the instantaneous dimensions.

True stress is obtained by dividing the axial load P by the actual instantaneous cross-sectional area A. After necking starts in a tension test, the area is measured at the minimum section of the neck; in this range of deformation, however, the "true" stress is only the average stress over the area, because the actual stresses vary across the section [74].

Natural strain, also called "true" strain, is the change in gage length with respect to the instantaneous gage length over which the change occurs:

$$\epsilon_{nat} = \int_{L_o}^{L} \frac{dL}{L} = \ln \frac{L}{L_o} = \ln (1 + \epsilon_o) \tag{2.1}$$

where L = length of a small element under a given load

L_o = original length of element before any load is applied

ϵ_o = conventional strain

From the hypothesis that during plastic deformation the volume of material remains constant, it may be shown that

$$\epsilon_{nat} = \ln \frac{L}{L_o} = \ln \frac{A_o}{A} \tag{2.2}$$

where A_o = original cross-sectional area before any load is applied

A = instantaneous cross-sectional area under a given load

This relation is particularly useful in obtaining the natural strain at the critical area of the necked section.

The general form of the true stress–natural strain diagram for a tension test is shown diagrammatically in Fig. 2.4. Observations indicate that the true stress is practi-

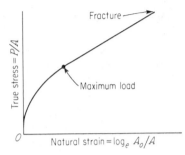

Figure 2.4 True stress–natural strain diagram for an axial-tension test.

cally linearly related to the natural strain from the maximum load point to the fracture load point. This type of diagram is much more significant than the conventional stress-strain diagram in studies concerned with metal-forming operations.

Nondimensionalization In the ordinary stress-strain diagram strain is dimensionless, but stress is dimensional. However, it is possible, and at times desirable, to make both stress and strain dimensionless. Ramberg and Osgood [77] developed a practical method for doing this.

Figure 2.5a shows an ordinary stress-strain diagram. If all the stress values are divided by a selected *base stress* σ_0 and all the strain values by a selected *base strain* ϵ_0, the diagram in Fig. 2.5b results. The point representing the two base values on the original stress-strain diagram will be located at the base point (1,1) on the new Ramberg-Osgood or *one-one diagram*.

To determine the base value σ_0, draw on Fig. 2.5a a straight line corresponding to the secant modulus E_s equal to some selected value such as 0.7E. The *secant yield stress* is at the intersection of the secant line and the stress-strain curve, and it deter-

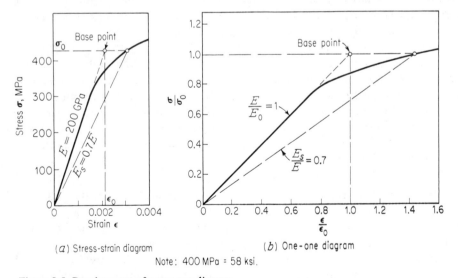

(*a*) Stress-strain diagram (*b*) One-one diagram

Note: 400 MPa ≃ 58 ksi.

Figure 2.5 Development of a one-one diagram.

mines σ_0. The base strain is the elastic strain corresponding to the base stress σ_0, or $\epsilon_0 = \sigma_0/E$. It can be determined as shown in Fig. 2.5a.

The base value of the elastic modulus (E_0) is E. Therefore E/E_0 will always be 1.0 for a one-one diagram. It is also evident that the secant yield stress corresponds to a value of $\epsilon/\epsilon_0 = 1/0.7 = 1.43$. After the location of the two common points $(1,1)$ and $(1,1.43)$ has been determined, all one-one diagrams are alike except for their shape.

It has been shown [77] that

$$\frac{\epsilon}{\epsilon_0} = \frac{\sigma}{\sigma_0} + \frac{3}{7}\left(\frac{\sigma}{\sigma_0}\right)^n \tag{2.3}$$

This is the Ramberg-Osgood equation, which simplifies the stress-strain diagrams so that the relation between ϵ/ϵ_0 and σ/σ_0 for a variety of materials is expressed solely by the exponent n, n being a function of the shape of the diagram. The equation is valuable in comparative studies of materials, since it permits the studies to be conducted on a more generalized basis.

The tangent-modulus ratio in dimensionless form is

$$\frac{E_t}{E} = \frac{1}{1 + \frac{3}{7}n(\sigma/\sigma_0)^{n-1}} \tag{2.4}$$

This ratio is useful in problems involving inelastic buckling. A method for determining the exponent n in the above equations from any stress-strain diagram has been developed [77]. Values of n for a few metals are shown in Ref. 81.

2.3 ELASTICITY

Elasticity is the property of a material by virtue of which deformations caused by stress disappear on removal of the stress. Some substances, such as gases, possess elasticity of volume only, but solids may possess in addition elasticity of form or shape. A perfectly elastic body is conceived to be one that completely recovers its original shape and dimensions after release of stress. No materials are known that are perfectly elastic throughout the entire range of stress up to rupture, although some materials, such as steel, appear to be elastic over a considerable range of stress. Some materials, such as cast iron, concrete, and certain nonferrous metals, are imperfectly elastic even at relatively low stresses, but the magnitude of permanent set under short-time loading is small, so that for practical purposes the material can be considered elastic up to reasonable magnitudes of stress.

Elastic action If a tensile load within the elastic range is applied, the elastic axial strains result from a separation of the atoms or molecules in the direction of the loading. At the same time they move closer together in the transverse direction. For a relatively isotropic material, such as steel, the stress-strain characteristics are closely similar irrespective of the direction of loading (due to the random arrangement of the many crystals of which the material is composed), but for anisotropic materials, such as wood, these properties vary with the direction of loading.

The degree to which elastic action is exhibited is often a function of the test conditions. Some materials, which are imperfectly elastic under virgin loading, appear to become elastic after having been prestressed, and overstressing in some metals appears to raise the limit of elastic action; thus, previous strain history has something to do with defining the limits of elastic action. The range of elastic action, which may be relatively great for some materials at normal temperatures, is usually reduced with increasing temperature. Also, the rapidity and/or duration of loading affect the apparent elasticity of some materials; for example, with wood and concrete, a load from which practically perfect recovery takes place in a short-time test may produce considerable permanent set if long sustained.

In accordance with the concept of elastic behavior as defined above, a quantitative measure of the elasticity of a material might logically be expressed as the extent to which the material can be deformed within the limit of elastic action. However, since engineers usually think in terms of stress rather than of strain, a practical index of elasticity is the stress that marks the (effective) limit of elastic behavior.

Elastic behavior is sometimes improperly associated with two other phenomena: the linear proportionality of stress to strain and the nonpermanent absorption of energy during cyclic variation in stress. These two phenomena are not necessarily criteria of the property of elasticity and are actually independent of it. Soft vulcanized rubber, for example, is elastic but does not exhibit a straight-line relation between stress and strain, even well within the limits of elastic action, and exhibits *hysteresis*[†] during a cycle of loading and unloading. The effect of permanent energy absorption under cyclic stress within the elastic range, called elastic hysteresis or frictional damping, is illustrated by the decay in amplitude of free vibrations of an elastic spring. It is a fact, however, that the stresses at which nonlinear stress-strain behavior or actual yielding begins are useful in designating practical limits of elastic action of the common constructional materials.

Yielding In tests of materials under uniaxial loading, three criteria of elastic strength or elastic failure have been used: the elastic limit, the proportional limit, and the yield strength.

The *elastic limit* is defined as the greatest stress a material is capable of developing without a permanent set remaining on complete release of stress. To determine the elastic limit within the strict interpretation of this concept would require successive application and release of greater and greater loads until a load is found at which permanent set is produced. The determination of the elastic limit in this sense is too arduous to be practical and therefore is rarely made.

The *proportional limit* is defined as the greatest stress that a material is capable of developing without a deviation from straight-line proportionality between stress and strain. Most materials exhibit this linear relation between stress and strain within the elastic range, and the values of the elastic limit for metals found by means of observations of permanent set do not differ greatly from the values of the proportional limit. The proportional limit is determined by use of a stress-strain diagram. The operations

[†]Hysteresis is discussed further in Art. 2.6.

are relatively simple; hence the proportional limit has been used as a measure of the elastic limit, and the terms have often been misused, one for the other. The proportional limit is sometimes called the "proportional elastic limit." The determination of the proportional limit can be subject to considerable imprecision because of the difficulty of detecting when the stress-strain curve departs from straight-line proportionality. Thus, for many practical purposes, a measure of elastic strength called the *yield strength* is used.

The yield strength, which uses as a criterion a specified degree of plastic yielding, may be a somewhat arbitrary measure, but it is more easily determined. The diagrams in Fig. 2.6 illustrate the idea. Figure 2.6*a* shows a hypothetical stress-strain diagram for a material loaded to a stress (YS) somewhat above the proportional limit (PL) and then unloaded. The distance *CA* represents the deviation or offset from Hooke's law at the stress (YS). The set after release of load is indicated as the strain *a* on the diagram. For several materials the ratio of stress to strain during release of load, from a stress somewhat above the proportional limit, is constant and closely approximates the ratio of stress to strain within the elastic range. That is, the line *AB* in Fig. 2.6*a* practically parallels the line *OC*, and the offset *CA* approximates the permanent set *OB*. The offset therefore approximates the inelastic deformation at a given stress. Although other methods have been used as the basis for determining yield strength, this concept forms the basis of the *offset method*, in accordance with the definition of yield strength given by ASTM. ASTM defines yield strength as the stress at which a material exhibits a specified limiting permanent set (ASTM E 6).

In Fig. 2.6*b*, *OX* represents a portion of the stress-strain diagram for a material that does not exhibit a marked yield at any particular stress but yields gradually after the proportional limit is exceeded. This may be considered to be the general case. Point *B* is marked on the strain axis at a distance *a*, the specified offset, from the intersection *O* of the stress-strain curve with the strain axis. The line *BA* is drawn parallel to the initial straight-line portion of the stress-strain diagram to intersect the curve at *A*, thus determining the yield strength (YS) as defined by the offset method.

The precision of the determination of the yield strength becomes less exact as the magnitude of the offset decreases. Therefore, too small a value of the offset should not be specified. Typical values for constructional materials range from 0.0001 to 0.001.

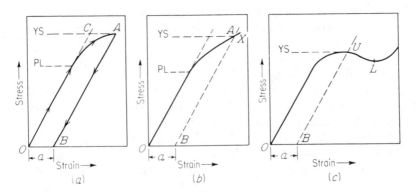

Figure 2.6 Determination of yield strength.

The term *proof stress*, found in British specifications, is very similar to the term yield strength as determined by the offset method.

The elastic limit and the proportional limit may be considered as special values of the yield strength, for which the limiting set is zero.

With a ductile material it is possible, if the test is conducted in an appropriate manner, to distinguish between two critical points in the yield range, the *upper yield point* (U) and the *lower yield point* (L), as shown in Fig. 2.6c. The upper yield point is the one usually reported, but it is highly sensitive to the rate of loading, and it appears that the lower yield point is of more real significance, as far as the fundamental properties of the material are concerned [86].

Another term frequently used, especially by structural engineers, is *yield stress*, which is understood to be the stress level of highly ductile materials, such as structural steels, at which large strains take place without further increase in stress. The yield stress is represented by the horizontal portion of a stress-strain curve and forms the basis of plastic analysis and design methods.

2.4 PLASTICITY

Plasticity is the property that enables material to undergo permanent deformation without rupture. A general expression of plastic action would involve the time rate of strain, since in the plastic state materials can deform under constant sustained stress; it would also involve the concept of limit of deformation before rupture. Evidence of plastic action in structural materials is called *yield, plastic flow,* or *creep*.

Plastic action Plastic strains are caused by slips induced by shear stresses, as illustrated in idealized form in Fig. 2.7. Such strains can occur in all materials at high stresses, even at normal temperatures; in Chap. 15 it is shown that plastic strains can also take place in materials at relatively low stresses provided ample time is allowed and favorably high temperatures are provided. Many metals show a *strain-hardening* effect when undergoing plastic deformations, because after minor shear slips have occurred they exhibit no further plastic strains until higher stresses are applied. There appears to be no appreciable change in volume as a result of plastic strains.

Although the maximum shear stress for tensile loading occurs on a 45° plane, the slip in a particular metallic crystal does not necessarily take place along that plane, as to do so would require a major rearrangement of the atoms. The overall plastic strain in a material depends on (1) the number of slip planes involved, which in turn depends on the atomic arrangement; (2) the overall effects of crystal orientation; and (3) the intensity of the shear stress.

Since slip does not involve any appreciable change in the spacing of the atoms, there is no tendency for them to return to their original position after the shear stress is removed. This indicates that plastic strains are irreversible and explains why they are permanent. The fact that the stress-strain diagram during unloading is commonly a straight line is also explained by the irreversible character of plastic strains.

Tests show that the yield stress is increased by a prior loading into the plastic range and is an indication of the strain-hardening effect. This results from the fact that

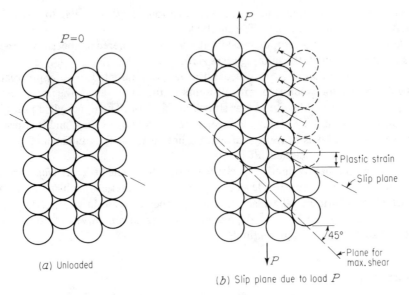

Figure 2.7 Plastic strain and slip plane.

any plastic strains produced by a prior tensile (or compressive) loading remain unchanged until a greater load of similar type is applied.

When a material is stressed into the plastic range, some individual crystals will undergo a permanent set while adjacent crystals, more favorably oriented, may deform only elastically. This will result in some *residual stresses* in the individual crystals of the highly stressed material.

Plasticity is of importance in forming, shaping, and extruding operations. This is discussed in Part Three.

Of particular importance in connection with mechanical testing is one manifestation of plasticity: ductility. *Ductility* is the property of a material that enables it to be drawn out to a considerable extent before rupture and at the same time to sustain an appreciable load. Mild steel is a ductile material. A nonductile material is said to be *brittle*; that is, it fractures with relatively little or no elongation. Cast iron and concrete are brittle materials. Usually the tensile strength of brittle materials is only a fraction of their compressive strength. The usual measures of ductility are the percentage elongation and the reduction of area in the tension test; the determination of these factors is discussed in Chap. 8 Ductility is also sometimes determined by a cold-bend test, which is discussed in Chap. 11.

Maximum strength   The term *ultimate strength* has to do with the maximum stress a material can develop. Ultimate strengths are computed on the basis of the maximum load carried by the test piece and the *original* cross-sectional dimensions; these may be referred to as the "nominal strengths." Ultimate strengths are usually stated in terms of the kind of stress producing the failure.

The *tensile strength* is the maximum tensile stress that a material is capable of developing, and in practice is the maximum stress developed by a specimen of the material during the course of loading to rupture. Figure 2.8*a* shows diagrammatically stress-strain relations for a ductile metal loaded to failure in tension. The solid line, as the diagram is customarily drawn, is based on the original cross-sectional area. The ultimate strength (US) is the stress at the highest point (A) on this diagram. Beyond that point, as the specimen contracts markedly or "necks down" to final rupture, the applied load decreases as a result of the decreasing resisting area, and if the test is conducted carefully, the nominal stress at failure (B) can be obtained. The stress at failure is sometimes called the "breaking stress" or "rupture stress."

The dashed line in Fig. 2.8a represents the true stress–conventional strain relation such as might be obtained if the load at any stage of loading were divided by the actual cross-sectional area, which decreases under tensile loading. The stress so obtained is sometimes called the "true" stress, although it is unlikely that it is the actual stress on the critical section in the range *C-D* on the diagram because the drawing of metal undoubtedly causes a complex stress distribution to develop. The characteristic form of the stress-strain diagram for a nonductile metal tested in tension is also shown in Fig. 2.8a. For a material of this type the breaking strength coincides with the ultimate strength (US).

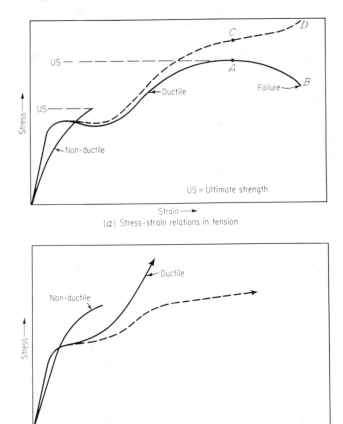

(a) Stress-strain relations in tension

(b) Stress-strain relations in compression

Figure 2.8 Stress-strain diagrams for ductile and nonductile materials.

The *compressive strength* is the maximum compressive stress that a material is capable of developing. With a (brittle) material that fails in compression by rupturing, the compressive strength has a definite value. In the case of ductile, malleable, or semi-viscous materials (which do not fail in compression by a shattering fracture), the value obtained for compressive strength is an arbitrary value dependent on the degree of distortion that is regarded as effective failure of the material. Figure 2.8*b* illustrates characteristic stress-strain diagrams for ductile and nonductile materials in compression, the dashed line again showing the true stress—conventional strain relation; in compression it is lower than the conventional stress-strain diagram owing to the increase in cross section of the specimen while under compressive loading.

The measures of ultimate strength in flexure and torsion, the *moduli of rupture* in flexure and torsion, are computed on the basis of certain assumptions that are discussed in Chaps. 10 and 11.

Hardness, which is a measure of the resistance to surface indentation or abrasion, may in general be thought of as a function of the stress required to produce some specified type of surface failure. In one type of test (the Brinell) a value of stress per unit area of indentation is computed when a ball indenter is pressed into the material under a given load. For most hardness tests, however, inasmuch as the stress conditions are complicated and cannot be evaluated, the hardness is simply expressed in terms of some arbitrary value, such as the scale reading of the particular instrument used. Details are discussed in Chap. 12.

Under repeated loads, failure may occur because of fatigue. The capacity of a material to withstand repeated application of stress has been termed its *endurance*. The *fatigue strength* is the maximum stress that can be applied a given number of times without causing failure. As the number of stress cycles becomes very large, the fatigue strength tends to reach a limiting value, called the *fatigue limit*. For ordinary steels the fatigue limit under reversed flexure is roughly half the static tensile strength. Methods of fatigue testing are discussed in Chap. 14.

2.5 FAILURE

When conducting tests to failure of materials and of structural parts or members, it is important to observe and to record the type of failure and the characteristic of the fracture. This observation should include not only the phenomena associated with final rupture but also all evidence of change of condition such as yield, slip, scaling, necking down, localized crack development, and so on. Although observations of failure are necessarily qualitative, much can be learned from a study of failures, and with experience it is possible to recognize from a break the kind of stress that caused failure and something about the kind and condition of the material. In this connection, it is important to be alert in order to discover the presence of flaws and defects, for premature failure is frequently caused by defects. Defects may be cause for invalidating the test; also, with any material it is desirable to form an estimate of the frequency of occurrence of defects. Further, distinction should be made between failure of the *material* and failure of the test piece owing to conditions inherent in the setup, such as instability.

The characteristic fractures under the various kinds of loading are discussed in detail in the chapters on particular types of tests or materials. Our interest in this article is in laying a basis for understanding some of the processes that lead to failure, as well as the nature of failure as a phenomenon of materials under load. We are, then, as much concerned with the nature of *resistance* to deformation as with the conditions under which an endpoint or instability is reached.

While complete rupture or disintegration may provide an obvious demonstration that some endpoint has been reached in the course of subjecting a material or a structure to some treatment, the destruction of the material is only one aspect of the question, What constitutes failure? In a sense, it is only a superficial aspect.

With respect to the performance of a material per se, an alteration in characteristic behavior governed by some basic physical property may be said to constitute failure. For example, if a material is stressed (or strained) beyond the elastic limit (that is, it does not recover its original shape or length on release of load), it may be said to have reached elastic failure, but this does not mean that the ability of the material to exhibit some elastic recovery has been obliterated. Rather, it means that relative to the material's initial condition a nonelastic or permanent strain has been induced. In a sense, then, the concept of failure is elusive.

Failure can occur in three fundamental ways: by slip or flow, by pulling apart or separation, or by buckling. A combination or succession of these actions may take place during the course of loading some test pieces or structures to final rupture.

Slip or *flow* occurs under the action of shearing stresses (see Art. 2.4). Essentially parallel planes within an element of a material move (slip or slide) in parallel directions. Continuous action in this manner, at constant volume and without disintegration of the material, is termed creep, or plastic flow. Slip may be terminated by rupture when the molecular forces are overcome. Tensile or compressive loadings, as well as direct shear, torsional, or flexural loadings, may induce states of stress in which the shearing stresses cause slip. Thus, in a simple tension test, failure may occur due to slip, which is a shear-induced action. Slip and its endpoint, shear fracture, may take place through intercrystalline material or within the crystal lattice along certain critical planes. The planes susceptible to slip depend on the arrangement of the atoms in the lattice.

Separation is an action induced by tensile stresses. It takes place when the stress normal to a plane exceeds the internal forces that bind the material together and is often called cleavage fracture. States of stress that involve tensile stresses sufficiently critical to cause cleavage fracture may be induced by loadings other than primary tensile loadings; for example, in torsional loading of a cylindrical piece of chalk the induced tensile stresses are responsible for the type of helicoidal fracture surface that normally is observed. Cleavage may occur at grain boundaries, through intercrystalline material, or across susceptible planes within the crystal lattice. The planes susceptible to cleavage depend on the arrangement of the atoms in the lattice.

Buckling is a compression phenomenon. It is exemplified by an end-loaded slender column; after some critical level of load the resistance of the composite mass drops and the system of forces is unstable. A buckling failure may be induced by loadings other than primary compressive loading; for example, the torsional loading of a thin-walled tube may result in buckling caused by the induced compressive stresses. Or, in a beam, failure may be initiated by the localized buckling of the wood fibers at the compression surface of the beam, as well as by lateral buckling of the compression flange as a whole.

2.6 ENERGY CAPACITY

The capacity of a material to absorb or store energy is of importance in connection with problems such as shock resistance and impact loading. The basic principle involved is that work or energy is equal to force times distance. Both are expressed in joules. In general, work done on an elastic body is stored as "strain" energy and can be recovered mechanically. Work done on a plastic body, however, is converted into nonmechanical energy, especially heat.

The amount of energy absorbed in stressing a material up to the elastic limit, or the amount of energy that can be recovered when stress is released from the elastic limit, is called the *elastic resilience*. The energy stored per unit of volume at the elastic limit is the *modulus of resilience*. For a unit volume the resilience is the product of average stress times strain:

$$\frac{\sigma}{2} \epsilon = \frac{\sigma}{2} \frac{\sigma}{E} = \frac{\sigma^2}{2E}$$

In terms of the stress-strain diagram, energy absorption is represented by the area under the diagram. In Fig. 2.9, which shows a typical diagram for mild steel, the elastic resilience is represented by area I. If the load is released from point A in the plastic range, the recovery diagram is approximately a straight line (AB) and the energy released is represented by area II; this has been called the *hyperelastic* resilience.

The modulus of resilience is a measure of what may be called the "elastic energy strength" of the material and is of importance in the selection of materials for service

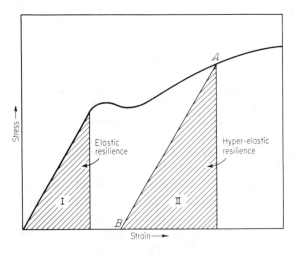

Figure 2.9 Resilience in terms of the stress-strain diagram.

where parts are subjected to energy loads but where the stresses must be kept within the elastic limit. It should be noted that for a high modulus of resilience a material should have a relatively high elastic strength, a low modulus of elasticity, or both. It is expressed in joules per cubic meter, which can easily be shown to equal pascals.[†] For various grades of steel it has values ranging from about 100 to 4500 kJ/m³. The higher the carbon or alloy content, the higher the modulus of resilience. For steel the modulus of resilience for shear is approximately equal to that for axial stress.

When a material is subject to repeated loading, during any cycle of loading and unloading, or vice versa, some energy is permanently absorbed or lost. This is true even in the elastic range, as is evidenced by the decay in the free vibrations of rods and springs. For metals such as steel it is true that the energy lost per cycle is small, but to this extent the idea that the modulus of elastic resilience represents recoverable energy is approximate. This phenomenon of lost energy is called, in general, *hysteresis*, and within the elastic range, *elastic hysteresis*. In terms of the stress-strain diagram, the hysteresis loss is represented by the area enclosed by the loop formed by consecutive segments of the diagram. In Fig. 2.10*a* a hypothetical diagram is shown for a material such as soft rubber, which is elastic but does not follow Hooke's law or release all the energy expended during loading. The shaded area represents the energy lost or the hysteresis loss. Figure 2.10*b* shows typical hysteresis loops for a material like steel when it is unloaded and reloaded to stresses above the elastic limit. The determination of hysteresis loops requires precise instrumentation. The most feasible means for determining the hysteresis loss in the elastic range is observation of the damping of vibrations.

Toughness involves the amount of energy required to rupture a material. It may be measured by the amount of work per unit volume of a material required to carry that material to failure under static loading, called the *modulus of toughness*. Using this criterion it may be represented by the area under the complete stress-strain diagram, as illustrated in Fig. 2.11.

[†]J/m³ = N · m/m³ = N/m² = Pa. In inch-pound units, lbf-in/in³ = psi. 1 kJ/M³ = 0.145 lbf-in/in.³

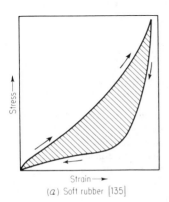

Strain ⟶
(*a*) Soft rubber [135]

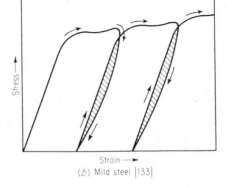

Strain ⟶
(*b*) Mild steel [133]

Figure 2.10 Hysteresis loops.

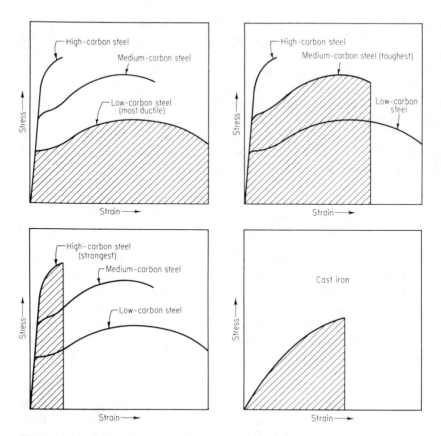

Figure 2.11 Toughness in terms of the stress-strain diagram.

Toughness is a measure of what may be called the ultimate energy strength of a material. It is important in the selection of a material for types of service where impact loads are applied that from time to time may cause stresses above the yield point. Toughness can be expressed in joules per cubic meter. For carbon steels, values range from about 35 to 120 MJ/m^3; medium carbon steels are the toughest, as may be noted from Fig. 2.11.

PROBLEM 2 Compression Tests of Small Wooden Columns

Object: To determine the relation of slenderness ratio to the ultimate unit load resisted by small pin-ended wooden columns.

Preparatory reading: Chapter on columns in textbook on mechanics of materials.

Specimen: Five or six small, clear, straight-grained wooden columns of uniform cross section, preferably about 20 by 40 mm and of varying lengths (200 mm to about 1.25 mm).

Special apparatus: Two pin-ended bearing plates.

Procedure:

1. Mount the pin-ended bearing plates on the ends of the column with the axes of the knifedges or pins parallel to the longer dimension. Measure the length of one column from center to center of

the pins and measure the cross-sectional dimensions of the column. Compute the probable maximum load by Euler's equation, assuming E to be about 10 GPa or looking it up in Chap. 20.

2. Center the end fixtures on the column and center the column in the testing machine. Apply load slowly and note the maximum load when a slight lateral deflection occurs. Release the load, shift both end fixtures slightly toward the convex side of the column as developed during the previous loading, and reload. If the buckling load is not higher than before, take the first value as the maximum load. If the second load is higher than the first, the column was not properly centered during the first loading. Repeat these operations until by trial the maximum buckling load is obtained.

3. Repeat the first two steps for each of the other columns.

4. Compute Young's modulus by substituting the maximum load carried by the longest column in Euler's equation for pin-ended columns. For each column compute the theoretical value of P/A. Specify the method of failure (buckling or crushing) for each column.

5. Plot a graph of the theoretical values of P/A against the slenderness ratio for each column. On the same sheet and using the same origin, plot a graph of the actual values of P/A against the slenderness ratio.

6. Determine the equation of a curve that is tangent to Euler's curve and that passes through the point corresponding to the ultimate strength at a slenderness ratio of zero.

Report: Submit an informal report that describes the test and compares the results with the theory. Discuss the following questions:

1. From your tests, what is the minimum slenderness ratio for which Euler's equation appears to be applicable? Is a straight-line equation satisfactory for lower values?

2. What effects do the end conditions of Euler columns have on the buckling load?

3. What property of a material determines the load that (a) a short compressive member and (b) a slender compressive member may carry?

DISCUSSION

1. A warning sign states a limit for the heaviest truck that may cross a bridge. Should the limit be given in kilograms or in newtons? Argue the case.

2. Distinguish among various indicators of elastic strength.

3. Distinguish among different moduli of elasticity.

4. Characterize the metal used in ordinary wire clothes hangers as to its resilience and toughness.

5. Speculate what the stress-strain curves for rubber, chalk, and glass might look like, and sketch them on one graph.

6. What are some possible failure criteria of a tennis racket?

7. What are some important properties of the metal used in a beam?

8. What characteristics would you specify for a material to be used to make saw blades?

9. Would you prefer a highly elastic or a highly plastic material to make a crash barrier? Explain in terms of energy.

10. Consider that you are conducting a series of flexure tests with "three-point" loading on beams of ductile material. Show how nondimensionalized curves can be drawn that plot load vs. deflection at the center. Assume that the specimens are identical except for the (constant) cross section.

READING

At this time it would be an excellent idea to review your mechanics of materials text and notes: you will need most of the subject matter during this course. If you have taken a materials science course, it will also be useful to refresh your memory on the microscopic structure of materials. In fact, it might be well to review your physics text—especially the parts on mechanics, optics, electricity, and magnetism.

Brady, G. S., and H. R. Clauser: *Materials Handbook*, 11th ed., McGraw-Hill, New York, 1977.

Brick, R. M., A. W. Pense, and R. B. Gordon: *Structural Properties of Engineering Materials*, 4th ed., McGraw-Hill, New York, 1977.

Eisenstadt, M. M.: *Introduction to the Mechanical Properties of Materials*, Macmillan, New York, 1971.

Gordon, J. E.: *The New Science of Strong Materials*, 2d ed., Pitman, Marshfield, Mass., 1979.

Guy, A. G.: *Introduction to Materials Science*, McGraw-Hill, New York, 1972.

Hanks, R. W.: *Materials Engineering Science*, Harcourt, Brace & World, New York, 1970.

Keyser, C. A.: *Materials Science in Engineering*, 3d ed., Merrill, Columbus, Ohio, 1980.

Popov, E. P.: *Mechanics of Materials (SI version)*, 2d ed., Prentice-Hall, Englewood Cliffs, N.J., 1978.

Shanley, F. R.: *Mechanics of Materials*, McGraw-Hill, New York, 1967.

Tottle, C. R.: *The Science of Engineering Materials*, Heinemann, London, 1966.

Van Vlack, L. H.: *Elements of Materials Science and Engineering*, 3d ed., Addison-Wesley, Reading, Mass., 1975.

THREE

ANALYSIS OF DATA

3.1 THE MANAGEMENT OF DATA

To be of value, data must be in a form that can be readily understood and easily used. The particular form in which the data should be summarized and the extent to which they should be interpreted obviously will depend on the intended purpose. In many cases, several commonly used procedures for analyzing and reporting data can be applied.

The tasks encountered in connection with the analysis of data include the following:

1. *Reduction of raw data.* The units in which the data are recorded are dictated by the measurement methods. Thus loads may be measured in kilonewtons and deformations in millimeters. Because most data have meaning only in *comparison* with similar data, the quantitative measures obtained from a test are reduced to values whose units are acceptable as a basis for comparison; therefore loads are reduced to stresses, say, in megapascals, and deformations to strains. The reliability of the data is conditioned by the errors of measurement. In reducing the data, corrections may have to be applied for systematic errors, and in order to express the final data in the appropriate number of significant figures, an estimate should be made of the accidental errors inherent in the measurements and of the effect of these errors on the accuracy of the reduced values. Errors are discussed more fully in Art. 5.6.
2. *Summary of data.* Although in the simple case the summarizing of test results may merely amount to setting down a few readily comprehensible facts and figures, in connection with large-scale operations there may be accumulated masses of data that are so numerous and variable it is practically impossible for the mind to digest and evaluate them in unassembled form. Advantage may then be taken of statistical procedures for summarizing the data.

3. *Study of relations between variables.* After the data have been reduced and assembled in manageable form, the final step in the analysis is usually to seek or to develop relations between the variables involved or between the data obtained from a particular test and previously obtained data or some theory. The skill with which this is done depends on the capacity, the ingenuity, and the background of the analyst. Common devices employed in studying such relations are tabulations, graphs, bar charts, and correlation diagrams; the procedure is usually to hold constant (insofar as is known or is possible) all variables except two, whose relation is investigated.

We can differentiate between two types of statistical methods. *Descriptive methods* help us present data in a comprehensible form. *Inference methods* help us generalize from the properties of a limited *sample* to those of the whole *population*, thus making testing more efficient.

A few statistical concepts useful in handling data from testing work are summarized in this chapter. Terms relating to statistics are defined by ASTM E 456, and examples of standards that are based on statistical methods include ASTM D 2915, which relates to grading of lumber, and ASTM E 206, which deals with fatigue testing.

3.2 VARIATIONS IN DATA

Practically all data derived from tests are subject to variation. The results of a test on a single specimen involve measurements that are subject to variation, and the results of a test on a series of similar specimens show variation among specimens. After the measurements have been corrected for the effects of systematic errors, it is usually found that the variations in adjusted or corrected measurements follow a chance distribution. For large numbers of data, variations in measurements and measures of properties have been found to coincide closely with variations computed from theoretical considerations. When the data are few, the coincidence is often not so good, but for convenience the concepts developed from the theory of probability (for many numbers) are applied and afford a fairly workable means of summarizing and utilizing data.

To illustrate, let us assume that for a sample of 80 sheets of galvanized iron the amount the coating is to be determined. The sheets are weighed before and after the galvanizing process and the resulting net masses are recorded in order. The results are listed in Table 3.1. It should be remarked that all data are assumed to be equally valid; any obviously faulty observations should have been discarded (see ASTM E 178). The results listed in this random order of testing are called *ungrouped data.* With some difficulty we can determine their general range and distribution, and we notice that some numbers appear more often than others. Beyond that it is hard to analyze the values.

An improvement is made by arranging the items according to magnitude, usually in ascending order. This is referred to as an array or an *ungrouped frequency distribution.* Merely by inspection, the minimum and the maximum values can be selected, and by a simple computation, the median or middle value and the range can be deter-

Table 3.1 Mass of coating on 80 galvanized iron sheets

Test	mass, g	Test	mass, g	Test	mass, g	Test	mass, g
1	38.81	21	37.63	41	40.56	61	38.67
2	40.02	22	42.61	42	43.14	62	41.06
3	42.03	23	45.18	43	41.62	63	42.41
4	43.06	24	39.39	44	44.17	64	42.88
5	41.58	25	42.84	45	42.34	65	42.98
6	42.58	26	43.24	46	44.63	66	40.12
7	42.80	27	43.70	47	45.01	67	43.03
8	43.34	28	45.44	48	42.93	68	43.40
9	43.72	29	41.25	49	46.63	69	41.80
10	38.12	30	43.64	50	39.61	70	44.50
11	40.63	31	44.06	51	46.21	71	42.14
12	44.28	32	44.33	52	40.38	72	39.51
13	44.45	33	42.45	53	47.22	73	44.47
14	43.21	34	44.90	54	41.36	74	41.44
15	43.79	35	44.51	55	44.77	75	43.91
16	43.55	36	41.74	56	45.61	76	42.73
17	42.26	37	45.30	57	39.84	77	42.19
18	43.03	38	43.84	58	40.88	78	43.48
19	41.29	39	42.67	59	38.98	79	41.13
20	46.52	40	46.89	60	41.92	80	40.58

mined. Although data in this form are still unwieldy, it is possible to study the array by dividing it into equal parts, such as quartiles (four parts), deciles (10 parts), or percentiles (100 parts), just as the median divides it into two parts. Table 3.2 shows the previous data in this form. Note that each of the columns in the table represents one quartile. We will use these data to illustrate some of the following discussions as well.

3.3 GROUPED FREQUENCY

In order to make the data more comprehensible, we must analyze them further so that the results can be presented in a condensed tabular or graphical form.

Grouping of data The first step in such an analysis involves grouping of data. When the time of occurrence is important, a chronological sequence is sometimes used and the data are presented as a time series. For example, a table (or graph) might show the amount of concrete placed on a project each day; the placement is a more or less continuous function, but it is grouped into amounts for each 24-h period. In materials testing, observations involving time of occurrence, such as determinations of plastic flow (creep) and determinations of progressive deterioration of materials subjected to alternate cycles of freezing and thawing, are sometimes presented as time series. Some data, such as results of test borings, may require geographical grouping.

Most data in materials testing are grouped according to *magnitude*. The symmetrical arrangement of data according to magnitude results in what is known as *frequency-*

Table 3.2 Ungrouped frequency table

Rank	value	Rank	value	Rank	value	Rank	value
1	37.63	21	41.36	41	42.88	61	44.17
2	38.12	22	41.44	42	42.93	62	44.28
3	38.67	23	41.58	43	42.98	63	44.33
4	38.81	24	41.62	44	43.03	64	44.45
5	38.98	25	41.74	45	43.03	65	44.47
6	39.39	26	41.80	46	43.06	66	44.50
7	39.51	27	41.92	47	43.14	67	44.51
8	39.61	28	42.03	48	43.21	68	44.63
9	39.84	29	42.14	49	43.24	69	44.77
10	40.02	30	42.19	50	43.34	70	44.90
11	40.12	31	42.26	51	43.40	71	45.01
12	40.38	32	42.34	52	43.48	72	45.18
13	40.56	33	42.41	53	43.55	73	45.30
14	40.58	34	42.45	54	43.64	74	45.44
15	40.63	35	42.58	55	43.70	75	45.61
16	40.88	36	42.61	56	43.72	76	46.21
17	41.06	37	42.67	57	43.79	77	46.52
18	41.13	38	42.73	58	43.84	78	46.63
19	41.25	39	42.80	59	43.91	79	46.89
20	41.29	40	42.84	60	44.06	80	47.22

distribution series. Sometimes these series are considered as being divided into two types, one consisting of *one* measurement of a given characteristic of each of *n* different pieces, the other consisting of *n* measurements of a given characteristic of a *single* piece. An example of the first type is some quality of the material itself, such as the tensile strength of wire in a given coil; an example of the second type is some statistic of a single piece, such as its diameter. Both types consist of homogeneous data. They come from a common parent population, and each is not a single-valued constant but rather a frequency-distribution function.

On first thought the diameter of a tensile specimen might not appear to be a distribution function, that is, a *variate*, but it should be remembered that the surfaces of test specimens cannot be given a costly ground and polished finish, and so finer measurements would tend to emphasize the variability due to a combination of indeterminate causes. This means each item is subject to accidental or chance variations that are composed of the inherent variability of the characteristic itself. In addition, errors of measurement are inevitable.

In the testing of materials the accuracy of machines, strainometers, gages, and other measuring devices is usually maintained between known limits, and the variations in measurements due to parallax, lost motion, and inertial effects are minimized by proper design and use. Furthermore, these limits (errors) are small compared with the usual variations in property measurements from specimen to specimen. When the errors become larger and systematic, as may be caused by temperature changes, either computed corrections are directly applied or a procedure is used to eliminate them. Fortunately, data on materials testing are predominately affected by the variability of the characteristic observed.

It is often useful to group the data according to subdivisions based on the variable being measured. These groups are called *cells*, or *class* or *step intervals*. After the length of the interval has been decided on, the number of items in each interval, called the *class frequency*, is determined. Often the *relative frequency*, which is the number in each interval divided by the total number of items, is used. This fraction of the total number is an important characteristic, especially when applied to the percentage below, outside, or beyond a specified limit, thus giving the "fraction defective." A rule should be established so that each item will lie within some interval and not indefinitely on a boundary line. The resulting grouped frequency table for the previous example of galvanized iron sheets is shown in Table 3.3. When there are a large number of items, 13 to 20 class intervals are recommended [117]. Too many intervals may give an irregular distribution. By using larger intervals (few divisions) the appearance of the distribution can often be improved. In this case we have chosen 10 class intervals. When the total number of items is less than 25, such a presentation is of little value.

Sometimes it is of interest to know the number of data that fall below (or above) a certain value. For this reason the *cumulative frequency* may be shown or the *fractional* or *relative cumulative frequency*. To obtain the relative cumulative frequency, the cumulative frequency is divided by the total number of data. For our example, these values are shown in Table 3.3. For a large number of data, of course, such tables are best generated by computer.

Table 3.3 also shows the product of frequency and midpoint for each class. Article 3.4 shows how the sum of these values can be used to find the approximate arithmetic mean.

Frequency diagrams Graphical representations usually help us visualize the nature of data. For graphs a cartesian coordinate system is established on which the x axis shows the variable studied. The frequencies, actual or relative, are plotted as ordinates to an arithmetical scale on the center line of each interval, and vertical bars equal in width

Table 3.3 Grouped frequency table

Mass of coatings, g		Class frequency	Relative frequency	Cumulative frequency	Relative cumulative frequency	Frequency × midpoint
Class interval†	Midpoint					
37.5–38.5	38	2	0.025	2	0.025	76
38.5–39.5	39	4	0.050	6	0.075	156
39.5–40.5	40	6	0.075	12	0.150	240
40.5–41.5	41	10	0.125	22	0.275	410
41.5–42.5	42	12	0.150	34	0.425	504
42.5–43.5	43	18	0.225	52	0.650	774
43.5–44.5	44	14	0.175	66	0.825	616
44.5–45.5	45	8	0.100	74	0.925	360
45.5–46.5	46	2	0.025	76	0.950	92
46.5–47.5	47	4	0.050	80	1.000	188
Total		80	1.000			3416

†Boundary values assigned to lower class.

to the class interval and in height to the frequency are drawn. A frequency bar chart is created, from which the vertical lines separating the bars may be omitted. Either way the diagram is called a *frequency histogram*, which is illustrated in Fig. 3.1*a*.

Instead of drawing bars, we may connect successive points by straight lines, creating a *frequency polygon*. The straight lines of the polygon are, of course, meaningless as ordinates: they do not show intermediate values. The concept is particularly useful when several frequency distributions are to be shown on the same diagram for comparison. Figure 3.1*b* shows a frequency polygon for our example.

The distribution commonly encountered in materials testing is a bell-shaped curve resembling the theoretical normal frequency curve. The normal frequency curve apparently was first used by Demoivre in 1733, but it is known as a probability curve, curve of error, gaussian error curve, Laplace-gaussian curve, or *normal distribution curve*. It is given by the function $y = ke^{-h^2x^2}$, where y is the probability of an error x occur-

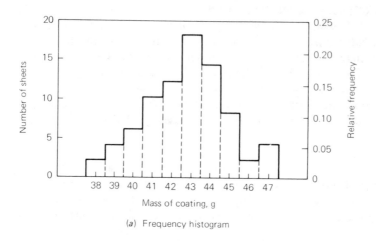

(a) Frequency histogram

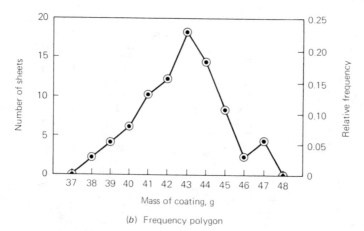

(b) Frequency polygon

Figure 3.1 Frequency graphs.

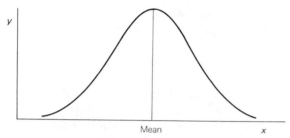

Figure 3.2 Normal distribution curve.

ring and *h* and *k* are constants that determine the spread and height of the curve, respectively. Figure 3.2 shows such a curve.

If the distribution curve has only one peak, it is said to be unimodal. Many times it is asymmetrical and is skewed to the left or right. Nonhomogeneous data, say from two different parent populations, might produce a bimodal distribution, but homogeneous data take other characteristic shapes such as J- or L-shaped curves, which are extremely asymmetrical, or possibly a U-shaped curve, which could be symmetrical in that it has a point of minimum frequency between higher frequencies at both ends of the range.

This discussion is restricted to unimodal frequency distributions, which are characteristic of data on the physical and mechanical properties of materials. The general recognition that the true magnitude of a property or quality of a material is a distribution function, thus involving the concept of probability (which in itself is considered one of the most important concepts in the field of statistics), is the modern viewpoint.

Cumulative frequencies, both actual and relative (or percentage), can also be plotted. The diagram, like the table, could be labeled to read number or percentage of items greater than or less than marked values. The variable under consideration is usually plotted on the *x* axis, and when both the *x* axis and the *y* axis are arithmetic,

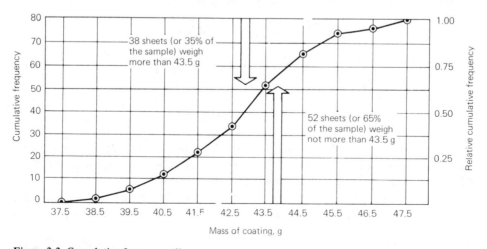

Figure 3.3 Cumulative frequency diagram.

the cumulative distribution takes a peculiar form commonly called an *ogiv* (or *ogee*) curve. This curve becomes a straight line for a normal distribution when the ordinates are plotted to a so-called probability scale.

In the case of grouped frequencies the lower or upper boundary, as appropriate, of the intervals is marked off on the *x* axis.

Figure 3.3 shows a plot of our example data. Note that the cumulative frequencies are plotted vs. the upper boundaries of the class intervals, not vs. the midpoints.

Since frequency diagrams are merely plots of the grouped frequency tabulations, they cannot yield more information than the tabulations themselves, but it is advantageous to show the relations in graphical form.

3.4 DISTRIBUTION CHARACTERISTICS

Often it is particularly helpful if a mass of data can be fairly represented by a few characteristic values. This is the idea behind averages and deviations, or measures of central tendency and dispersion.

Central tendency A measure of central tendency, or the tendency to be grouped about a central value, is called an *average*, which purports to summarize the data by locating this typical value. The most significant averages are the arithmetic mean, the median, and the mode. Two other averages, the geometric mean and the harmonic mean, are not often used in materials testing.

The *arithmetic mean* is the most widely used of all criteria of central tendency. It is the "average" used in everyday speech. The customary symbol for the arithmetic mean is $\overline{X}$, and the formula is

$$\overline{X} = \frac{1}{n} \sum_{i=1}^{n} X_i = \frac{X_1 + X_2 + \cdots + X_n}{n} \tag{3.1}$$

For the five items 2, 3, 5, 8, and 9, the arithmetic mean is

$$\overline{X} = \frac{2 + 3 + 5 + 8 + 9}{5} = \frac{27}{5} = 5.40$$

When the frequency f of various discrete observations is greater than unity, the calculation can be simplified to

$$\overline{X} = \frac{\Sigma f_i X_i}{\Sigma f_i} \tag{3.2}$$

This may even be done for grouped data such as those in Table 3.3, with only a slight loss in accuracy. For our example, the mean calculated in this manner is 3416/80 = 42.70 g, whereas the precise mean, based on Table 3.1, is 42.69 g.

The arithmetic mean is a calculated average (in contrast to the median) and is affected by every item, but it is greatly distorted by unusually large values at the extremes.

The *median* is the value of the middle item in an array. In a distribution where the central values are closely grouped, the median is typical of the data since it is not affected by unusual terminal values. It is not as well known, however, as the arithmetic mean. For the previously given array (2, 3, 5, 8, 9), the median is 5. When the number of observations or classes is even, the median is set equal to the arithmetic mean of the two middle values. Thus for the data in Table 3.2 the median is (42.84 + 42.88)/2 = 42.86 g.

In much the same way that the median divides the sample into two parts, values may be found that divide the sample into four, 10, or any number of parts. We may then speak of the "lower quartile," the "upper decile," or the "15th percentile."

The *mode* applies only to grouped data and is the value that occurs most frequently. In an ideal (smooth) frequency distribution the modal value is located by the maximum or highest ordinate of frequency. When a large number of items are used, it is the most typical average. It is an average *position* and is unaffected by large items near the ends of the range. The mode may be given as the midpoint of the modal interval (class or cell) with the highest frequency. It may also be computed by some method such as moments of force or moving averages or by smoothing the frequency distribution when a limited amount of data is available. In the array 2, 3, 5, 8, 9 no items are repeated; hence there is no mode. In the example of Table 3.3, the class occurring most frequently is that with a midpoint of 43, which occurs 18 times; thus the mode is approximately 43.

The median falls between the arithmetic mean and the mode. For a symmetrical unimodal distribution curve the mean, the median, and the mode all have the same value, of course.

Dispersion   An inspection of the frequency diagram will furnish a qualitative indication of an important characteristic—the dispersion, scatter, or variation about the average value. A crude measure of this deviation from the average is the *range* or *spread*. Although it is easily determined, since it is merely the actual difference between the maximum and minimum items in the distribution, it is dependent on only two values that are of unusual occurrence (low frequency), particularly when n is large. So the range may give misleading indications of distribution.

A criterion that considers the location of every item, rather than only the two extremes, would be more meaningful. An obvious measure would be the average distance of the items from some measure of central tendency. If the arithmetic mean and plus and minus distances were used, such an average would be zero; however, this difficulty can be overcome by disregarding the signs of these distances, which are called deviations.

One measure of dispersion, then, is the *average deviation*, called the *mean deviation* when it refers to the arithmetic mean. Mathematically,

$$\text{Average deviation} = \frac{\sum_{i=1}^{n} |X_i - \overline{X}|}{n}$$

The quantity $|X_i - \overline{X}|$ is the absolute value of a deviation from some average. If the median is used, the average deviation will always be smaller than when the arithmetic mean ($\overline{X}$) is used.

The most widely employed measure of dispersion is the *standard deviation*, which is the square root of the average of the squares of the deviations of the numbers from their arithmetic mean $\overline{X}$. The standard deviation is a special arrangement of the average deviation and is usually designated by the symbol σ. The following equation gives its most general form:

$$\sigma = \sqrt{\frac{\sum\limits_{i=1}^{n} (X_i - \overline{X})^2}{n}} = \sqrt{\frac{\sum\limits_{i=1}^{n} X_i^2}{n} - \overline{X}^2} \tag{3.3}$$

A glance at the general form makes obvious the reason for the often used designation, root-mean-square deviation. When the mean of a sample, rather than the usually known mean of the entire population, is employed, it is more nearly correct to divide by the "number of degrees of freedom," usually $(n - 1)$, instead of by the total number of items n. This is known as Bessel's correction, and the estimate of σ becomes

$$\sigma \simeq s = \sqrt{\frac{\sum\limits_{i=1}^{n} (X_i - \overline{X})^2}{n - 1}} = \sqrt{\frac{\sum\limits_{i=1}^{n} X_i^2}{n - 1} - \overline{X}^2 \frac{n}{n - 1}} \tag{3.4}$$

where $\overline{X}$ is the mean of a sample. For large masses of data a computer should be used.

The calculation of the standard deviation for our example problem, using Table 3.1 or Table 3.2, yields $\sigma = 2.089$ g.

The standard deviation, since it uses second powers of all items, places more emphasis on widely dispersed items than the average deviation does. For a normal distribution the average deviation is 0.7979σ. This relation can be applied with fair approximation to slightly skewed distributions.

An important characteristic of the standard deviation is the number of items or the area included between ordinates erected on each side of the center of the distribution curve at a distance of one, two, and three standard deviations. These numbers or areas for the normal (theoretical) probability curve are 68.26, 95.44, and 99.73 percent, respectively. Even in the actual analysis of materials-testing data it is unlikely that less than 96 percent of the values for a frequency distribution of more than 100 items will be within the range of three standard deviations on each side of the arithmetic mean. Figure 3.4 illustrates the relation.

Since significant comparisons of dispersions cannot be made by using absolute measurements, a variation expressed as a ratio or percentage must be determined. The most commonly used is the standard deviation, which so expressed is usually called v, the *coefficient of variation:*

$$v = \frac{\sigma}{\overline{X}} \tag{3.5}$$

It should be noted that when $\overline{X}$ is very small or even zero, as it could be in the case of temperature measurements, if degrees Celsius rather than kelvins are used, the resulting v would be quite misleading. For our example, $v = 2.809/42.69 = 0.0489 = 4.89$ percent.

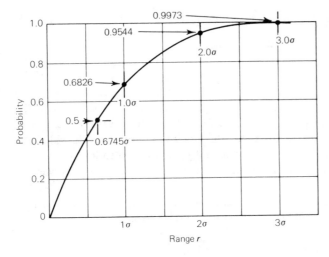

(a) Probability of observation
falling within given range

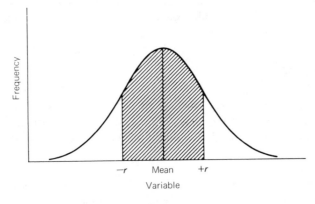

(b) Normal distribution curve
showing range of deviations

Figure 3.4 Probabilities of deviations (errors).

Skewness When a frequency distribution departs from a theoretical normal shape, it is said to be skewed or asymmetrical. Absolute and/or relative measures of this lack of symmetry can be based on the values or locations of the several measures of central tendencies, all of which are identical for symmetrical distributions. Symmetry is naturally measured about the high point of the frequency polygon—the modal value. The relative location of the arithmetic mean, since it is affected by the magnitude of extreme items, offers an excellent criterion of the *skewness factor k* in either the simple form

$$k = \frac{\text{mean} - \text{mode}}{\text{standard deviation}} \qquad (3.6)$$

or an alternate form, which gives a more exact value

$$k = \frac{\displaystyle\sum_{i=1}^{n} (X_i - \overline{X})^3}{n\sigma^3} \tag{3.7}$$

For symmetrical curves k is obviously zero. Since the scale of the variable is usually arranged so that the values increase toward the right, then for curves skewed to the right, that is, with the excess tail to the right, the larger values on that side will cause the arithmetic mean to be larger than the mode and to be located to its right with a resulting plus sign for k. Hence a negative coefficient k means the curve is skewed to the left.

Unless a sample contains more than 250 data or so, k is not significant. For our example, the skewness factor k can be estimated from Eq. (3.6) as

$$k = \frac{42.70 - 43}{2.089} = -0.144$$

or calculated as $k = -0.175$ by using Eq. (3.7), but it may not be very meaningful because there are only 80 data. The values do indicate that the curve is skewed to the left. Compare these results with Fig. 3.1, in which there are five nonzero frequencies to the left of the mode, but only four to the right.

3.5 SAMPLING AND ERRORS

The distribution characteristics pose some very practical questions in materials testing. One question concerns the nature and the size of a sample and their implications on the population. Another concerns the probability with which we can expect certain errors to occur in a test program.

Samples Samples should be taken in a random manner, so that each specimen has an equal chance of being selected every time a choice is made. Sampling may be done with or without replacement: the chosen specimen may be returned to the population before the next choice is made, or discarded. For destructive tests the latter method must be used, and it is usually more efficient in any case.

The size of the sample is important, as the mean of one sample is likely to differ from that of another. If in the example problem we had made only four observations instead of 80, we feel that we would have obtained a less accurate representation of the population, but we do not know how much less accurate. Several interesting facts can be demonstrated:

1. If we have a population of size N, then the number of possible samples of size n is $N!/[n!\,(N-n)!]$.
2. The mean of all the individual sample means equals the mean of the population.
3. If N is very large compared to n, the standard deviation of the sample means σ_s from the population mean σ_p is related to σ_p by the expression

$$\sigma_s = \frac{\sigma_p}{\sqrt{n}}$$

We call σ_s the *standard error of the mean*. If σ_p is unknown, as is usually the case, it may be estimated, for example, by using the standard deviation of the sample as an approximation.

Two significant conclusions can be drawn from these relations:

1. The distribution curve for the sample means is narrower (and hence higher) than the curve for individual observations, and increasingly so for larger samples.
2. The standard error of the mean varies inversely with the square root of the number of observations or specimens. For example, taking four times the number of specimens only doubles the precision.

In our examples of 80 galvanized sheet specimens we had calculated the mean to be 42.69 g and the standard deviation to be 2.089 g. If we assume the standard deviation of the entire population to be equal to this value, then the standard error of the mean is $2.089/\sqrt{80} = 0.234$ g. If we had chosen only four specimens, the corresponding value would be 1.045 g.

Errors Other measurements courses, such as surveying, usually provide some familiarity with the theory of errors. Here it is desirable to point out two things: (1) the relation of certain terms often used in connection with the discussion of errors to terms used in the general field of statistics, and (2) the method of determining the error of a computed quantity.

The criterion of central tendency for a series of measurements of like quantities, which differ because of accidental errors only, is taken as the arithmetic mean, which is the most probable value of the measurements. In the theory of errors, the deviations from the mean are usually referred to as the "residuals."

Commonly used criteria of dispersion are the *average error* (which is the same as the average deviation) and the *probable error*. The probable error of a single measurement in a series of measurement (of equal weight) is computed from the expression $0.6745 \sqrt{\Sigma (v^2)/(n-1)}$, where the v's are the residuals and n is the number of observations. The quantity under the square-root sign is seen to be the standard deviation for a limited number of observations. The probable error is *not* the error most likely to occur, but simply marks the limit within which the chances are 50–50 that any error taken at random will fall (see Fig. 3.4). The probable error of the mean of a series of measurements is found by dividing the probable error for a single observation by $\sqrt{n}$. The standard deviation provides just as useful and significant a measure of dispersion as the probable error and might just as well be used. For comparative purposes, the *relative error* or *precision ratio* is commonly taken as the ratio of the probable error to the value of the quantity measured; this corresponds to the coefficient of variation.

The probable error of a quantity calculated from the combination of several independently measured quantities is derived from the theory of least squares. If $u = f(x, y, z, \ldots)$, and if R_u is the probable error of a quantity u, R_x of quantity x, and so on, then

$$R_u^2 = \left(\frac{\partial u}{\partial x} Rx\right)^2 + \left(\frac{\partial u}{\partial y} R_y\right)^2 + \left(\frac{\partial u}{\partial z} R_z\right)^2 + \cdots$$

This may be applied to several specific cases as follows:

1. Product u of x and a constant C: $R_u = CR_x$
2. Area of a circle, u, from measured diameter x:

$$R_u = \sqrt{\left[\frac{\partial(\pi x^2)}{\partial x} R_x\right]^2} = \frac{\pi D}{2} R_x$$

3. Sum u of like quantities $x, y, z, \ldots$:

$$R_u = \sqrt{R_x^2 + R_y^2 + R_z^2 + \cdots}$$

If the probable error of each quantity is the same,

$$R_u = R_x \sqrt{n}$$

where n is the number of quantities

4. Modulus of elasticity E from measurements of P, the applied load; L, the gage length; A, the cross-sectional area; and e, the deformation:

$$R_E = \sqrt{\left(R_P \frac{L}{Ae}\right)^2 + \left(R_L \frac{P}{Ae}\right)^2 + \left(R_A \frac{-PL}{A^2 e}\right)^2 + \left(R_e \frac{-PL}{Ae^2}\right)^2}$$

The relative error may be used instead of the probable error in the above equations.

3.6 CORRELATION

In order to study a relation of a group of paired measurements, the obvious procedure would be the construction of a chart with arithmetic rectangular axes. This is
known as a *scatter diagram*, and the line representing the best fit is the *regression* line;
if the line were straight, its general form would be $y = mx + b$, where m would be the
regression coefficient. If all points were on the regression line, that is, if there were no
spread or variation normal to the line, the correlation would be perfect and the *coefficient of correlation* would be unity, or 1, the sign depending on the slope of the line.
For a straight regression line a wide scatter would decrease the coefficient of correlation. The letter r is used to designate the coefficient of simple linear correlation and is
sometimes referred to as the Pearsonian coefficient. There are appropriate procedures
for computing the location of the regression line, such as the method of least squares,
which makes the squares of the deviations of the points from the line a minimum. An
easier procedure is to sketch it in freehand and let the diagram tell the story without
the computation.

Well-known scatter diagrams in materials testing are the strength-moisture and
strength-density diagrams for wood and the strength-hardness relation for steel. The
latter is probably most widely used in inspection and control work and wherever

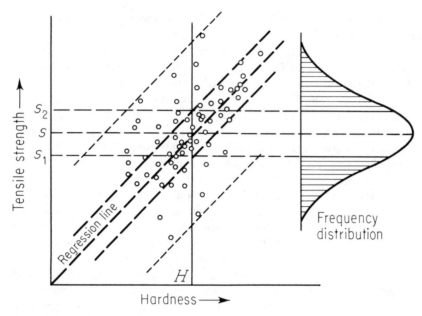

Figure 3.5 Scatter diagram.

destructive tests are not feasible. Strength is usually plotted on the y axis and hardness, the independent variable, on the x axis, as in Fig. 3.5. The heavy dashed lines equally spaced on both sides of the regression line can be placed so as to indicate any desired probability limits. For the example given, a hardness of H indicates that the chances are even (1 to 1) that the tensile strength will be between s_1 and s_2 because the limits are placed $\pm 0.6745\sigma$ (the probable error) on each side of the central value s. In the frequency distribution shown to the right, the open area is equal to that shown cross-hatched, each being one-half the total. The frequency polygon shows, however, that the most likely deviation (error) is zero, or, saying it another way, the most likely or probable strength, indicated by H, is the central (modal) value s. It must be remembered, of course, that the hardness of a given specimen is in itself subject to errors of measurement.

3.7 CONTROL CHARTS

The pioneer writer on the subject of scientific control of engineering materials and products was W. A. Shewhart. Little was written previous to the appearance of his publications, which have formed the basis for widely accepted techniques of measuring and predicting the control of the quality of manufactured products. The value of these methods is in the detection of assignable causes for variations in quality of the final product. Only a brief statement is feasible in this book.

It is practically impossible to attain a given value of a quality in each successive manufactured article because the quality is a variable and the change in its magnitude is a frequency distribution.

The variation in the magnitude of some statistic of a measurable property such as tensile strength can be used as a criterion of quality. Most frequently used criteria are the arithmetic mean, standard deviation, fraction defective, and range.

Values of a given function of quality, such as the arithmetic mean of the tensile strength of successive samples, each containing an equal number of items, say five, are plotted as ordinates against a scale of abscissas that gives the numerical sequence of samples increasing in the customary way from left to right, as shown in Fig. 3.6.

The control chart presents the data so that their consistency and regularity can be seen at a glance. The important feature of the chart, however, is the controls that appear as lines parallel to the abscissas on the chart and show the limits of the variability in the quality being measured that should be left to chance. These limits are commonly set at three standard deviations on both sides of the central value. With a normal distribution, 99.73 percent of the samples will then satisfy the criterion. Stating this in terms of probability, there is only one chance in about 370 that a single sample will fall outside these limits. The limits have been found to be satisfactory in industrial applications. The limits of three standard deviations vary with the number of items in each sample, being larger—that is, farther from the central value—for small numbers. Tables are available that give the limits for the different statistics such as average, standard deviation, and range for each sample size from 2 to 25 items. Formulas and coefficients for the computation of control limits, when samples contain more items, are also available in many texts on statistics.

When the control chart is used in connection with a standard, the limits are established with respect to the specified value, but if no standards are given, the limits are determined on the basis of the data themselves as they are accumulated. In both of these cases the location of the plotted points outside the established limits is taken as an indication of the existence of a causative factor.

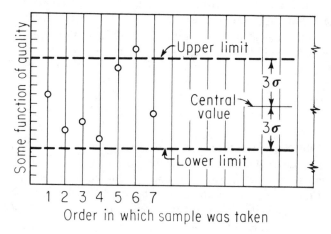

Figure 3.6 Control chart.

3.8 STATISTICAL SUMMARIES

The meaning and significance contained in a mass of measurements are made clear by their interpretation, which in turn is facilitated by a well-arranged presentation. However, the accuracy of the data itself is in no way altered by the method of presentation or by subsequent study and interpretation.

Since the investigator has no way of knowing what use may later be made of the data, supporting evidence regarding the test specimens themselves should first be presented in brief, clear, unequivocal statements. These should include even more information than at the time may seem relevant, since later developments may prove the value of all collateral evidence presented regarding the history of the specimens, including their selection and preparation, and the grade and character of the material. In the use of control charts such extraneous data as the day of the week and the weather should not be ignored.

The next information should deal with the test itself and the conditions under which it was made. This includes the measurements together with an adequate description of the methods used to eliminate constant errors and to reduce the size of chance or accidental errors. Observations on the general and specific test conditions and procedures should be documented, giving particular attention to the matter of special difficulties and their treatment. The aim is to present evidence of all the precautions taken to establish controlled conditions and to secure reliable, trustworthy data.

In the presentation of materials-testing data itself, the essential information is generally contained in four statistics, namely, the number of items n, the arithmetic mean $\overline{X}$, the standard deviation σ, and the coefficient of variation v. The skewness factor, the range, the median, and the mode may also be of interest. The percentage outside a certain limit, such as the defective fraction, is of importance. However, it is obvious that any single statistic cannot present a complete picture of the manner in which the original data were distributed; hence, the statistics employed depend on the purpose of the investigation and the use to be made of the test results.

For the example problem of this chapter, the statistics are summarized as follows:

Number of specimens, n:	80
Arithmetic mean, $\overline{X}$:	42.69 g
Standard deviation, σ:	2.089 g
Coefficient of variation, v:	4.894%
Skewness factor, k:	0.175
Range:	9.59 g
Median:	42.86 g
Mode:	43 g

Of the various graphical representations, a statistical summary normally should at least include a frequency histogram or polygon. Sometimes a cumulative frequency diagram is helpful.

PROBLEM 3 Sieve Analyses of Aggregates

Object: To determine the gradation of coarse and fine concrete aggregates.
Preparatory reading: ASTM C 136, D 75, and E 11.
Special apparatus: Set of sieves and shaker, sample splitter, scales.
Procedure:
1. By means of quartering or with the aid of a sample splitter, obtain one sample of about 5 kg of coarse aggregate (gravel) and one of about 2 kg of fine aggregate (sand). Weigh these.
2. Select an appropriate set of sieves for each sample and perform a sieve analysis, determining the mass retained in each sieve. Sieve sizes are listed in App. B.
3. Tabulate the results, obtaining grouped frequency tables, and compute the cumulative frequencies. Note that in this exercise the frequencies are given in units of mass rather than as numbers of items.
4. Draw a percentage frequency histogram for each sample, as well as a percentage cumulative frequency diagram. Plot the sizes to logarithmic scale on the x axis. Label the diagrams so that their meaning is clear to an intelligent layperson.
5. Compute the measures of central tendency and those of dispersion.
Report: Submit an informal report of your results and findings. Describe the sampling and test methods. Comment on the following:
1. Are the aggregates "well" or "poorly" graded? Explain the meaning of these adverbs with the aid of your charts.
2. Why is a logarithmic x axis used, and what are some of the implications of this? How would the curves change if the number of particles, rather than their combined mass, were plotted on the y axis?
3. How does the frequency diagram compare with a normal distribution? Is there obvious skewing, right or left?
4. What is the mode, and how does it compare with the median and the mean? Is that as expected?
5. What factors influence the accuracy and the precision of a sieve analysis? Is there an inherent bias? What is the effect of the shape of the aggregate particles?

DISCUSSION

1. If you were told that the average grade point average at your university was 2.65 last semester, how significant would you consider this figure? What other measures might you be interested in?
2. Give two examples of sampling, one with and one without replacement. Under what conditions would each type be preferable?
3. If the arithmetic mean of the strength of some bolts were 40 kN and the standard deviation 2.0 kN, between what limits would you expect half the bolts to fall?
4. Measure the three dimensions of a box or a brick, then repeat this several times. Let someone else do it too. Comment on the precision of the results.
5. What is the meaning of "probable" in the term *probable error*?
6. Think of some correlation problems with which you have become acquainted. Consider, for example, the statement, "If you get good grades, you will get good job offers."
7. Suppose there is a population of 200 megapeople. If you ask 1.5 kilopeople, selected as random, what their ages are, what deviation of the sample mean from the population mean can you expect?
8. If people's ages are given in years, as they normally are, comment on the implied grouping. Draw a distribution curve for your university's student population, as you imagine it, and discuss it statistically.
9. State various methods of computing a standard deviation and compare and relate them to one another.
10. If you were to make quality-control tests of the concrete mix and the reinforcing bars used in a dam, of which material would you take a larger sample? Why?

READING

There are many rather general books on mathematics (perhaps one or two that you own) that contain parts or chapters on statistics and probability.

Alder, H. L., and E. B. Roessler: *Introduction to Probability and Statistics*, 5th ed., Freeman, San Francisco, 1972.

Averill, E.: *Elements of Statistics*, Wiley, New York, 1972.

Breiman, L.: *Statistics*, Houghton Mifflin, Boston, 1973.

Campbell, S. K.: *Flaws and Fallacies in Statistical Thinking*, Prentice-Hall Englewood Cliffs, N.J., 1974.

Ehrenberg, A. S. C.: *Data Reduction*, Wiley, London, 1975.

Freund, J. E.: *Modern Elementary Statistics*, 4th ed., Prentice-Hall, Englewood Cliffs, N.J., 1973.

Kennedy, J. B., and A. M. Neville: *Basic Statistical Methods*, IEP, New York, 1976.

Lipson, C., and N. J. Sheth: *Statistical Design and Analysis of Engineering Experiments*, McGraw-Hill, New York, 1973.

Mandel, J.: *The Statistical Analysis of Experimental Data*, Interscience-Wiley, New York, 1964.

Mikhail, E. M.: *Observations and Least Squares*, Harper & Row, New York, 1976.

FOUR

PRESENTATION OF RESULTS

4.1 ARRANGEMENT OF REPORTS

Even the best test is valuable only if it is reported accurately and effectively. A report as a whole should be planned to meet the needs of the individual or group for whom it is intended. It should be clear in meaning and in a readable form. Attention should be given to (1) the format or mechanics of makeup, that is, the method of reproduction, kind of type, and size of page; (2) style and composition; and (3) the way in which the data are presented.

It is a good idea to consult a reference work on the preparation of technical reports, especially if one has never taken a course in the subject. It would also be valuable to study some reports prepared by engineering organizations. Yet there is much room for critical judgment and creative thinking.

The laboratory reports to be prepared in this course will be relatively short, but they will include most of the elements found in reports written by professionals for their clients.

A workable arrangement of the subject matter of a test report is as follows:

Statement of the problem
Materials, apparatus, and/or methods of testing
Summary of test results—including tables, diagrams, and some discussion of detailed results
Discussion of the general features of the test and the findings
Appendixes, to include complete or sample computations, data sheet, and other special supporting evidence if available

The detailed arrangement and the headings used may vary somewhat with the particular type of problem on which a report is made. The report should be as brief as possible, consistent with the inclusion of all important facts, observations, and discussions.

The *statement of the problem* describes the objectives of the test, the general method of attack, and the type of material (or apparatus) tested.

In the section on *materials, apparatus, and methods of testing*, the material tested is described and identified, important pieces of apparatus are listed and briefly described, and the principal features of the test procedure are stated. Sketches are usually helpful here. Standardized and well-known methods and apparatus may be identified by reference to ASTM standards.

In the *summary of test results*, the results obtained are shown, preferably in tabular and/or graphical form. All tables and diagrams are introduced by appropriate statements in the text, which point out the nature of the particular data. The arrangement of the tables must be clear and readable, and the diagrams must be accurate, forceful, and easily understood. The tables and figures are numbered and referred to by number in the text. As a rule, each should be sufficiently complete to show the intended relations without the necessity of searching through the text for essential explanations.

The *discussion of findings* section includes in general terms the results of the test and draws attention to salient facts discovered. Figures and tables are referred to as necessary. It is helpful to compare the results with data found in the technical literature, identifying the references as footnotes or in the appendix. The behavior of the material tested should be described, and the test procedures and results evaluated. Definite conclusions concerning the quality of the material tested and its acceptability under given specifications are also essential.

An *appendix* includes information that supports the main part of the report but is not essential to understanding it. Such information may include sample calculations and other explanations of the data reduction. The sources of any formulas used and the meanings of symbols should be stated. The appendix may also include copies of data sheets, printout records, computer programs, and other supporting documentation.

In writing all parts of the report, it is important to consider the interests of potential readers, who may be looking for some particular information but will frequently not be inclined to read through the entire report in order to find it. Most report readers are busy people.

Many types of modern office machines and graphic aids can be used to good advantage in producing an attractive report. Normally the original is kept on file and copies are distributed, which should be kept in mind during the preparation. Finally, the report is submitted with a cover letter or memorandum.

These various aspects are discussed in a little more detail in the following articles, but a full treatment of technical report writing is beyond the scope of this text. Some useful references are listed at the end of the chapter.

4.2 NARRATION

The reader of a test report will normally be interested in various facts besides tabulated data. Why was the test conducted? What apparatus was used? Did anything unusual occur? How reliable are the results obtained? Which conclusions can be drawn? There is much information that is best communicated verbally.

Sentences also tie the mathematical and numerical information together. They are as necessary to the report as mortar is to a stone wall. The quality of the sentences, like that of the mortar, greatly affects the worth of the whole project. Good writing creates confidence and, more importantly, understanding.

It is nice to write elegantly, but it is essential to do so clearly and correctly, in complete sentences and with proper punctuation. Spellings and meanings of words should be checked, especially in cases of common misuse. For example, "on either side of the specimen" and "on both sides of the specimen" are not synonymous.

In mathematical developments, equations are tied together with suitable phrases: "Substituting for E in Eq. (3) yields ..." or "... which after rearranging becomes ...." All tables and figures should be referred to in the appropriate place: "These results are shown in Table 4" or "Figure 5 shows the arrangement used."

The style should be consistent. In many instances there is a choice of abbreviations, spellings, styles, symbols, and the like. For example, one may write "Eq. (3)" or "Eq'n (3)" or "equation (3)." Unless one option is chosen and consistently used, the report will look disjointed. For correct S1 usage, see App. A.

Another virtue is brevity. Using more words than are necessary wastes the reader's time and is never appreciated. To make the report readable, clear, complete, and concise, the writer must edit it several times and, preferably, let someone else read it too. The final product will make the time spent worthwhile.

4.3 MATHEMATICS

Equations (and inequalities) are useful in reports, as elsewhere, because they can express some relations more precisely and succinctly than words can. They should be included whenever they are needed for this purpose, and not otherwise.

Some equations are particularly important or are referred to for some other reason. These should be numbered (in parentheses) near the right-hand margin, at the same level as the equal sign.

Symbols for various physical quantities are an important part of equations. They must be handled with discretion and care. Some symbols, for example a for some length, are invented for a particular problem. The same basic symbol can be used for similar quantities by adding appropriate subscripts; strains in the x and y directions for example might be called ϵ_x and ϵ_y. All symbols should be defined either in a list of symbols or where they first appear, or both. Under no circumstances may the same symbol be used to express two different quantities or two different symbols be used to express the same quantity.

Symbols expressing quantities (as opposed to units) should be *italic* (*slanted*). If a typewriter does not have italic type, one may use roman (vertical) letters. If there are no Greek letters on the typewriter, these may be lettered by hand. Try to be consistent, however, because a handwritten letter may not look exactly like its typed counterpart. The same holds true for various mathematical symbols, such as

The Euler equation for pin-ended columns gives the critical
load P as

$$P = \frac{\pi^2 EI}{L^2} \qquad (3)$$

where E is Young's modulus and I and L are the minimum moment
of inertia and the length, respectively. By expressing I as
Ar^2 in terms of the cross-sectional area A and the radius of
gyration r and dividing both sides of Eq. (3) by A, the average critical stress

$$\sigma = \frac{\pi^2 E}{(L/r)^2} \qquad (4)$$

is obtained. The slenderness ratio, L/r, becomes

$$L/r = \pi \sqrt{\frac{E}{\sigma}} \qquad (5)$$

For Douglas fir, E is about 12 GPa and σ_c about 50 MPa. Thus
at an average stress of half the ultimate compressive strength
Eq. (5) yields a slenderness ratio of 68.8.

Figure 4.1 Example of mathematics.

integration and root signs. In some cases it may be best to hand-letter an entire equation. After writing the first draft, one can decide on a consistent style.

SI symbols for prefixes as well as for units are always in roman type and are not
followed by periods. Other rules for SI symbols are explained in App. A (see also
ASTM E 380).

Figure 4.1 shows an example of the kind of mathematical text that might appear
in a report for this course.

4.4 TABLES

The presentation of factual material in tabular form economizes space and facilitates
the comprehension of the scope and range, as well as the significance, of the data.
Often more varied information can be condensed into a table than can be shown in a
chart or diagram. Sometimes it may be desirable to evaluate an equation and tabulate
the results for the convenience of the reader. Most test results, including all raw test
data, must be expressed in numerical form. Thus tables are always necessary, and they
ought to be clear and complete.

Tables are essentially two-dimensional arrays, and both the rows and the columns
must be properly identified. Most text results are shown in tables in which the rows
pertain to the various trials, load increments, or some other controlled quantity (in-

TABLE 4 - SUMMARY OF RESULTS

No.	Length, mm	L/r	Load, kN	σ, MPa
1	80.2	13.9	21.4	53.5
2	159.4	27.6	20.0	50.1
3	241.0	41.7	17.9	44.7
4	320.7	55.5	14.2	35.5

Figure 4.2 Example of table.

dependent variables) and in which the columns represent various observed quantities, such as dial readings, and computed values, such as strains (dependent variables).

Whether tables are used in the text or in an appendix, they should have adequate and clearly stated titles, preferably so brief they can be understood on the first reading. The column headings should be complete and contain units for the values in each column; these units should not be attached to items in the body of the table. In tables of some length, ease of reading is improved by grouping the rows, such as by the use of horizontal dividing lines or by increasing the spaces between every third or fifth row. If particular items in the table, such as a column heading or perhaps an unusual result, need further explanation, footnotes may be added immediately below the table (not separated from the table at the bottom of the page).

Numerical values normally should range from 0.1 to 1000, but tables may justify an exception. All numbers in a column ought to have the same unit, the prefix being determined by the majority of the values. In typing the table, the decimal markers are aligned vertically. As a rule, all values are rounded to the same number of digits after the decimal marker.

A typical table for a report is shown in Fig. 4.2.

4.5 FIGURES

Illustrations used in connection with an accompanying text are referred to as *figures* (often abbreviated Fig.). In reports on materials tests these figures may be sketches or photographs of apparatus or specimens, or schematics of processes, or plots of data (graphs). Since the purpose of illustrations is to aid in the presentation of information, the matter of appearance and clearness should be given special attention in the preparation of charts and graphs.

The size of the reproduction may be greater or less than the original, and the size of lettering and weight of line on the original must be proportioned in accordance with any contemplated change in size of the reproduction so as to make it clear and easily read.

The line drawings most essential to reports on materials testing are usually referred to as graphs, charts, or curve sheets, although the designation *diagram* is frequently employed, as in "stress-strain diagrams."

Special cross-section, graph, or coordinate sheets are available in many varieties of rulings, printed from accurately made plates, and are not to be confused with ordinary quadrille sheets. A widely used paper is millimeter ruled, with centimeter lines emphasized.

When it is not intended that numerical values be read from a graph, it may be best to omit rulings or to have only a few.

The choice of *scales* depends on the range in the data and the purpose of the graph. Sometimes the overall picture is given by a small-scale drawing, often called a thumbnail sketch or key drawing, and the significant or important part of the curve is drawn to a larger scale, magnifying or emphasizing the importance part of the relation. The approximate scale is the range in data to be shown, divided by the number of principal divisions, which should preferably be subdivided into 5 or 10 lightly marked scale divisions. The exact scale should be selected so that the smallest division will be some simple part of the scale and not an awkward fractional part. Well-chosen scales make the lines on the cross-section paper helpful in reading values of random points on a curve. Beginners often overlook the rather obvious necessity of placing the principal scale divisions on the heavy lines of the graph paper. The principal values should be lettered in the margins.

Stresses and forces are usually scaled on the y axis as ordinates and strains and deflections on the x axis as abscissas. The plotted points should be fine pencil dots enclosed in some appropriate symbol such as a circle, triangle, or square. Only the symbols, however, are inked (using drawing instruments), and the inked curve or graph is extended to and not through them. If the diagram is to be used for computing (reading values off the chart instead of from tables), the experimental points are not shown, and the curve is drawn as a continuous line of lighter weight than that used when merely showing a general relation; in the latter case the curve should be heavy so as to stand out clearly from the grid, because it is the important part of the illustration. When experimental points are shown, the use of a smooth averaging curve is recommended rather than the broken line that would result from joining points by straight lines. Some mechanical aid, either a flexible ruler or a template, called an irregular or french curve, should be used in drawing the line.

If the entire drawing is specially prepared, the grid lines are not drawn through any symbols or lettering, and usually only the principal lines, corresponding to the heavy lines on prepared graph paper, are shown.

Possible distortion and unintentional overemphasis can arise by using only a partial scale, namely one that does not run continuously from zero. If it is inexpedient to show the entire scale, a conspicuous note or a break in the scale will indicate that it is not complete.

The lines on cross-section paper usually have equidistant spacing; that is, they are on an arithmetic scale. Many types of graph paper are available not only in different arithmetic rulings but in *semilog*, which is logarithmic on one axis and arithmetic on the other, and in *logarithmic* rulings on both axes, in either one or more cycles, that is, units of 10. Log rulings are also known as ratio rulings. Other special rulings, such as *probability* and reciprocal, may sometimes be advantageously employed. If special rulings are used, this fact should be stated, for example, "Age, days (log scale)."

Emphasis can sometimes be more effectively accomplished by plotting deviations from an assumed or theoretical relation, especially if the relation is linear, instead of increasing the scale of one of the variables. Some special hints apply particularly to the construction of stress-strain diagrams.

Materials such as soft steel undergo very little strain up to the proportional limit but develop large strains beyond it. For materials of this type two stress-strain diagrams using the same data should be plotted, one diagram including the entire data and the other diagram covering only the elastic range. The second diagram should be plotted so that the line makes an angle of approximately 60° with the strain axis. This diagram permits a more accurate determination of certain properties (see, for example, Fig. 8.18).

Some materials, such as steel and wood, follow Hooke's law of the proportionality of stress and strain over a considerable portion of their strength, whereas the stress-strain diagrams for such materials as cast iron and concrete may depart from a straight line even at very low stresses. Thus, for ductile metals and wood, the lower part of the diagrams should be made to consist of straight lines even though these lines appear to depart somewhat from the plotted points. To determine quickly the approximate location for a straight line through average values, use an ink line on a separate piece of white paper placed underneath the transparent cross-section paper.

The first two or three observed strains may be erratic, and the straight line of the lower portion of the diagram may therefore not pass through the origin. The origin may be shifted so that the curve will pass through it, but this is not usually done. However, in determining the modulus of elasticity of the material, it is convenient to draw a line through the origin parallel to the straight portion.

Each graph should have an appropriate title, and the material, specimen(s), testing machine, investigator, and date should be identified. If symbols or lines of more than

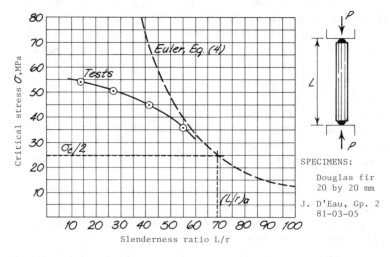

Figure 4.3 Example of graph.

one kind are used, they should be identified next to each curve or shown in a legend or key to the symbols.

Figure 4.3 gives an example of one type of figure, namely a graph of some experimental data. Compare this figure with Figs. 4.1 and 4.2.

4.6 MECHANICS OF REPORT PREPARATION

A few other aspects of final report preparation deserve some attention. There is considerable latitude on these, but certain characteristics are normal and will be expected.

Odds and ends The *cover page* ought to convey all the information necessary to identify the report: what, who, when, and where. The cover page gives the title of the report, identifies the group that performed the experiment and the author, states the dates of testing and submission, and includes the firm and its address or the university and course. Commercial reports also state for whom the report was prepared. The design is a matter of choice, but it should be standard for an organization. Figure 4.4 shows an example.

The pages must be numbered. Preliminary material should have lowercase roman numerals, e.g., "iv," and the appendix may be numbered separately, e.g., "A-4." There ought to be a *table of contents* giving the main headings, and the headings in the table of contents and in the text must agree. Sometimes lists of symbols, illustrations, and tables are also included, immediately following the table of contents.

One should be consistent in the use of *headings* and subheadings. Various choices exist: capitals or lowercase; centered, left or right justified; underlined or not; roman or italic. An outline will help the author decide on a suitable hierarchy. A numbering system for the headings will facilitate reference to the various parts of the report.

Whenever information is produced from another source, honesty demands that the reference be cited. This can be done by footnote if there are only a few. Otherwise, it is preferable to collect the references in a list in the appendix, ordered according to some logical scheme. Each reference is numbered and referred to in the text by inserting the reference number in parentheses. There are standard formats for listing the references. Incidentally, such a list of references is *not* a bibliography, since a bibliography would include a comprehensive listing of the pertinent literature.

For longer reports, or whenever the client requires it, an *abstract* (or summary or synopsis) is included among the preliminary material. This is a brief (usually one-page) digest of the purpose, nature, and findings of the test. The main purpose of the abstract is to tell the reader whether the report will be useful to him or her. The abstract should be headed with a full bibliographical description so that it can be separated from the report. If the report is to become accessible by an information retrieval system, *key words* are listed, usually with the abstract. These words or phrases (e.g., *aluminum, hardness, testing*) are taken from the retrieval system's master list.

Assembly and duplication The reader will assume the makeup of the report reflects the care that went into the entire project. Like the package of a product, it helps to sell the work.

```
                    ALMA MATER UNIVERSITY
                       Pedantry, USA




                    TEST OF WOODEN COLUMNS
                    A laboratory report
                       submitted by
                      Jeanne D'Eau
                           to
                    Prof. I.M. Picky




                         Course:   CE 232
                         Group 2,  Problem 7
                         Date of test:    81-03-05
                         Date of report:  81-03-15
```

Figure 4.4 Example of title page.

The pages should be of standard size, normally typed on one side only, double-spaced, and with adequate margins to allow for binding. If a larger sheet is needed, it should be easy to fold out, and of the same height as the other pages if at all possible. If a page has to be turned 90 degrees, perhaps to accommodate a large table or illustration, it should be read from the right side.

Many typewriters have interchangeable elements with various type styles and a full complement of mathematical symbols. These options are helpful, but they require consistency in their use. Clean type and a carbon ribbon will ensure that the copy is crisp and black.

In lettering equations (when necessary) and figures, a fine-point pen (felt-tip, ball-point, or stylus) can be used, but it must have reproducible ink. Whether a lettering guide, transfers, or freehand lettering is used depends on the skill of the drafter.

Figures, especially those on graph paper, can advantageously be cut out and pasted on the page, which can then be duplicated. It is best to confine graphics to black-and-white and shadings: color duplication is expensive and not always available. Reducing copying machines are generally available and can profitably be used to decrease the size of some figures and tables. Computer printouts present a special problem, which can be alleviated by appropriate formatting as well as by reduction.

The whole report should be securely bound in a folder. The kind with a transparent front cover is best; otherwise a label giving the report title, the author's name, and the submission date must be attached. The folder should be flat so that it can be conveniently filed.

Covering correspondence   When a report is submitted, it normally is accompanied by a letter or memorandum of transmittal. The purpose of this correspondence is primarily to record the date of submission, the name of the person responsible, the name of the recipient, the circumstances of the submission, and the distribution of the report. In addition, the covering letter presents an opportunity to make less-formal observations and recommendations than the report proper should contain.

A *letter* is customarily used when the report is submitted to a person outside one's own organization, such as a customer or client. Typed on letterhead stationery, a letter shows the date[†] at the top, followed by the name-and-address block of the recipient. The name should be accompanied by the addressee's title and organization. Rarely, the name is omitted and the letter addressed by title. Sometimes the subject matter is indicated after the word "Re:" (the Latin for "thing"). The text is normally preceded by a salutation (e.g., "Dear Dr. Schmidt") and followed by a complimentary close (e.g., "Sincerely yours,") and the signature. Under the signature is the writer's signature block, that is, his or her typed name and title. The word "Enclosure" (or "Encl.") should appear to show that the report is enclosed. Finally the names of any copy recipients should be shown and identified by "Copies to:" or "cc:"; if any copy recipient is not sent a report, the phrase "(w/o encl.)" follows the name. In some modern letter-writing practice the salutation and the complimentary close are omitted as essentially meaningless, particularly if the recipient is addressed by title or if a group is addressed. To avoid sexist phrasing, one may write "Dear Sir or Madam:" or "Ladies and Gentlemen:" as a salutation.

In communications within an organization the *memorandum* (or memo) offers a more concise form. The address block, salutation, close, and signature block are omitted and replaced by "To:," "From:," and "Subject:" lines at the top. The subject is indicated by a short descriptive phrase, e.g., "Submission of Report on Fatigue Test for Job 84-123." The memorandum may be signed or initialed, and in all other respects its form follows that of a letter.

For both the letter and the memorandum the text may be very brief, perhaps simply stating that the subject report is submitted as an enclosure: the main purpose is to create a record for the file. Here is the opportunity, however, for some further

---

[†]The standard form of writing a date adopted by ISO is year-month-day (in descending order), e.g., 1984-04-20.

observations or opinions of the writer, which would be particularly appropriate if he or she had not personally conducted the tests or written the report, but were, perhaps, in charge of the testing laboratory. Still, it is well to keep such remarks brief.

4.7 ORAL PRESENTATIONS

Occasionally a testing group may be asked to report orally on the results of some project. Compared with a written report, the advantages include the possibility of some "show and tell" and the ability to respond to questions.

The preparation of an oral presentation should parallel the preparation of a written report, except for the final phases. Because the length of the report will be strictly limited, however, a speaker must prepare even more carefully.

The narrative becomes the spoken word. A detailed outline of what is to be said is necessary; a full text is not. The outline, plus data, quotes, and the like, might be written on file cards for easy reference during the talk. The spoken word should be illustrated with *visual aids*. At least one rehearsal is a must.

Mathematical, graphical, and tabular material may be shown on slides or transparencies. The latter are cheaper, faster, and easier to make. They can be marked on during the presentation, and it is possible to use colors, overlays, and other creative tricks. Whichever type is used, the graphics should be kept simple: large tables and complicated derivations are difficult to comprehend and will detract attention from the talk.

Sketches of devices and specimens can be shown in the same way. Depending on their size and the number of listeners, however, it may be better yet to show the actual object rather than a picture. Passing items around the room, however, will split the audience's attention. Another possibility is charts taped to the wall, provided these are large enough to be seen by all and displayed only when referred to.

On the day of the presentation, the presenter should review the material, show up early, check the room and the visual aids (difficulty in locating a light switch or a slide out of order would be embarrassing), and relax. If all is prepared, all will go well.

A presenter should speak clearly, confidently, and in complete sentences. There is no need to worry about a pause now and then, or to rush: the points need time to sink in. Another advantage of overhead projectors becomes apparent now—the speaker can continue to face the audience while pointing out features on the transparency with a pencil or a knitting needle. Finally, questions should be invited and answered as simply and briefly as possible.

PROBLEM 4 Report Preparation

Object: To prepare written and oral reports on test results previously obtained.
Special apparatus: Copiers, overhead projector, tape recorder.
Procedure:
1. Assemble the results of Problems 2 or 3 and outline a full report on them.
2. Write the rough draft, incorporating the tables and figures.
3. Revise, edit, and produce a smooth original report, complete in all respects.

4. Proofread and correct the report and copy it. Bind the copy into a folder.
5. Outline a short (10-min) oral presentation, that is, a briefing.
6. Select significant aspects of the test apparatus and the results, and prepare on ordinary paper visually effective displays.
7. Make transparencies (or slides) of your displays.
8. Select other visual aids that you think helpful.
9. Compose the text and prepare notes.
10. Rehearse and revise (if necessary, shorten) your presentation.
Report:
1. Submit the bound report with a letter of transmittal.
2. Present a briefing. Record the briefing on tape (videotape, if available).

DISCUSSION

1. Obtain an actual report and criticize it. Look for inconsistencies, unclear passages, awkward language, deficient illustrations, improper SI usage, and the like.
2. Imagine that you conducted tensile tests on various brands of fishing line. Prepare a table of contents for your report.
3. Use the information in the example title page of Fig. 4.4 and put it into a form artistically more pleasing to you.
4. Use the computer to generate a table that gives the cross-sectional area of pipe for various outside diameters and wall thicknesses.
5. Prepare a graph showing the temperature variation in your city during a 24-h period. Include the high and low predicted.
6. Write a short paper that explains the relation between bending moment and extreme fiber stress in a beam.
7. Write a proper bibliographical entry for this book, one for an article found in a current engineering periodical, and one for a handbook.
8. Explain, in words only, what a helical spring looks like. Would an equation or a sketch make this easier?
9. Make some sketches that show how a bicycle pump works.
10. We mentioned "either" as an example of a word that is often misused. Can you think of any other words whose meaning or usage is widely misunderstood? How about "dilemma," "due," "data," or "disinterested"? Can you find examples of poor or incorrect usage in this text? Of course you can.

READING

You can probably find a lot of half-forgotten information in your old English textbooks. Look through dictionaries, thesauri, and other tools of the trade. Above all, enrich your life as well as your language by reading thoughtfully and critically.

Andrews, D. C., and M. O. Bickle: *Technical Writing: Principles and Forms,* Macmillan, New York, 1978.
Berry, T. E.: *The Most Common Mistakes in English Usage,* McGraw-Hill, New York, 1971.
Campbell, W. G., and S. V. Ballou: *Form and Style,* 5th ed., Houghton Mifflin, Boston, 1978.
A Manual of Style, 12th ed., The University of Chicago Press, Chicago, 1969.
Souther, J. W., and M. L. White: *Technical Report Writing,* 2d ed., Wiley, New York, 1978.
Walpole, J.: *Writer's Guide,* Prentice-Hall, Englewood Cliffs N.J., 1980.
Turabian, K. L.: *A Manual for Writers of Term Papers, Theses, and Dissertations,* 4th ed., The University of Chicago Press, Chicago, 1973.
Van Leunen, M. -C.: *A Handbook for Scholars,* Knopf, New York, 1978.

TESTING PROCEDURES

5.1 PURPOSES OF MECHANICAL TESTS

One of the important considerations in the design of a structure or a machine is *strength*, a characteristic that enables the device to serve its function safely and well. The boom of a crane, for example, must support the design loads without breaking, and also without bending so much that control becomes difficult for the operator. In a broad sense, strength refers to the ability of a structure or a machine to resist loads without failure, which may occur by rupture due to excessive stress or may take place owing to excessive deformation. The latter cause of failure, in turn, may be the result of a limiting stress having been exceeded or the result of inadequate stiffness. The properties of materials that are of significance in relation to this general problem are the *mechanical properties*.

Mechanical properties may be specifically defined as those having to do with the behavior (either elastic or inelastic) of a material under applied forces. Mechanical properties are usually expressed in terms of quantities that are primarily functions of stress or strain, but they are occasionally expressed in terms of other quantities such as time and temperature.

Mechanical testing is concerned with the determination of measures of mechanical properties. The primary measurements involved are the determination of load and of change in length. These are translated into terms of stress and strain through consideration of the dimensions of the test piece.

Fundamental mechanical properties are strength, stiffness, elasticity, plasticity, and energy capacity. The strength of a material is measured by the stress at which some specified limiting condition develops. The principal limiting conditions or criteria of failure are termination of elastic action and rupture. Hardness, which is usually indicated by resistance to indentation or abrasion at the surface of a material, may be considered as a particular type of measure of strength. Stiffness has to do with the magnitude of deformation that occurs under load; within the range of elastic behavior, stiffness is measured by the modulus of elasticity. Elasticity (but not the modulus of elasticity) refers to the ability of a material to deform without permanent set on release of stress. The term plasticity is used in a general sense to indicate the ability of a material to deform in the inelastic or plastic range without rupture. Plasticity may be expressed in a number of ways; for example, in connection with tension tests of ductile metals, it is referred to as ductility. The capacity of a material to absorb elastic energy depends on both strength and stiffness. Energy capacity in the range of elastic action is termed resilience, the energy required to rupture a material is taken as a measure of its toughness. These concepts were more fully described in Chap. 2.

In this chapter we discuss the general features of mechanical testing associated with the mechanical properties of materials mentioned above. The definitions given are based on the ASTM definitions of terms relating to testing (ASTM E 6), although they differ in some respects. For this introductory discussion, only the general procedures of mechanical testing are outlined; details of particular types of tests are given in separate chapters in Part Two.

It is not to be inferred that in particular cases all the mechanical properties are determined. For reasons of economy, the number and difficulty of tests are kept at a minimum. For a particular problem, only a few pertinent tests need ordinarily be made, and for control work a single type of tests of the simplest sort, selected because it appears to be a significant indicator of the required quality, is often sufficient. For example, in the commercial production of some steel products, simple hardness tests made at suitable intervals are often sufficient to indicate whether or not the quality of the steel is being maintained within specification limits.

5.2 TYPES OF MECHANICAL TESTS

In order to approximate the conditions under which a material must perform in service, a variety of test procedures are necessary. The relation between different test procedures can be made evident by an orderly classification of test variables, the principal types of which are (1) those having to do with the manner in which load is applied and (2) those having to do with the conditions of the material or specimen itself and of the surroundings (ambient conditions) during the test.

Loading methods The method of loading is the most common basis for designating or classifying mechanical tests. There are three factors involved in defining the manner in which load is applied: the kind of stress induced, the rate at which the load is applied, and the number of times the load is applied.

In the mechanical testing of prepared specimens there are five primary types of loading, as governed by the stress condition to be induced: tension, compression, direct shear, torsion, and flexure. In tension and compression tests, one tries to apply an axial load to a test specimen so as to obtain uniform distribution of stress over the critical cross section. In direct shear tests, one attempts to obtain uniform distribution of stress, but this ideal condition is never satisfied in practice because of the way the shear stresses are developed within the body under direct shear loads, and because of incidental stresses set up by the holding devices. Pure shear can be developed in cylindrical bars subjected to torsion, although the intensity of shearing stress varies from zero at the center to a maximum at the periphery of the cross section. Torsion tests have an advantage over direct shear tests in that strains can be determined by measurement of the angle of twist. In flexure tests, both tension and compression are involved (and also shear, if other than pure bending is induced), and composite effects are studied; for example, deflections are measured directly, and the modulus of rupture can be calculated.

In certain special tests, a complex stress condition may be induced by superposition of the primary types of loading; for example, a triaxial compression test involves compression in three directions, or a test may be made in which, say, torsion, is combined with tension. In some cases, a direct primary stress may be combined with secondary effects of bending such as when buckling occurs in a column.

A complex stress condition also occurs when the intensity of stress varies from point to point in a piece of material as the result of localized load application or abrupt changes in the shape of the piece. This stress condition is an inherent and significant condition in such tests as indentation hardness tests and notched-bar impact tests. It may be noted in passing, however, that insofar as routine testing is concerned, tests of this sort involve only the observation of simple phenomena (for example, resultant deformation) and do not involve determination of stress distribution.

With respect to the rate at which load is applied, tests may be classified into three groups. If the load is applied over a relatively short time and yet slowly enough that the speed of testing can be considered to have a practically negligible effect on the results, the test is called a *static* test. Such tests may be conducted over periods ranging from several minutes to several hours. By far the majority of tests fall into this category. If the load is applied very rapidly so that the effect of inertia and the time element are involved, the tests are called *dynamic* tests; in the special case where the load is applied suddenly, as by striking a blow, the test is called an *impact* test. If the load is sustained over a long period, say months or even years, the test is a *long-time* test, of which *creep* tests are a special type.

With respect to the number of times load is applied, tests may be classified into two groups. In the first group, which includes most of all tests made, a single application of load constitutes the test. In the second group, the test load is applied many times, millions of times if necessary. The most important category of tests in this group is the *endurance* or *fatigue* tests, whose purpose is to determine the endurance or fatigue limit of a material (of which specimens are made) or of an actual part.

Test conditions In addition to loading conditions, one must consider the condition of the material at the time of the test and the ambient conditions if they affect the test results.

Depending on the temperature at which the tests are conducted, three general classes of tests are recognized. In the first class, comprising the majority, are tests carried out at normal atmospheric or room temperatures. In the second class are tests made to determine properties of materials, such as the brittleness of steel, at very low temperatures, as might be required in the development of high-altitude aircraft or of cryogenic devices. In the third class are tests carried out at elevated temperatures, as in the development of rockets, jet engines, and gas turbines, to evaluate the strength, ductility, and creep of materials under such conditions.

The mechanical properties of some materials are affected by moisture conditions. For example, the strength of materials such as concrete, brick, stone, and wood is markedly influenced by the moisture in the material. The standard tests on concrete are made while the material is in a saturated condition, whereas those on brick are made on oven-dried specimens. Tests of wood may be made on specimens either in a green or in an air-dry condition, but the moisture content at the time of the test is always determined. Long-time tests of these materials may require controlled humidity conditions. Known moisture conditions are required for a standard test so that the test results obtained by different operators will be comparable.

For special purposes, tests may be conducted that involve the use of corrosive atmospheres, salt sprays, or baths containing substances designed to ensure neutral or corrosive reactions.

In planning or specifying the details of a test, the makeup of the specimen in relation to the physical condition or nature of the material requires consideration. Some of the factors involved in the selection and preparation of specimens are discussed in Art. 5.4. In conducting a particular test, the manner of holding, gripping, supporting, or bedding the specimen should receive attention, as well as the stability of the specimen or its parts.

In research investigations, the procedure in finding the effect of given variables is to maintain all conditions constant except those under investigation. In designing, conducting, or reporting tests, significant test conditions must be specified, controlled, or known.

5.3 DESIGN OF TESTS

In the design of a test, the following are suggested as fundamental questions to be considered:

What is the nature of the answer sought?
What test can be made to provide an answer?
How will the test results be related to performance?
What are the limitations of the type of test selected?
How should the precision of the work be adjusted to be in accord with the limitations
    so as to achieve economy of effort and consistent reliability of results?
What type of specimen is best suited for the test?
How many samples are necessary to obtain representative results?

The ideal test should be meaningful, reliable, reproducible, of known precision, and economical. The selection of a procedure should be controlled by the significance of the test, guided by economy of effort, and influenced by a sense of proportion.

The following observations bearing on the design of tests are abstracted from an early NBS handbook on materials testing [94]:

An adequate measure of a given property is possible when (1) the property can be defined with sufficient exactness, (2) the material is of known composition or purity, (3) the attending conditions are standard or are known, (4) the experimental methods are theoretically correct, (5) the observations and their reductions are made with due care, and (6) the order of accuracy of the results is known. This ideal is rarely if ever reached, but as it is striven for the results pass from the qualitative to the quantitative stage and are called *constants* because redeterminations will not yield sensibly different results. Approximate results are improved upon steadily as more precise instruments and methods are devised. The degree of accuracy to be sought becomes a very practical matter in a testing laboratory. The time and labor involved in tests may well increase out of proportion as the limits of attainable accuracy are approached. For the determination of physical constants or fundamental properties of materials the degree of accuracy sought may be the maximum possible. In general the degree of accuracy striven for should be that which is strictly good enough for the purpose in hand.

5.4 TEST SPECIMENS

The success of any test depends to a decisive degree on intelligent choice and suitable preparation of the specimens.

Selection Specimens must be selected and prepared so as to give a reliable indication of the properties of the materials or parts they represent. There are two problems involved in the selection of specimens for testing: (1) the setting up of physical procedures for securing samples and (2) the determination of the number of specimens or the frequency of tests necessary. As regards the first problem, standard specifications for sampling have been especially prepared by ASTM for a number of engineering materials, for example: cement, C 183; lime, C 50; brick, C 67; concrete masonry units, C 140; concrete aggregates, D 75; bituminous materials, D 140; and wood, D 143. In addition, many specifications for particular materials contain sampling requirements. Some of the considerations involved in the selection of specimens are outlined below. As to the second problem, a guide is furnished by sampling theory, which is outlined in Art. 3.5. Still, in many instances the number of specimens or tests to be used is based on custom or experience. For routine testing and inspection the sampling procedure is usually definitely prescribed, yet it is desirable that the inspector have sufficient background knowledge of the production of a particular material to know what constitutes a really representative sample. Above all, good sense should be used in both the selection and the preparation of specimens.

In testing material from metal plate, due regard should be given to the direction of rolling; sometimes tests are made both on specimens cut parallel to the direction of rolling and on specimens cut perpendicular to the direction of rolling. Directional effects are particularly important in wrought iron but are not as important in steel; apparently there is little directional effect in rolled brass and copper. The strength and ductility of metal cut from rolled structural shapes appear to be influenced to some extent by the working under the rolls; the thinner parts tend to be slightly stronger and less ductile. The properties of metal cut from castings are influenced by the rate of cooling and by shrinkage stresses at changes in section; generally, specimens near the surface of iron castings are stronger. If specimens are cast separately, the size of casting requires consideration. With some materials (such as bronze) the composition of the molten mixture may change during the time required to cast the metal, so that a specimen taken at the beginning may not correspond to the average composition of the final product. To obtain representative specimens from forgings or heat-treated parts may offer real difficulties; each case must be treated independently.

With specimens that are molded (such as mortar, concrete, rubber, and plastics), care must be taken that molding conditions do not cause defects in the specimen; for example, restraint to shrinkage in the molds may induce cracks in brittle materials. Attention should be directed to the maintenance of known or standard "curing" conditions (moisture and/or temperature) if they tend to influence the results of the tests.

In the selection of specimens of brick, tile, and other ceramic products, consideration must be given to variation in degree of burning, defects, and the like. Specimens of stone should be selected with regard to homogeneity of the deposit and the piece, as well as with regard to the direction of bedding planes.

Specimens of wood should be selected with due regard for the direction of the grain, the density of the wood as influenced by the rate of growth, the proportion of earlywood to latewood, the degree of seasoning, and the presence of defects.

Preparation In preparing specimens of metal, if a rough "blank" is taken by shearing, punching, or flame-cutting, make sure that the finished specimen does not contain any of the damaged metal. The finished surface of specimens from sheared blanks should be at least 3 mm from sheared faces and at least 6 mm from flame-cut faces. Care should be taken not to bend the piece, because working of the metal tends to change its properties; in tests of specimens cut from tubing, flattening of specimens is sometimes prohibited for this reason.

The finish cut on machined metal specimens should be made by turning, planing, or milling and should give a surface fine enough not to influence the break. If the ends of a specimen are to be threaded, slightly rounded threads, rather than sharp V threads, should be cut, particularly if the specimen requires heat-treatment.

Specimens of concrete, mortar, and several other types of materials have to be molded while in the plastic state. Special consideration must be given to each type of material. For specifications for molding concrete, see ASTM C 31.

The sawing, coring, or grinding of stone, concrete, and ceramic materials should receive attention with respect to flatness and squareness of bearing surfaces and adequate smoothness of lateral surfaces. The axes of stone specimens should be so oriented that bedding planes do not form planes of weakness for the particular test specimen.

Wood specimens should be sawed, planed, or turned in such a way as to avoid sharp saw cuts, nicks, or split fibers on critical surfaces.

The size of the finished specimen is in general governed by the size of the piece or product from which it is taken and by the capacity of the testing machine available to test it. In many materials, the degree of homogeneity or uniformity of structure of the material may dictate the size of specimen that can be used. For example, the diameter of concrete specimens should be three or four times the diameter of the largest aggregate particles.

Dimensions and tolerances for standardized specimens should be noted and adhered to; these are discussed in connection with particular types of tests.

Finally, attention should be given to the marking and identification of test specimens and the method of relating test samples to the lot or lots of material they represent.

5.5 TESTING APPARATUS

The selection of apparatus for a particular test involves considerations of (1) the purpose of the test, (2) the accuracy required, (3) convenience or availability, and (4) economy. In a number of instances, the final choice represents a compromise between the last three.

For routine testing, the accuracy and precision required should be readily attainable with normal equipment, but this accuracy and precision should be known and maintained. The usual requirement for ordinary testing machines is that they be accurate to within 1 percent in the loading range. If the final result is to have the same degree of accuracy, then measurements of strains must also be accurate to at least 1 percent. Note that in order to obtain a 1 percent accuracy for a cross-sectional area, the diameter must be measured to within 0.5 percent. For work of high precision, consistent accuracy in all measurements is needed. It may require some study of the performance of various pieces of apparatus in order to select the right equipment for a particular test.

Considerations of convenience and economy are generally dictated by the equipment available in a particular laboratory. Certainly, however, within the limits of required accuracy and precision, the procedure that is simplest and least time-consuming should be selected.

5.6 MEASUREMENTS

Although the determination of quantitative measures of the many properties of materials calls for a wide diversity of observations, the basic quantities actually to be measured are relatively few; some of the more important are lengths (including changes in length), angles, volumes, masses, forces, pressures, time intervals, temperatures, electric currents, voltages, and resistances.

In mechanical testing most measurements ultimately have to do with the determination of stress and strain. Although direct comparison with known weights and distances often is used as the means for determining force and length, in general a variety of physical principles and phenomena are employed in the numerous types of apparatus used to determine load and deformation. In addition to mechanical devices that multiply or magnify load and length changes, there are instruments that take advantage of phenomena such as elasticity, reflection of light, interference of light waves, electric resistance, magnetism, inductance, and sonic vibrations. Electric instrumentation is especially powerful.

In order to control the accuracy of numerical data, it is necessary to know the accuracy, or the limit of error, of the contributing measurements. The error (that is, the difference between an observed value and what is believed to be the true value) in the indicated readings of a measuring instrument is normally determined by a process of calibration. The true or correct value is obtained through some more fundamental method of measurement or by comparison with observations on apparatus of known accuracy. It can also be inferred from statistics, as described in Chap. 3.

Intimately related to the *accuracy* of an instrument are the *sensitivity* and the *least reading* of the instrument. These are indications of precision. The sensitivity is expressed in terms of the smallest value of the quantity to be measured that induces a response on the indicating device of the measuring instrument; an instrument that requires a relatively large change in magnitude of the quantity being measured in order to be actuated is said to lack sensitivity. The least reading is the smallest value that can be read from an instrument having a graduated scale. Except on instruments provided with a vernier, the least reading is that fraction of the smallest division which can be conveniently and reliably estimated; this fraction is ordinarily one-fifth or one-tenth, except where the graduations are very closely spaced. On vernier instruments the least reading is the least count of the vernier. Obviously, the precision of a measuring device is limited by its sensitivity and least reading, but even a sensitive instrument is not necessarily accurate.

All measurements, except the counting of individual objects, are subject to accidental variation, which must be controlled or known if the final results of a test are to be of known *precision*. The difficulty with measurements, including those made in connection with materials testing, is that successive readings usually are not identical; that is, *discrepancies* occur. It can be inferred that we do not ordinarily obtain the *true* measurement, and when perchance we do, we do not know that we do. Thus *errors* are unavoidable and must be considered. They fall into two distinct classes, systematic and accidental errors, which we will discuss briefly.

Systematic errors are illustrated by the speedometer of a car that indicates 90 km/h whenever the car is actually traveling at a speed of 87 km/h. Some are "natural errors," such as will occur when corrections for the influence of thermal expansion on the length of a steel tape are not made, or when abnormal humidity influences an instrument or specimen. "Instrument errors" are caused by constructional imperfections in an instrument, such as poor optics in a microscope, or such as failure to level a balance. Finally, there are "personal errors" brought about by some sensory limitation on the observer's part, such as a slowness of reaction that might cause the observer to push the button on a stopwatch a bit late. Systematic errors are often cumulative in that they tend to fall on one side of the true value. Alertness and care are needed to detect them, and reasoning ability to correct or reduce them.

Accidental errors, in contrast, tend to fall randomly on the plus or minus side of the true value and thus tend to be self-compensating if a series of measurements is taken. They are generally caused by the inability of the observer to match the precision of an instrument. Even if we had a perfectly accurate ruler, for example, it would be impossible consistently to obtain the same length of a brick with a precision of 0.1 mm. Note, incidentally, that if in this example we were satisfied with a precision of 1 cm, no accidental errors would occur: discrepancies become more likely as the precision is increased.

In contrast to systematic errors, accidental errors cannot be corrected. Instead, statistical methods are used to analyze them. In the simplest case, for example, we might measure the brick 10 times and average the lengths, expecting thereby to obtain a more reliable result. A summary of basic statistical methods, useful in the interpretation of testing data, was presented in Chap. 3.

In Chap. 7 we will describe some of the more common instruments for determining load, length, and deformation and the principles on which they operate. The discussion is confined primarily to the subject of measurement of mechanical properties, since most of the effort in the ordinary materials-testing laboratory is concerned with these. However, the same principles apply directly to measurements of many other physical properties; for example, the determination of thermal expansion or of contraction due to drying involves chiefly length-change measurements.

In conjunction with mechanical tests a number of auxiliary measurements and tests may be necessary, usually for furnishing data on dimensions, area, volume, mass, density, moisture content, and so on. These are mentioned here for the purpose of directing attention to the desirability of making such measurements with accuracy and precision consistent with that of load and strain measurements. The calibration of the test apparatus may be considered an essential example of auxiliary tests.

5.7 CONDUCTING TESTS

Once a test has been started, a variety of functions must be performed, often in rapid succession or even simultaneously and always reliably. Loads must be adjusted, readings taken, behavior observed, data recorded, and preliminary calculations made. Automated recording equipment will help in many instances, but it is essential in any case that the testing team be well organized and fully prepared. This applies to the laboratory exercises in this course just as much as to any commercial testing facility. A team might consist of a recorder, an operator, an observer, and a computer. The following instructions, directed to each team member, would apply.

1. *Recorder*. You are the leader of the party, since you are in a position to supervise the general progress of the work. Before the work begins, you should understand all that is to be done and should so plan the work that the problem will be completed within the allotted time. Although each member of your party will be held responsible for adequate preparation before the work begins and for the performance of his or her duties in the laboratory, the responsibility for the satisfactory execution of the problem rests on you. You will give directions to other members of the party and see to it that all the required data are obtained. Do the following personally:

   Prepare a sheet with appropriate headings for the entry of all data if data sheets are not provided. On the data sheet show such general information as the number in the party, names of party personnel, the problem number, special equipment used, and the date. Before the test is begun, determine all pertinent details of the test setup. Record all data as soon as they are observed. Do not record any data on other sheets with the intent of transferring them later to the regular data sheet. Record the speed of the movable head to the testing machine or the rate of load application, if applicable. Draw a sketch of any special apparatus and describe its operation. Draw a sketch of any ruptured specimens and describe the character of the fracture.

2. *Operator.* You check over all equipment, except the measuring devices used by the observer. Become familiar with the equipment so that no delays or difficulties will arise during the test.

Adjust the equipment, select an appropriate load range, and zero the machine.

For tension tests see that the proper grips are ready for use and make sure that they bear for their full length against the specimen and also against the head of the testing machine. For compression tests see that a compression plate and spherical bearing block are in position. During the test, do not apply load so rapidly that the load and strain observations are of doubtful accuracy.

In using a hydraulic machine move the piston slightly out of the cylinder before positioning the movable crosshead. Then, if the crosshead should accidentally stall against the specimen, it can be released by lowering the piston.

If the grips remain stuck at the end of a test, use a hammer to strike the sides of the specimen. Do not strike the grips or grip handles. After the test, clean up the machine and its surroundings. Check all tools and equipment, and leave them in proper order. Report any lost or damaged equipment. Before leaving the machine, make certain that the motor has stopped. A mechanical machine should be left out of gear, and a hydraulic machine should be left with the unloading valve open and the loading valve closed.

3. *Observer.* You check all measuring apparatus and see that it is in proper order. Study the action of any strainometer to be used and become familiar with its operation. Note its range. Determine how to convert strainometer readings to unit strains. Attach the strainometer to the specimen and center the latter in the testing machine. Set the strainometer to read zero at zero load, or at some predetermined small load, and check to see that most of its range is available for use. Report all measurements and observations to the recorder. Keep the specimen constantly under observation in order to determine critical points in the test.

4. *Computer.* You make any preliminary computations that may be necessary. Determine the probable maximum critical values to be observed and select suitable increments for the development of adequate stress-strain or other curves. Assist the observer in taking readings during the test.

Testing and common sense Scientific experimentation and testing, as well as mathematics, have come to be an important tool of the engineer. Testing should not be used as a substitute for thought, although an appropriate experiment may aid in an analysis.

Before a test is undertaken, its purpose should be well understood, and the general character of the results should be envisioned. The magic of tests does not lie in turning them on and hoping for the best, rather it results from careful, intelligent planning and the slow, painful process of overcoming difficulties.

It is important for engineers concerned with the performance of tests to have developed the ability to visualize what goes on behind the physical operations of tests—the paths of stress and deformation, reactions, the movements of component parts, circuits of flow, and so on. They must be aware of opportunities for error and be quick to see where mistakes might occur. They should be alert to note the unusual, for herein lies the embryo of new discovery. They should be the *first* to test their results by the "seem-reasonable" criterion and be ready to check them if they do not make sense.

The following remarks, paraphrased from the closing article of the late George Fillmore Swain's book *Structural Engineering—Strength of Materials* [87], are pertinent:

> The point to remember is that in testing, as in the use of mathematics, common sense should always be in command. Young engineers will do well to cultivate the habit of distrusting what does not seem reasonable. At times the "seeming" may be wrong—the result may be correct; in such case their senses need cultivating. Finally, it may be appropriate to refer to a remark of Aristotle who observed that it is the mark of an instructed mind to rest satisfied with that degree of precision which the nature of the subject admits, and not to seek exactness where only an approximation of the truth is possible.

An experiment or test is unfinished until it is summarized, checked, and interpreted. It should be the pride, as it is the duty, of the engineer to present the results of his or her findings in a clear, forceful, understandable, and pleasing way. The nature of a report should be adjusted to fit the needs of the audience. Nontechnical persons and uninformed users of materials have a tendency to think of tests, especially acceptance tests, as precise, infallible, and of general application. Tests are always made subject to limiting conditons, and the results are not properly reported unless accompanied by a practical interpretation.

PROBLEM 5 Withdrawal Test of Nails

Object: To determine the pulling resistance of nails embedded in wood.
Preparatory reading: ASTM D 143, D 1761, D 1037, F 547.
Specimens: About 40 common nails, 1–3 mm in diameter. A piece of "2 by 4" lumber (38 by 89 mm), approximately 1 m long.
Special apparatus: Hand tools, heavy-duty scales, weights.
Procedure:
1. Devise an apparatus that will allow a weight to pull a nail partially driven into wood. Perhaps a lever arrangement is feasible. Weights from the consolidation machine in your soil mechanics laboratory might be used, as might a sack containing gravel. Do not use a testing machine.
2. Drive a nail a certain distance (mm) into a wide side of the piece of wood and measure the force (N) required to pull it out. Repeat the test two or three times for the same penetration. Then repeat the whole procedure several times for different amounts of penetration.
3. Take notes of all pertinent facts, and make a sketch of the apparatus.
4. Determine the density of the wood.

Report: Submit your tabulated results as well as such pertinent information as the type of nail and species of wood used. Graph the results. Analyze your data, and try to determine a relation between nail penetration and withdrawal force. Comment on the following questions:

1. In retrospect, how could your apparatus have been improved?
2. How much confidence do you have in the accuracy of your measurements? What was the precision? Exactly how did you measure penetration?
3. How did failure occur? Did it usually happen suddenly? What are the implications for safety?
4. How do you imagine that the nails resisted being withdrawn? Draw a free-body diagram of a nail.
5. Try to outline a standard specification for a test of this kind. How does your specification compare with the procedure described in ASTM D 1761?

DISCUSSION

1. What are some of the important strength characteristics of materials?
2. Make some simple sketches to illustrate the five primary loading types.
3. How would you test the shock absorber in the landing gear of a lunar lander that might touch down during either the day or the night?
4. What are the factors involved in designing a test?
5. If you had to select a representative number of test specimens from a container of bolts, how would you go about it?
6. Suppose that the volume of a brick is to be accurate within 5 percent. What does this imply as to the necessary accuracy of its three dimensions?
7. What is the sensitivity of your watch, your car's speedometer, your ruler, and a room thermometer? What do you estimate the accuracy of these instruments to be?
8. If your testing team has three members, how do you divide the duties?
9. Make a general outline for a report of a typical test, such as a tension or compression test.
10. Devise a test method to determine the coefficient of restitution of, say, tennis balls.

READING

Make a trip to the library to inspect its holdings in the materials testing area, including periodicals. What is the subject code in the classification system your library uses? To which journals does your library subscribe? Ask about indexes and information retrieval systems.

Barry, B. A.: *Engineering Measurements*, Wiley, New York, 1964.
Fenner, A. J.: *Mechanical Testing of Materials*, George Newnes, London, 1965.
Holman, J. P., and W. J. Gajda: *Experimental Methods*, 3d ed., McGraw-Hill, New York, 1978.
Sandor, B. I.: *Experiments in Strength of Materials*, Prentice-Hall, Englewood Cliffs, N.J., 1980.
Schenk, H.: *Theories of Engineering Experimentation*, 3d ed., McGraw-Hill, New York, 1979.

TESTING MACHINES

6.1 TYPES OF TESTING MACHINES

In laboratory testing, loads are applied to test specimens by means of testing machines. Depending on their purpose, these vary in configuration, size, capacity, versatility, and all manner of detail.

Two essential parts of a testing machine are (1) a means for applying load to a specimen and (2) a means for measuring the applied load. Depending on the design of the machine, these two parts may be entirely separate or they may be superimposed one on the other. In addition to these basic features, there are a variety of accessory parts or mechanisms, such as devices for gripping or supporting the test piece, the power unit, controllers, recorders, speed indicators, and recoil or shock absorbers.

The load may be applied by mechanical means, through the use of screw-gear mechanisms, in which case the machines are referred to as "screw-gear" or "mechanical" machines. When the load is applied by means of a hydraulic jack or press, the device is called a "hydraulic" machine. The power may be supplied by hand or by some prime mover (usually an electric motor) to a pump or gear train, depending on the design of the machine and its capacity.

Some machines are designed for one kind of test only, such as tension machines made for testing chain and wire and compression machines for testing concrete specimens. If a machine is designed to test specimens in tension and compression, it is called a *universal testing machine*, or sometimes UTM for short. By means of appropriate attachments, universal testing machines can be used to perform flexure, shear, hardness, and other tests. There are also special machines for torsion, hardness, impact, fatigue, cold-bending, and other tests. In some of these special machines, load is not measured.

Sometimes it is advantageous to have the specimen horizontal, as when the testing chain or long specimens of wire rope. On the other hand, vertical machines are preferable when testing columns to prevent the column from bending due to its own weight, which would occur if it were in a horizontal position.

Some machines are designed for the application of static loads only, while others permit cyclic load applications to test fatigue strength. Most machines provide only axial loading, but some are capable of torsional loading as well.

The various features mentioned above, together with the type of load-indicating mechanism and the size or capacity, serve to classify a testing machine.

Excepting the direct use of weights, in its elementary (and earliest) form the testing machine consisted of a single lever, which was used both to apply and to balance the load (see Fig. 6.1a). In such a machine, the lack of means to compensate for the deformation of the specimen and the movement of the machine parts was a serious disadvantage, so that the next step in the development of the testing machine was to provide a means of loading independent of the means of weighing (see Figs. 6.1b and 6.1c). In Fig. 6.1b the load is shown applied hydraulically, whereas in Fig. 6.1c a mechanical means is employed. Such machines are of interest in demonstrating fundamental action, although single-lever machines are now rarely used.

Two principal types of power-driven universal machines are now in common use: (1) screw-gear machines with multiple-lever and movable-poise or pendulum weighing devices or with electronic load-measuring devices, and (2) hydraulic machines, the more accurate types of which use the Emery capsule and a modified bourdon† tube, or a bourdon tube in combination with an isoelastic spring or an electronic device, to measure and indicate the load. The principles of operation of the two types of machines are illustrated schematically in Fig. 6.2.

Some of the general requirements for testing machines are as follows:

1. The required accuracy must be sustained throughout the loading range; errors are ordinarily required to be less than 1 percent, but 0.5 percent or less is desirable.
2. The machine should be sensitive to small load changes, that is, precise.
3. Jaws in the crossheads should be in alignment.
4. Movable heads should not rock, twist, or shift laterally.

†Pronounced '*boor*-don.

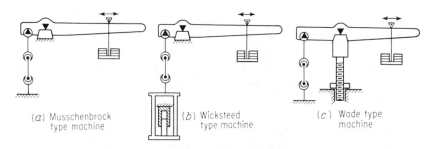

(*a*) Musschenbrock type machine   (*b*) Wicksteed type machine   (*c*) Wade type machine

Figure 6.1 Early testing machines. (*From Gibbons, Ref. 28.*)

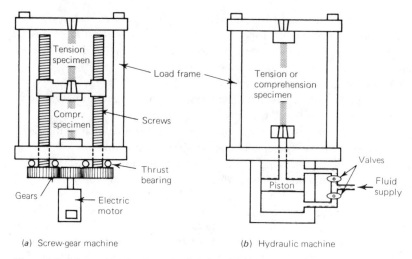

(a) Screw-gear machine   (b) Hydraulic machine

Figure 6.2 Schematics of universal testing machines.

5. Load application should be uniform, controllable, and capable of considerable range in speed.
6. The machine should be free from excessive vibration.
7. The recoil mechanism should be adequate to absorb the energy of rupture of test pieces breaking suddenly, so as to avoid injury to the machine when loaded to capacity.
8. The machine should be capable of easy and rapid manipulation and adjustment and should permit easy access to specimens and strainometers.

Sometimes autographic or semiautographic stress-strain recorders become part of the testing machine ensemble. Closed-loop control systems allowing the selection of various command inputs are also available. In machines of small to moderate capacity the recoil energy may be absorbed by rubber pads, but on large machines hydraulic recoil cylinders are used. Gripping devices, bearing blocks, and specimen supports are essential accessories.

This chapter deals mainly with universal testing machines. Specialized machines are discussed as part of the tests or in connection with the materials to which they apply.

6.2 DETERMINATION OF LOAD

The following paragraphs describe some of the methods of measuring load in materials-testing practice. Certain of these methods may be used alone or in combination with other methods.

Weights When weights of known magnitude are used directly as the means of applying load, they also serve to measure the load. The procedure is limited in application, but it may be used for nondestructive testing of full-scale structures.

By means of a horizontal lever, the arms of which are of fixed but not necessarily equal length, a given load on one arm may be balanced by some combination of weights on the other arm. This principle is sometimes used to bring a lever weighing system within a desired *range* of load, but because the process of balancing by continually adding separate weights is a slow one, it is rarely used alone in testing machines.

For testing purposes one of the useful principles of weighing is that of the steel-yard, whereby the load applied to the short arm is balanced by a weight of constant magnitude placed at the appropriate point on the long arm. The long arm, or scale beam, is graduated to indicate the load corresponding to the position of the movable weight, which is sometimes called a rider, poise, or jockey weight (see Fig. 6.1). Another form of the variable-lever principle is the pendulum. The steelyard method of weighing load requires manual operation for balancing, but the pendulum method together with the use of a suitable scale is self-balancing.

The actual load to be balanced by the elementary weighing device is sometimes reduced or stepped down from a given full load by an intermediate compound- or multiple-lever system. This is necessary when large loads are to be measured, in order to keep the weighing device within convenient, usable proportions.

Hydraulic devices Liquid pressures are commonly measured by means of manometers or bourdon tubes. A manometer is simply a glass tube, usually placed vertically, in which a liquid (say, mercury) can rise to the level at which it balances the applied pressure; the level of the liquid is read from a graduated scale. Obviously the manometer is limited to the measurement of relatively low pressures, so that its use for large loads would require an intermediate transmission device to step down the load.

The bourdon tube is essentially a closed-end curved metal tube that tends to straighten out as the pressure is increased in the liquid in the tube. In the usual bourdon gage the motion of the end of the tube is mechanically magnified to rotate a pointer over a scale, as indicated schematically in Fig. 6.3. The accuracy of the ordinary bourdon gage may be considerably affected by temperature changes, hysteresis, and the friction of its moving parts.

The load to be weighed may be transmitted hydraulically either by a hydraulic

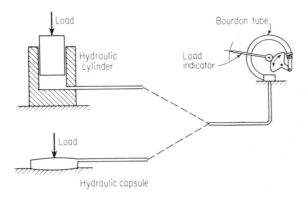

Figure 6.3 Weighing by hydraulic pressure. (*From Gibbons, Ref. 28.*)

cylinder and piston or by a closed flexible capsule, both of which are shown schematically in Fig. 6.3.

Interconnected hydraulic devices of different piston areas may be used in place of an intermediate lever system to step down the load, and the small piston can be made to actuate a steelyard or pendulum weighing device; this is simply the reverse of the usual hydraulic-jack principle.

The hydraulic cylinder has two marked disadvantages when used in load-weighing systems: the leadage of the liquid in loosely fitted pistons and the variable friction on the piston when packing is used. The friction can be reduced by the use of cylinders fitted with carefully ground and lapped pistons, and it can be still further reduced by a mechanism that rotates the piston during operation of the unit; these devices do not, however, fully eliminate the difficulties, and they complicate the manufacture of the apparatus.

The hydraulic capsule, which operates without appreciable friction, has proved to be a very satisfactory means of transmitting load and has come into wide use. Figure 6.4 shows a schematic cross-sectional view of the Emery capsule, which essentially consists of a thin, flexible metal diaphragm and a heavy recessed block tightly clamped together to seal in the liquid, usually oil, to form a closed hydraulic system. Because the liquid is practically incompressible, the load on the block is transmitted to the oil in the shallow chamber without much movement of the diaphragm. The oil shown is only in the load-measuring system. Another, separate oil system is used in testing machines that apply the load hydraulically.

Mechanical devices In general, "dynamometers" are transducers by means of which power output or power transmission can be measured. Because the mechanical measurement of power usually resolves itself into the determination of a force (along with other quantities), the term dynamometer is often applied to self-contained load-measuring instruments (usually portable).

Many dynamometers (in the restricted sense of a load-measuring instrument) utilize the deformation or deflection of an elastic member as the basis for determining the force applied to the device. In use, a dynamometer is inserted into the force circuit and the force to be measured (or a known fraction of the force to be measured) is transmitted through the dynamometer. By calibrating under known forces the

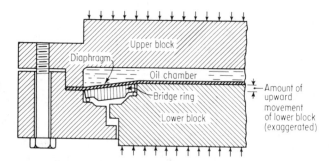

Figure 6.4 Partial section of Emery capsule.

Figure 6.5 Load cell. (*Courtesy of MTS Systems Corp.*)

deflection of the elastic element can be translated directly into terms of force transmitted, either by using a suitably graduated scale or by applying a calibration factor to the indicated deflections.

In materials testing, two types of dynamometer are used. One type is a spring balance made with a closely wound helical spring that can be used directly to measure the loads on a small specimen or can be used in conjunction with a multiple-lever or hydraulic transmission system. In the other type, instead of a helical spring, the elastic deflection of a beam, frame, or ring may be used to measure load. The "calibration ring" is such a device; it is described in Art. 6.6 in connection with the calibration of testing machines.

Electric devices Quite often, the load on a specimen must be translated into an electric signal capable of actuating some kind of electronic readout or control. This is accomplished by using "load cells" consisting of an elastic device to which electric-resistance strain gages (described in Chap. 7) have been permanently attached. Such load cells have the additional advantage of being able to monitor high-frequency dynamic-load applications. Figure 6.5 shows an example.

6.3 STATIC TESTING MACHINES

Most testing machines are designed to impart a constant or gradually varying load to a specimen. They can be used for "static" tension, compression, shear, bending, hardness, and creep tests. Although there are special-purpose machines, these tests can be performed with the universal testing machines described here.

In a mechanical machine the load ordinarily is applied to a specimen through the "movable crosshead," which in most machines moves downward during a test. In the case of a specimen in tension, the load is resisted by a "fixed" crosshead, which may, however, be placed in various positions. In a compression or cross-bending test the load is resisted by the "bed" or "platen" of the machine. In screw-gear lever-type machines, the fixed head or the platen then transmits the load to the compound-lever weighing system.

In a hydraulic machine the load ordinarily is applied by the movement of the piston of the hydraulic system, which is connected either to the platen of the machine or to a movable crosshead. The crossheads may have hydraulic lifts and, if so, may also be clamped to the columns by hydraulic locks. The load-weighing mechanism may originate in either the fixed or the movable portions of such machines.

Figure 6.2 shows schematically on what principles the two types of machines operate. In some hydraulic machines the load application is one-directional, as in mechanical machines.

The capacities of screw-gear machines are generally less than 1500 kN†. The difficulty of providing adequate knife-edge fulcrums, which must carry the full load developed by the machine, places a practical upper limit on the size of machines using levers to measure the load. Screw-gear machines in general tend to be rather noisy.

The hydraulic machine offers a means of attaining very large capacities. The largest machine, used for compression only, is the 10 000 000-lbf (45-MN) machine at the National Bureau of Standards [12]. A number of universal machines in the range from 5 to 20 MN are in use. There are also small, hand-pumped testers with capacities as low as 5 kN. In the modern hydraulic machine the load can be applied quickly, easily, with little noise or vibration, and with good control of the loading rate. Some cheaper hydraulic machines that use the pressure in the loading cylinder as an indication of the load on the specimen may be susceptible to larger inaccuracies, but the better machines, especially those that incorporate the Emery capsule or electric load cells, can be very accurate.

Screw-gear machines In some universal machines, a motor-driven screw-gear mechanism actuates the movable head, which transmits the load through the test piece directly to the platen or to the fixed head and then indirectly to the platen. The load on the platen may in turn be balanced by a multiple-lever system that terminates with the scale beam and poise weight. However, some screw-gear machines weigh the load by a direct-reading pendulum weighing system. In some machines, such as the one shown in Fig. 6.6, the load is measured by means of an electric load cell that electronically operates the load indicator.

In some testing machines the screws themselves rotate within bearing nuts mounted in the movable head as shown in Fig. 6.2; in other machines the screws are fixed to the movable head and the bearing nuts are in the gears below the bedplate. Either system serves satisfactorily to move the head.

Machines having two, three, or four screws are used. The two-screw machines

† 1 kN = 225 lbf.

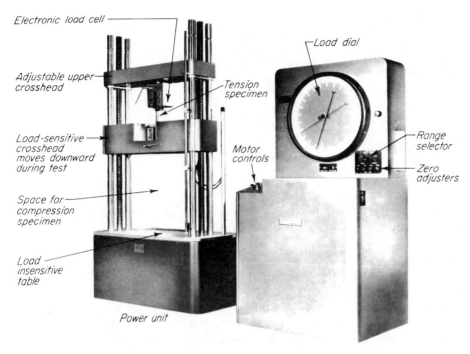

Figure 6.6 Mechanical universal machine with electronic load cell. (*Courtesy of Wiedmann Machine Co.*)

are well adapted to tension and transverse tests, but when they are used for compression tests, the specimen must be carefully placed in the plane of the screws and midway between them to avoid bending the screws. The specimen is not so accessible in the three- and four-screw machines as it is in those having only two screws, but the former are not so readily damaged by accidental eccentricity or off-center loads.

Single-screw machines are sometimes used for tension tests of wire, rubber, fabric, or cement briquets.

The knife-edges of a machine should be checked from time to time, because if they become dulled, chipped, clogged with debris, or displaced on the seat, the accuracy and sensitivity of the machine may be appreciably reduced.

To prevent the platen from jumping off the knife-edges when a specimen ruptures suddenly, hold-down bolts with rubber-insulated recoil nuts are used. These nuts should *not* bear on the platen and should always be checked at the beginning of a test to see that they are just barely loose.

At the beginning of any new test or series of similar tests, the scale beam or indicating mechanism should be balanced at zero load, with all the equipment on the platen that is to be used in the test but that will not stress the specimen. The scale beam or indicating mechanism is usually balanced by adjusting small counterweights.

Some screw-gear machines are designed so that different poise weights or pendulum weights may be used. By using a weight smaller than that required for full-capacity

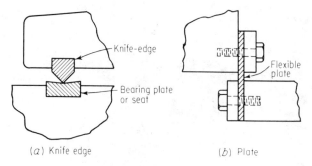

(a) Knife edge   (b) Plate

Figure 6.7 Fulcrums.

loads, the machine can be used for small loads with a precision somewhat comparable to that for large loads, provided adequate sensitivity can be maintained.

In all mechanical machines, *fulcrums* or pivots are important details. It is necessary that they operate with a minimum of friction and without lost motion and that they maintain a constant position (lever arm). Further, they should be designed so as to be stable and remain in alignment under load.

In testing machines using a lever weighing system, the fulcrums are usually hardened-steel "knife-edges" in which two ground surfaces meet at a 90° angle to produce a straight line, which is the bearing edge. In small instruments, the angle between the surfaces meeting at the bearing edge is often much less than 90°. The bearing plate or knife-edge seat as shown in Fig. 6.7a which is usually made of hardened steel, also has ground surfaces commonly meeting at a reentrant angle somewhat less than 180°. In testing machines the allowable compressive load on knife-edges range up to 2MN/m.†

An alternative design, shown in Fig. 6.7b, places the fulcrum in tension by using thin sheets with a short unsupported length. Allowable tensile loads of 10 MN/m have been used in plate fulcrums. Translatory motion of the joint can be prevented by using two adjacent fulcrum plates at right angles to each other.

Hydraulic machines In one type of hydraulic machine, the load is applied by a hydraulic press and is measured by the pressure developed within the hydraulic cylinder. The main piston is usually carefully fitted and lapped; to reduce the friction of the small piston used in the weighing system, this piston is rotated during operation of the machine. The load is weighted by a pendulum device or by a bourdon tube.

In another type of hydraulic machine, the load is applied by a hydraulic press independently of the weighing system, which is actuated by a hydraulic capsule. In some machines, such as that shown in Fig. 6.8, the very slight motion of the tip of the bourdon tube operates electronic units, which in turn operate the load pointer. In other machines the direct use of the bourdon tube has been replaced by a mechanism in which the bourdon tube operates the pointer indirectly by an air pressure system and an isoelastic (linearly elastic) spring. These methods overcome the well-known disadvantage of the ordinary bourdon tube, namely, that it does not give a

† 1 MN/m = 5700 lbf/in.

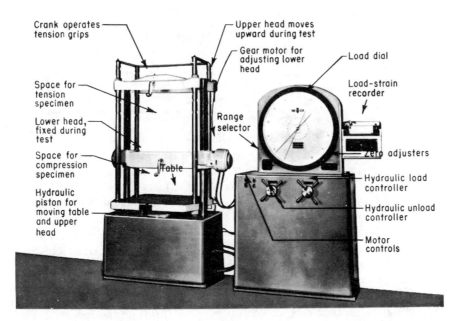

Figure 6.8 Hydraulic universal machine with electronic load indicator. (*Courtesy of Tinius Olsen Testing Machine Co.*)

straight-line relation between pressure and motion of the tip. In some machines the hydraulic capsule is located in the movable crosshead.

In a closed hydraulic load-indicating system, the presence of air in the system causes erratic and inaccurate measurements. Whenever repairs have been made, or if there is any possibility of leaks, the reliability of the load-indicating system should be checked.

Readout devices Most hydraulic machines are equipped with two or more load-indicating dials to serve for different ranges of load, or they have one dial with a mask that can be rotated to expose different groups of figures and thus permit the one dial to serve for various ranges of load. Suitable load-measuring mechanisms are used for each load range so that small loads can be observed with a precision comparable to that for large loads.

The loads may also be displayed by a digital indicating system. These readout devices normally provide four to six digits of readout and may be adjusted for several load ranges. The maximum load attained may be stored and recalled.

Many machines have a built-in *recorder* to generate a plot of load vs. deformation or strain. Such a recorder can be seen on the right side of the machine pictured in Fig. 6.8 and is diagrammed in Fig. 6.9. It detects deformation as well as load.

At the top right of Fig. 6.9, a deformation of the specimen causes the core of a "differential transformer" in the strain instrument (1) to move proportionately and

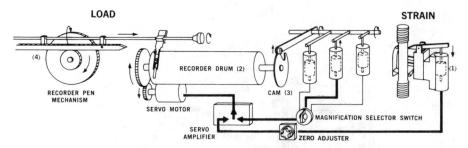

Figure 6.9 Load-strain recorder. (*Courtesy of Tinius Olsen Testing Machine Co.*)

produce a displacement voltage and direction signal. After amplification, the signal activates a servomotor that rotates the recorder drum (2) and a cam (3). As the cam turns, it moves the cam follower and the attached cores of three differential transformers (one for each magnification range). The core motion produces an opposite and equal signal. The signal for the selected magnification range travels to the servo amplifier and restores the null balance when the opposing signals are of equal intensity. Thus, the pen line drawn by the rotation of the recorder drum indicates the specimen strain. While the recorder drum is being rotated at a rate proportional to the specimen strain rate, the load-activated pen mechanism moves the pen point across the drum in direct proportion to the increasing or decreasing stress (load) on the specimen. The recorder pen assembly accommodates several test ranges by means of a double-cut rack (4), which provides a full recorder chart scale for either full- or one-half-range capacity. The strain magnification can also be changed. Strain detection is discussed further in Chap. 7.

Speed adjustment The driving mechanisms for screw-gear testing machines are usually made to operate the head at four or more speeds. The several speeds may be obtained by the selective use of different gear ratios, by the use of various fixed motor speeds, or by electronically controlled drives, which make possible any speed of testing desired. In most hydraulic machines, any rate of load application can be obtained by the use of an appropriate pump speed or valve setting controlling the flow of oil from the pump to the loading cylinder. In such machines, the load rate is often controlled by an auxiliary pacing arm or disk on the load-indicating dial; to apply load at a desired rate, the operator sets the pacer at the given rate and then adjusts the motor or pump controls to make the load-indicating pointer follow the pacer.

Some more sophisticated controls will be discussed in the next article in connection with dynamic test systems.

6.4 CYCLIC LOADING MACHINES

For some tests, machines must be able not only to exert a constant or gradually varying load but also to change the load according to some predetermined pattern. Such tests include the determination of various types of fatigue strength, both for material specimens and for manufactured components. Machines that can cycle a load are called dynamic testing machines. They are normally of the hydraulic type, although mechanical methods of varying the load, such as by means of an eccentric flywheel, exist (see Chap. 14). Machines that apply impact loads are discussed in Chap. 13; while these are also dynamic machines, they do not cycle the load.

Hydraulic machines As in static machines, in a hydraulic machine the load is transmitted from a piston that is acted on by pressurized fluid. In fact, a test is often started by applying a static load, which is then cyclically varied. This variation is made possible by a servo valve that controls the flow of fluid from one side of the piston to the other. The piston moves within the "hydraulic actuator." Its movement can be measured by an actuator displacement transducer, about which more will be said in Chap. 7.

If rather flexible materials are to be tested at high frequencies, pressurized hydraulic fluid must be supplied to the actuator at a high rate, possibly several liters per second, at a pressure of, say, 20 megapascals.[†] This is the function of a performance unit, which consists of an electric pump, a fluid reservoir, and related components. In high-capacity testers the unit is physically separated from the load unit, with hoses providing the connection.

In addition to load and performance units, a complete system, as shown in Fig. 6.10, includes a control unit. A control unit consists of electronic circuitry that allows the determination of test programs and provides for the readout of test data. Either ramp or cyclic functions can be generated, and data relating to load, displacement, and strain, as well as to cycle count, can be displayed. Numerous options are available, so that the manufacturer's literature should be consulted for details.

Control systems The performance of many types of tests requires a control of the rate of loading or of strain that is more accurate than a human operator's reflexes allow. Various mechanical, hydraulic and especially electronic devices provide such controls, at greatly varying levels of sophistication.

Especially in the case of dynamic testers, closed-loop systems permit tests to follow various preselected command inputs such as ramp or sinusoidal load or displacement functions. A closed-loop system is shown diagramatically in Fig. 6.11. Force is measured by a load cell and displacement by a displacement transducer, while strains may be monitored by a deformeter as described in Chap. 7. A selected feedback variable is continuously compared with the command input by the controller, which then transmits a correction signal to the servo valve. Simultaneously the monitored variables are transmitted to the readout devices, that is, to digital displays, X-Y plotters, and the like. In this way rigorous test control as well as full test documentation can be achieved.

6.5 ACCESSORIES

In addition to the basic machines described, various accessories are needed for testing. These include devices that grip or support all manner of specimens, and fixtures that

[†]1 MPa = 145 psi. Note that $(L/s) \cdot MPa = (10^{-3} m^3/s) \cdot (10^6 N/m^2) = kW$.

Figure 6.10 Dynamic test system. (*Courtesy of MTS Systems Corp.*)

allow other than axial loading conditions. Displacement and strain-measuring instruments, which are discussed in Chap. 7, are needed. Many options exist in the addition of control, readout, and recording devices.

Grips are needed to support flat or round, threaded or button-ended specimens, as well as special items such as the compact tension specimens used in fracture mechanics. The grips may be accommodated in closed or semiopen crossheads and may be operated by levers or by cranks with a rack and pinion arrangement. Levers may be operated manually or hydraulically. There are self-aligning grips and water-cooled versions for high-temperature testing.

Special fixtures include those for flexure tests, which utilize a base with two supports and a loading device to provide one or two concentrated loads. There are fixtures that allow various direct shear tests, and others that permit a compression machine to be used as a hardness tester. Other devices accommodate special tests such as compression tests on concrete cylinders and on cubes of various sizes; cylinder-splitting tests; tests on bricks, blocks, and other masonry units; guided weld tests; tension tests of wires and cables; cold-bending tests of reinforcing bars, and the like. Environmental chambers are available that surround the specimen and subject it to extremely low or high temperatures or to other environmental conditions.

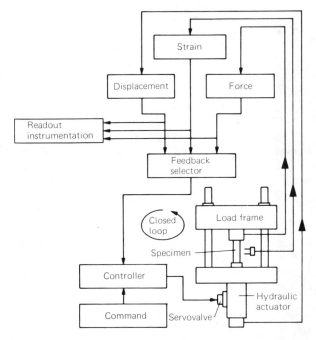

Figure 6.11 Closed-loop test system. (*Courtesy of MTS Systems Corp.*)

A wide range of *electronic accessories* can also be added. There are remote control units, for example, and automated amplitude controllers that permit unattended testing. Readout options include oscilloscopes (discussed in Chap. 7) to observe signals, dual digital indicators to display upper and lower peaks of a test parameter, strip-chart and X-Y recorders to plot one function vs. another, and card printers to produce reports. Of course, many variations and combinations are possible. Figure 6.12 shows an X-Y recorder combined with digital readout, and Fig. 6.13 shows a strip-chart recorder.

6.6 CALIBRATION

Periodically a testing machine must be *calibrated*, that is, adjusted so that the load readings conform to a standard. Three methods commonly used to calibrate testing machines are (1) use of weights alone, (2) use of levers and weights, and (3) use of elastic calibration devices (ASTM E 4). Weights, of course, are awkward to transport and handle. The simplest and most common method of calibrating large-capacity machines is by use of an elastic calibration device, which consists of an elastic metal member or members combined with a mechanism for indicating the magnitude of deformation under load. Two forms of this device are (1) a steel bar together with an attached strainometer and (2) a *calibration ring*, which is a steel ring or loop combined with some type of deflection indicator. The steel bar is suitable principally for use in ten-

Figure 6.12 Flatbed X-Y recorder. (*Courtesy of Tinius Olsen Testing Machine Co.*)

sion, although some bars are used in compression. The ring or loop devices are made for use either in tension or in compression. A calibration ring for use in compression is illustrated in Fig. 6.14. A compressive load shortens the vertical diameter, and the change is measured by the micrometer. From this change and the calibration data for the ring, the applied load can be determined. Calibration rings of this sort are available in capacities up to 1.5 MN,[†] but compression bars having capacities approaching 15 MN, which are equipped with electronic strain gages, are available at NBS. Also, for calibrating very large machines in compression, several calibration rings or bars can be used in parallel.

The following are three important requirements for an elastic calibration device (ASTM F 74):

1. It should be so constructed that its accuracy is not impaired by handling and ship-

[†]1 MN = 225 kip.

Figure 6.13 Strip chart recorder. (*Courtesy of Tinius Olsen Testing Machine Co.*)

ping and that parts subject to damage or removal can be replaced without impairing the accuracy of the device.

2. It should be provided with shackles or bearing blocks so constructed that the accuracy of the device in use is not impaired by imperfections in the shackles or blocks.

3. It should be calibrated in conjunction with the strainometer that is to be used with it, and the strainometer should be used in the same range as that covered by the calibration.

Care must be taken to minimize any temperature changes during the use of an elastic calibration device. Furthermore, the actual temperature at the time of use and at the time of its own calibration must be known, since the elastic properties of the device change with temperature. In general, the reading of a ring-type device changes by about 0.025 percent for each 1 K change in temperature from the standard.

Devices of this sort may be calibrated directly by standard weights. A testing machine having a capacity of 500 kN, produced by the use of standard weights, is available at NBS. For loads in excess of this, calibration is performed by the use of combinations of previously calibrated devices. ASTM requires that the calibrating loads be accurate to 0.02 percent for loads up to 450 kN and to 0.1 percent for loads from 450 to 1350 kN, but allows somewhat larger errors for greater loads (ASTM E 74). Before calibrating an elastic device it is desirable to subject it to a series of cyclic loadings over a range about 20 percent greater than that of its proposed use. The usable loading range for ordinary use is defined as that range in indicated loads for which the percentage error does not exceed specified tolerances (ASTM E 74).

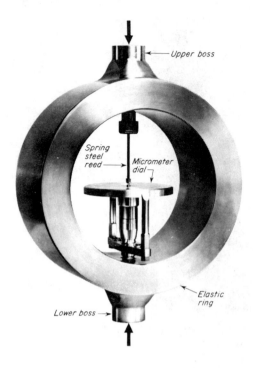

Upper boss

Spring steel reed

Micrometer dial

Elastic ring

Lower boss

Figure 6.14 Calibration ring for compression. (*Courtesy of Morehouse Machine Co.*)

In all ordinary calibration work, the calibrating loads should be applied so that the resultant load acts as nearly as possible along the axis of the loading. In special instances, calibrations may be made with the load applied at known eccentricities.

There is a difference between the *calibration* of testing machines, or the procedure of determining the magnitude of error in the indicated loads, and what ASTM (ASTM E 4) calls the *verification* of testing machines. Verification is a calibration to ascertain whether or not the errors are within a stated permissible range, and it implies certification that a machine meets stated accuracy requirements. The "permissible variation," or maximum allowable error in indicated load of a testing machine, is 1 percent. The "loading range" is the range of indicated loads for which the machine gives results within the specified permissible variation. The allowable loading range should be stated in any verification certificate. It is recommended that no correction be used with machines tested and found to be accurate within the limits prescribed (ASTM E 4).

It is specified that calibration corrections shall *not* be applied to indicated loads to obtain values within the required range of accuracy. Obviously this implies that the machine must be adjusted or modified until the calibration shows it to be within specified limits.

The adjustment of machines having a poise weight or a pendulum is readily accomplished by changing the weight of these elements. For hydraulic machines having an isoelastic spring, adjustment is accomplished by changing its effective length; for those using an electric gage at the tip of the bourdon tube, adjustment is effected in the tube connection; for those having a simple bourdon-tube load gage, adjustment must be made in the linkage of the gage.

Temperature changes do not affect the accuracy of a mechanical machine but do have a slight effect on all hydraulic machines using a bourdon tube. However, for normal temperature changes, the errors so introduced are usually less than about 0.1 percent.

PROBLEM 6 Study and Calibration of a Testing Machine

Object: To determine the errors in the loads indicated by the weighing system and the sensitivity for each load applied.
Preparatory reading: ASTM E 4.
Special apparatus: Proving ring or other calibration devices.
Procedure:
1. Study the construction of the testing machine and note how it applies tensile or compressive loads. Draw a sketch to show how the laod is applied and measured.
2. Before placing the proving ring in the tesing machine, turn the dial of the ring at least six revolutions away from contact with the reed mechanism. This is done so that when the preload is applied, the reed will not contact the anvil and bend, causing a large change in the zero reading.
3. Place the proving ring in a vertical position, with the lower boss resting on a hardened steel plate and a soft steel plate bearing on the upper boss. After adjusting the machine to read zero load, preload the ring to the limit specified and then unload.
4. Obtain the initial reading of the ring by turning the dial until the anvil almost touches the reed hammer. Vibrate the reed by deflecting the hammer about 10 mm to one side and releasing it; then slowly turn the dial until the anvil touches the hammer, which will be indicated by a buzzing sound. Read the dial to 0.1 division and observe the temperature. Apply load in 10 approximately equal increments, but before each increment withdraw the anvil from the reed. For each increment record the reading of the calibration device and the load indicated by the machine.

The difference between the initial and subsequent dial readings, after correcting for the temperature deviation and multiplying by the ring factor, gives the true load.
5. Observe the sensitivity of the machine for each load applied. The measure of sensitivity is the increase in the applied load required to produce an observable movement of the load-indicating device. Small weights may be placed on the platen of a mechanical machine and some types of hydraulic machine for this determination.
Report: Submit a report as you would if the university had engaged you to perform an independent verification of the machine. State all pertinent facts.
1. Show the error in each indicated load as a percentage of the true load. For indicated loads that are too high, the error is positive. Plot a curve of errors as ordinates against indicated loads as abscissas. On this graph show the maximum error permitted by ASTM. Also plot a curve showing the sensitivity at each load.
2. Indicate in which portions of the loading range covered by the test the machine meets ASTM requirements for accuracy. Draw conclusions regarding the suitability of the machine for ordinary testing insofar as accuracy and sensitivity are concerned. If the machine was calibrated for only a portion of its capacity load, discuss the probable inaccuracies for loads above the calibration limit.

DISCUSSION

1. Name in order of relative accuracy several methods of calibrating a testing machine. State the limitations of each method.

2. What may cause a change in the calibration of (*a*) a hydraulic machine and (*b*) a screw-gear machine?

3. Is any error introduced by the use of heavy base plates inserted below the specimen in a compression test? Explain and qualify your answer.

4. How does the Emery hydraulic machine differ from the other types of hydraulic machines? What are its advantages and disadvantages?

5. Does ASTM permit a blanket acceptance or rejection of a testing machine as a result of its calibration? Explain.

6. Is a sensitive machine always accurate? Can an accurate machine be sluggish?

7. What is the most simple method of correcting the error in (*a*) a lever machine and (*b*) a hydraulic machines so that the indicated loads will meet the ASTM requirements for accuracy?

8. Discuss the effect of temperature changes of, say, 10 K on the accuracy of testing machines.

9. Can you think of any natural phenomena based on the same principle as the bourdon tube?

10. Explain the principle of closed-loop control systems. Can you illustrate the idea with a familiar example?

READING

Few books deal primarily with the subject of testing machines. Look for opportunities, however, to inspect some current catalogs and other manufacturer's literature. That will help you keep up to date with a constantly changing technology.

SEVEN

INSTRUMENTS

7.1 LINEAR MEASUREMENTS

Laboratory testing work often requires fairly precise linear measurements, such as in determining the dimensions of specimens. Various instruments are available that improve the accuracy and precision of the readings by reducing human error.

The major obstacle to attaining accuracy is parallax, a problem familiar to photographers who do not have a single-lens reflex camera. Reading a graduated scale consists of estimating the position of some mark (an index line, a pointer, or the like) along the scale. Unless the eye, the mark, and the correct position line up, a parallax error results. Parallax can be eliminated in one of two ways, by "edge coincidence" or by use of the "mirror-scale" principle. In the first method, the mark is made to lie in the plane of the scale graduations as in Fig. 7.1a. One form of a mirror-scale device is shown in Fig. 7.1b. When the cross wire or pointer appears to coincide with its image, the line of sight through the cross wire is perpendicular to the scale and mirror.

Precision is subject to the limitations of eyesight. The least reading of a scale depends on the spacing of the graduated marks, and wherever possible it is desirable to estimate fractions of divisions. For greater refinement in reading fractions of a division, a *vernier* may be used.

A direct-reading single vernier, perhaps the most common type, is shown in Fig. 7.1a. A distance equal to nine divisions of the scale is divided into 10 equal divisions on the vernier, so that each division on the vernier equals nine-tenths of a division on the scale. Hence, if the first mark of the vernier beyond the zero or index mark matches up with any mark on the scale, the index is one-tenth of a division beyond the preceding scale mark; if the second mark on the vernier matches up, the index is two-tenths of a division beyond the preceding scale mark, and so on.

Whereas measurement of the distance between two points can be made directly by comparison with a graduated steel scale or tape, the distance between opposite surfaces of a solid object is best determined by using a caliper, the separation of the points of which can be measured directly with a scale. The direct use of a graduated

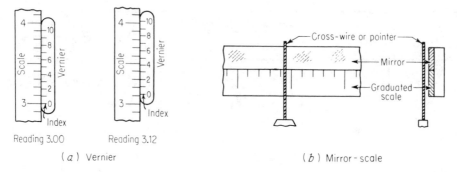

Reading 3.00   Reading 3.12

(*a*) Vernier   (*b*) Mirror-scale

Figure 7.1 Measurement aids.

scale yields results of limited precision, because the practical least reading of a scale with the unaided eye is about 0.25 mm. With a *vernier caliper*, which is shown in Fig. 7.2, the precision can be improved to about 0.05 mm.

Micrometers If finer measurements are needed, a micrometer[†] may be used. By definition, a micrometer is simply an instrument for giving a magnified indication of small distances. In many micrometers, the distance is in effect traversed by some moving part, and the resulting movement is magnified and measured. The determination of distances greater than the range of motion of the micrometer device requires that measurements be made with respect to some fixed point whose position is accurately known.

Perhaps the simplest form of micrometer is the screw micrometer. A common illustration of the screw micrometer is found on the micrometer caliper, shown in Fig. 7.3. If the screw has a pitch of 0.5 mm and the barrel has 25 divisions, each division corresponds to 0.02 mm, and readings with a precision of 0.01 mm (10 μm) can easily be estimated. Some micrometers are more precise, and some have measuring lengths of 500 mm or more.

In many uses of the screw micrometer, the end of the spindle or screw must contact the piece with reference to which measurements are being made. Some method of

[†]Note that the pronunciation of micrometer (mi-'*crom*-et-er), meaning "small-distance measurer," differs from that of micrometer ('*mi*-cro-met-er), the unit of length.

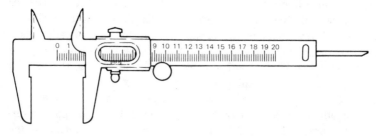

Figure 7.2 Vernier caliper.

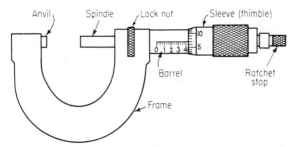

Figure 7.3 Screw micrometer caliper.

controlling the contact pressure is necessary if consistent results are to be obtained. The machinist's micrometer caliper (Fig. 7.3) is often equipped with a spring ratchet that releases at a definite contact pressure. In some applications of the screw micrometer (e.g., to measure deflection in a proving ring), the spindle is positioned accurately by the use of a vibrating reed arranged as indicated in Fig. 7.4. The reed is set in motion, and the screw is advanced until it comes in contact with the reed and alters its tone.

An even more precise device is the *micrometer microscope*. In one form, the magnified image of the points or marks with respect to which the measurement is to be made is superimposed on a scale within the instrument. In another form, the magnified distance to be measured is traversed by a movable cross hair, the motion of which is indicated by a screw micrometer. The practical least reading of an ordinary micrometer microscope is about 1 μm[†] with a range of about 2 to 5 mm.

Interferometer The most precise known means of measuring small movements is an adaptation of the phenomenon of interference of light waves. An interferometer is constructed in such a way that when monochromatic light is reflected from two nearly

[†]μm = 0.000 039 in.

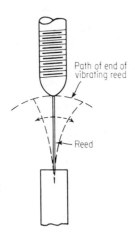

Figure 7.4 Vibrating reed principle.

parallel plane surfaces, reflected rays that are out of phase by one-half wavelength of the light interfere, causing a series of light and dark bands to be seen (see Fig. 7.5). As the distance between the reflecting surfaces changes, the bands appear to move across the field of view. The shift of one band across a reference mark corresponds to a movement of the plates of one-half wavelength of the light used.

The interferometer carries its own calibration, since the wavelengths of light are known to a high degree of accuracy.[†] Such an instrument is extremely sensitive and is suitable for very fine measurements. Under good conditions, measurements to one-tenth wavelength are easily possible, which for green light from a mercury-vapor lamp, of wavelength 546 nm, corresponds to about 50 nm. The total range of an interferometer is small, being about 1000 bands (often less), or 0.25 mm.

7.2 INSPECTION GAGES

In connection with the inspection of materials, it is often required that not only the quality of the material but also the dimensions and tolerances be checked. In the case of manufactured machine parts, this is done by the use of "gages." Although gaging is not an essential part of materials testing, it is important in inspection. A brief outline of common gaging procedures is given in this article.

Gages are devices used for determining whether one or more dimensions of a single part or assembly are within permissible limits. A simple gage is an individual instrument and can be used to check only one dimension, whereas a micrometer or scale, where it can be used, will serve for any measurement within its range. The gage, however, ensures reliability and rapidity of measurement not obtainable with a micrometer and, when proper limits are established, ensures interchangeability of manufactured parts, so that any of those accepted will fit into the same place without adjustment. Reference gages may be used to check service gages that are used in manufacturing operations or inspections. In many cases, manufacturing or inspection gages are calibrated directly by comparison with gage blocks of known accuracy or by the use of an interferometer instead of using check or reference gages.

[†]Note the definition of the meter: "The *meter* is the length equal to 1 650 763.73 wavelengths in vacuum of the radiation corresponding to the transition between the levels $2p_{10}$ and $5d_5$ of the krypton-86 atom." These wavelengths thus are about 606 nm long.

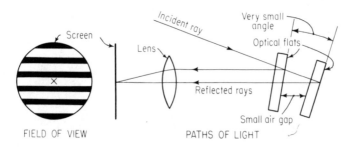

Figure 7.5 Principle of the inferometer.

There are many types of gages, varying from the simplest snap gage to the most complex arrangement, which corresponds to a fixture or jig in the manufacturing process made to check simultaneously such items as parallelism of surfaces, squareness, and distances between holes.

Simple gages are made in "go" and "no go" (or "not go") sizes. The go gage will fit over or enter into the part being measured as shown in Fig. 7.6, which illustrates two simple types of gages: the receiving and the entering or plug gage, and an example of a snap gage.

Indicating gages employ a device such as a lever and graduated scale or a dial indicator to show the variations in the dimensions being checked. The zero setting of the dial can be adjusted by a gage block or by some other unit, and go and no go marks can be made on the dial or actual plus-minus readings from the zero position observed.

Gaging is closely related to the concept of *tolerance*, which was developed during the nineteenth century and made interchangeable parts practical. Consider Fig. 7.6*a* and *b*. In the case of a shaft and bearing, a set of go gages might bring together a shaft of maximum diameter (one that will just pass through the go ring gage) and a bearing having the minimum or smallest allowable diameter (one that will just fit over the go plug gage). The mating of these parts will produce the minimum clearance. In order to control the maximum clearance, which is encountered when the smallest allowable shaft (one that just fails to enter the no go ring gage) is mated with the largest allowable bearing (one that just fails to be entered by a no go plug gage), the size of the no go gages must be based on the desired fit of the mating parts.

"Clearance" may be defined as the difference in dimensions between mating parts, but when clearance is specified as a definite amount, it is designated "allowance." We might say that allowance is intentional clearance. Since it is impossible to make a part intentionally to exact size, a definite leeway or variation must be permitted. This difference or range between maximum and minimum acceptable sizes is known as tolerance, or more briefly, tolerance is the allowable variation in a dimension. Thus the dimensions of many standard test specimens have given positive and negative tolerances.

7.3 DETERMINATION OF DISPLACEMENT

Various devices called *deformeters* are used to measure deformations or displacements, that is, changes in a length (or an angle) rather than the length (or angle) itself. Devices

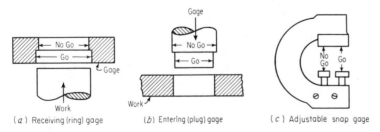

(*a*) Receiving (ring) gage   (*b*) Entering (plug) gage   (*c*) Adjustable snap gage

Figure 7.6 Simple gages.

used to determine changes in length resulting from linear strain are commonly called strainometers or strain meters and include extensometers and compressometers. Devices used to measure deflection due to bending are called deflectometers, and those used to measure angle of twist due to torsional loading are often called detrusion indicators. The instruments described in this article apply principally to the measurement of linear strains; deflectometers and detrusion indicators are described in later chapters in connection with the types of tests to which they apply. In essence we are concerned with applying the principles and techniques discussed in Art. 7.1 to more complex apparatus suited to the measurement of specimen deformations in testing practice.

Mechanical devices A type of micrometer in very wide use is the dial micrometer or dial indicator. In these instruments, the motion of a spindle actuates a lever or gear train, which in turn operates a pointer on a graduated dial. The dial indicator has the great advantage of being self-indicating. Figure 7.7 shows the principle of one type that uses a gear train mechanism driven by a spring-loaded spindle. Many dial indicators have divisions corresponding to movements of 10 μm, but more precise indicators are available. Ranges vary from 0.5 mm to over 100 mm, 10 mm being quite typical.

To obtain linear strains, dial indicators or other devices for measuring small displacements may be attached to fixtures that allow the measurement of deformation along a gage line over a specific gage length. Most of these devices, ingenious as they are, have been overtaken by electric instrumentation; however, some fixtures with continuing value are described below. Observe that their mechanisms improve the precision of the measurement by magnifying the deformation.

One type of strainometer is the averaging type of *collar extensometer* or *compressometer*, shown schematically in Fig. 7.8. This type of instrument gives the average strain in a specimen very closely, even though slight bending may occur. The deformation is first magnified (usually about twice) by the lever action of the collars and then is measured by some type of micrometer. This type of apparatus is frequently used on concrete cylinders in compression, but may also be used on metal bars in tension.

To obtain Poisson's ratio, lateral strains are determined by means of "lateral

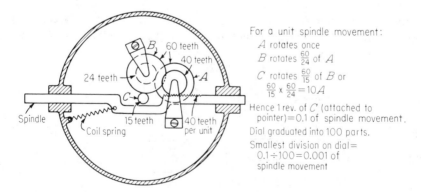

For a unit spindle movement:

A rotates once

B rotates $\frac{60}{24}$ of A

C rotates $\frac{60}{15}$ of B or

$\frac{60}{15} \times \frac{60}{24} = 10A$

Hence 1 rev. of C (attached to pointer)=0.1 of spindle movement.

Dial graduated into 100 parts.

Smallest division on dial= $0.1 \div 100 = 0.001$ of spindle movement

Figure 7.7 Typical dial-indicator mechanism.

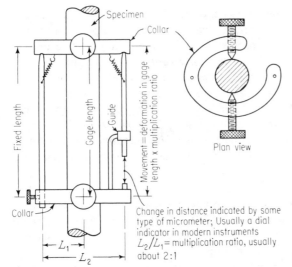

Figure 7.8 Collar deformeter.

strainometers." Many lateral strainometers are constructed along the lines indicated in Fig. 7.9a, with dial indicators used as a means of measurement. However, to achieve precision, extremely fine measurements are necessary. Vose's adaptation of the interferometer principle to measuring lateral strains is illustrated in Fig. 7.9b [98].

Optical devices Mechanical-lever systems are limited by friction in the joints and by the weight of the levers. By using jeweled bearings, plate fulcrums, or knife-edges, the bearing friction can be greatly reduced, but the levers still have weight; this difficulty can be removed by the use of beams of light for magnification. One such application, as used in *Marten's* extensometer, is illustrated in Fig. 7.10a. A plane mirror is attached to a double knife-edge or lozenge so that the mirror is rotated as the specimen changes length. The degree of magnification of this movement depends on the length of the long lever, which is the distance between the mirror and the scale, and the length of the short lever, which is the distance between opposite edges of the lozenge to which the mirror is attached. Sometimes a roller is used instead of knife-edges. The complete strainometer consists of two such units attached to opposite sides of the specimen.

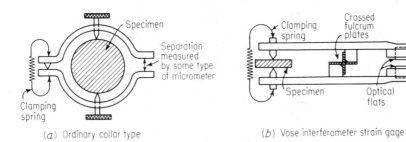

Figure 7.9 Lateral strainometer.

One of the most precise strainometers is *Tuckerman's* optical strain gage, which is illustrated in Fig. 7.10*b*. One surface of the lozenge, which has been polished to optical flatness, is used as a rotating or rocking mirror. Its rotation with respect to another mirror fixed in position with respect to the frame is measured with an optical device called an *autocollimator*. Strains as small as 2 μm/m can be measured.

The principal features of a *comparator* device are shown schematically in Fig. 7.11. Two microscopes are mounted on a fixed base or bar. One or both of the microscopes are fitted with micrometer eyepieces to measure the movement of the targets on the specimen. Except for the fact that the microscopes are replaced by telescopes, this general type of device is often used for observing the deformations in specimens tested at elevated temperatures.

Electric transducers Electrical principles may also be employed in measuring deformations. Devices based on these principles are particularly useful in that they are responsive to dynamic deformations and allow automatic recording of results. In much of testing they have replaced both mechanical and optical devices.

The *linear variable differential transformer (LVDT)* transducer is an electromechanical device that provides an output voltage proportional to a displacement. It consists of a transformer with a primary and two secondary coils wound on one cylindrical form and of a core that is oriented axially within the hollow body of the transformer. The arrangement of the transformer and the connection of the secondaries in series opposition are such that if the core is placed in a position that induces equal electromotive forces in the two secondaries, the net output voltage will be zero; the core is then said to be at electrical null. If the core is displaced toward one end of the unit, mutual inductance increases between the primary and one secondary while decreasing between the primary and the other secondary; the output emf's of the two secondaries are no longer equal and a net output voltage proportional to the displacement results. This output can be amplified and used to display a signal on a readout device.

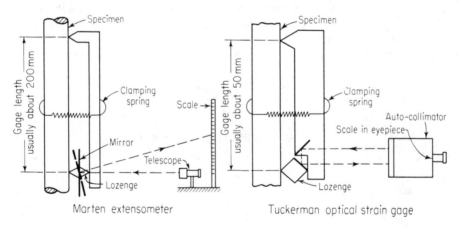

Figure 7.10 Mirror or "light-lever" extensometers.

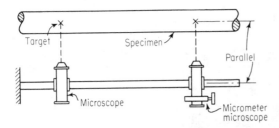

Figure 7.11 Optical comparator.

We previously alluded to LVDT applications in connections with recorders (Art. 6.3) and control systems (Art. 6.4) of testing machines. Now it is possible to visualize how LVDT deformeters, such as the one pictured in Fig. 7.12, work. They are used for measuring and controlling strain in tension, compression, static, and dynamic (up to about 300 Hz) tests. They also measure displacements such as crack openings in fracture-mechanics tests. Gage lengths range from 5 to 250 mm and more. In some models a mechanical breakaway arrangement safeguards the device against damage when a specimen ruptures.

Other devices are based on the electric resistance strain gages described in the next article.

7.4 ELECTRIC STRAIN GAGES

As we have seen, deformeters can be used to obtain strains by dividing deformations by gage lengths. There are devices, however, that allow strains to be obtained directly. One such device, described in this article and the next, is the electric resistance strain gage. This type of instrument has additional advantages, such as the capacities for remote sensing and automatic recording of strain, that make it highly convenient, flexible, and powerful. A less common device is the photoelastic gage, which is based on principles explained in Chap. 16.

Strain and resistance Lord Kelvin discovered that the electric resistance of a given wire is a function of the strain to which it is subjected, tensile strains usually increasing the resistance and compressive strains decreasing it. For strain-gage work it is common to express the change in resistance in terms of the change in strain, which gives a ratio called the strain sensitivity or gage factor K equal to $(\Delta R \cdot L)/(R \cdot \Delta L)$, where ΔR represents resistance change in the total gage resistance R and ΔL is the corresponding change in length of the total length L of the conductor. The strain ϵ then equals $\Delta R/(R \cdot K)$.

The strain sensitivity is markedly influenced by the type of resistance wire, as is shown in Table 7.1.

The various points to be considered in the selection of a resistance alloy, in order of their importance, are (1) gage factor (the higher the better); (2) resistance (the higher the better); (3) temperature coefficient of resistance (the lower the better so that the gage will not be too sensitive to temperature); (4) coefficient of linear expansion (the lower the better); (5) thermoelectric behavior, or the tendency to generate a

Figure 7.12 Clip-on LVDT extensometer. (*Courtesy of Tinius Olsen Testing Machine Co.*)

thermal emf at the connections (the lower the better); (6) physical properties (soft wire easily formed and soldered preferable to hard, springy wire); and (7) hysteresis behavior (undesirable).

Of the kinds of metal shown in Table 7.1, wire of the Constantan type is preferred for most gages because it has a good enough gage factor, a negligible temperature coefficient of resistance, and works easily. For dynamic measurements, when a high temperature coefficient of resistance is not of much importance, it is desirable to use isoelastic wire because of its high strain sensitivity. This provides a higher output, lowers the amplification required for the measuring instruments, and results in a lower cost of instrumentation.

Bonded resistance strain gages The original resistance-wire gage used in the United States was made by the Baldwin-Lima-Hamilton Corporation and was known as the SR-4 gage. A few common types of resistance-wire gages are shown in Fig. 7.13. In

TABLE 7.1 Characteristics of resistance wires

Trade name of wire	Composition	Strain sensitivity (gage factor)	Temp. coef. of resistance
Nichrome	80% Ni; 20% Cr	2.0	High
Manganin	4% Ni; 12% Mn; 84% Cu	0.47	Very low
Advance, Copel, or Constantan	45%; Ni 55% Cu	2.0	Negligible
Isoelastic	36% Ni; 8% Cr; 0.5% Mo	3.5	High
Nickel	Ni	−12.1	Unstable

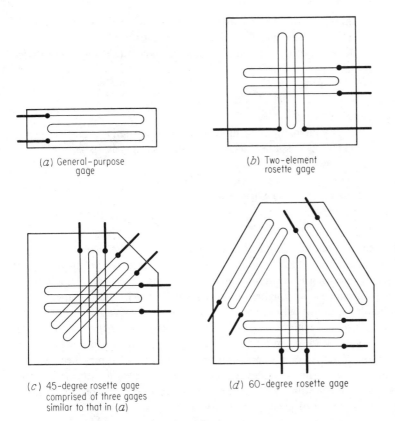

(a) General-purpose gage

(b) Two-element rosette gage

(c) 45-degree rosette gage comprised of three gages similar to that in (a)

(d) 60-degree rosette gage

Figure 7.13 Principal types of electric strain gages.

some cases the sensitive elements are made up of a continuous length of wire looped back and forth so that all the loops are in the same plane. The wire is then cemented to paper, plastic, or another type of carrier matrix (backing material). In other cases the sensitive element is made up of a continuous length of wire wrapped in a helical pattern around a thin, flat paper core. This sensitive element is then sandwiched between two cover papers for protection. Most often an etched alloy foil is used. The effective gage lengths range from less than 3 mm to more than 200 mm, but this range is not available in all types of gages.

A few special types of gages are made for particular service conditions. One such gage, called a post yield strain gage, is used for measuring strains up to 20 percent, which is far beyond the 1 to 2 percent strain range of other gages.

Some gages are temperature-compensated to cancel out unwanted temperature effects, which allows measurement of only the strains due to stress. Other gage patterns have been developed to eliminate the effect of transverse stresses from readings—these are known as stress gages. A special gage made of nichrome foil can be used at temperatures up to about 650°C if it is suitably bonded to the test piece. Another gage is encased in a plastic cover so that it can be embedded in fresh concrete without danger of interference from the moisture in the concrete. Resistance gages can also be used to sense temperature, and some are specially suited for this purpose.

The outstanding advantages of bonded electric resistance strain gages, compared with mechanical gages, are

1. Ease of installation
2. Relatively high accuracy
3. Adjustable sensitivity (by changing the gain of the amplifier used)
4. Remote indication (making possible the observation of strains at distant, inaccessible points)
5. Very short gage lengths
6. Measurement of strain at the surface of the test member
7. Response to dynamic strain

Attachment of gages The surface to which a bonded electric resistance strain gage is to be attached must be thoroughly clean and even, but not too smooth or highly polished. Satisfactory cleaning is accomplished with acetone or carbon tetrachloride. Very smooth surfaces should be roughened with a medium-grit emery paper. Many adhesives are available, most of which are an epoxy type. It is well to consult the manufacturer's literature to achieve a good match of gage and adhesive. A liberal amount of cement should be used in bonding the gage to avoid the formation of air pockets in the cement. The use of moderate weights, or spring clamping pressure applied through a rubber pad, will squeeze out excess cement. Thorough drying is essential to avoid yielding of the bonding material; drying can be aided by careful heating.

A high degree of stability of gage resistance is essential since the measurement of strain involves the determination of very small changes of resistance. Errors due to instability or drift should not exceed a small fraction of the resistance change due to the strain being measured.

Gage instability is caused primarily by moisture absorbed by the gage. Moisture produces (1) changes in conductivity and (2) dimensional changes of the bonding cement, which result in strains in the gage wires and thus in turn cause resistance changes. Also, long-time aging of the bonding cement may cause some instability. This is largely corrected by thorough drying or baking of the gage, and the effect of moisture is minimized by thorough moistureproofing. However, only a few types of service require this protection; in most applications the effect of moisture changes is not significant during the course of a test.

Before moistureproofing any gage it is necessary that the cement be thoroughly dried or cured and that the gage be free of all absorbed or surface moisture. The gage may then be waterproofed by coating with grease, wax, or polymeric compounds.

Stress determination When using a 45° rosette (Fig. 7.13c), if strains ϵ_1, ϵ_2, and ϵ_3 are observed at 0°, 45°, and 90° angles respectively with the x axis, then

$$\epsilon_x = \frac{(\epsilon_1 + \epsilon_3)}{2} + \sqrt{\frac{(\epsilon_1 - \epsilon_2)^2}{2} + \frac{(\epsilon_2 - \epsilon_3)^2}{2}}$$

where ϵ_x equals the principal strain most nearly in the direction of the x axis. The other principal strain occurring at right angles to ϵ_x is

$$\epsilon_y = \frac{(\epsilon_1 + \epsilon_3)}{2} - \sqrt{\frac{(\epsilon_1 - \epsilon_2)^2}{2} + \frac{(\epsilon_2 - \epsilon_3)^2}{2}}$$

Converting strains to stresses, based on the relation $\sigma = E\epsilon$, and correcting for the transverse strain with Poisson's ratio μ, the principal stresses are

$$\sigma_x = (\epsilon_x + \mu\epsilon_y)\frac{E}{1 - \mu^2}$$

$$\sigma_y = (\epsilon_y + \mu\epsilon_x)\frac{E}{1 - \mu^2}$$

for plane (two-dimensional) stress conditions. These relations hold for an isotropic material only (E the same in all directions) and for stresses below the proportional elastic limit. In using these equations a tensile strain is positive and a compressive strain is negative.

Special arrangements The clip gage shown in Fig. 7.14 is a special arrangement of ordinary strain gages to permit the measurement of large elastic or inelastic strains. It consists of two gages cemented to a U-shaped clip of spring bronze. When the test piece to which the clip gage is attached is subjected to strain, one gage is elongated while the other is compressed, resulting in a large change in resistance between the two. Each clip gage must be calibrated, but that is a relatively simple procedure. One advantage of this gage is that it may be used repeatedly on many different specimens.

In another application strain gages can be attached to a block of steel. This load cell, as shown in Fig. 6.5, can then be calibrated to transduce compressive forces and can be inserted into a load train. There are many mechanical and structural applications.

7.5 STRAIN-GAGE INSTRUMENTATION

The strains sensed by the gages described in the previous article must be interpreted. Various instruments are designed to transduce the electric signals into intelligible readouts.

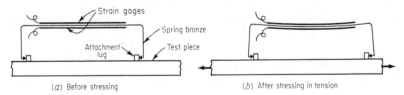

(a) Before stressing   (b) After stressing in tension

Figure 7.14 Clip gage.

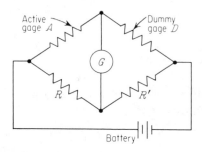

Figure 7.15 Wheatstone-bridge hookup.

Wheatstone-bridge hookup Static strains may be determined by placing a gage A in a four-arm Wheatstone-bridge dc circuit, as shown in Fig. 7.15. When the gage resistance is changed by deformation of the gage, the bridge circuit is unbalanced. To compensate for strains caused by temperature and humidity variations, a so-called dummy gage D is connected into the Wheatstone-bridge circuit. The dummy gage is a duplicate of the active or strain-measuring gage A. It is attached to a piece of unstressed material of the same kind and subject to the same temperature and humidity conditions as the member to which A is attached. The active gage measures strain due to stress plus deformation due to temperature and humidity effects, while the dummy gage measures deformation due to temperature and humidity only. The circuit is arranged so that the resultant bridge unbalance is a function of strain due to stress, free from deformation due to temperature or humidity.

The Wheatstone bridge consists of resistances (A plus D) and (R plus R') in parallel, with a means of determining points of zero potential by moving the lower lead of the high-resistance exploring galvanometer G along the resistances R and R' until it reads zero, thus using R and R' as a slide-wire resistance. In this way the resistances are divided between the four arms so that $A/D = R/R'$. The resistances R and R' are commonly made variable to avoid the need for moving the galvanometer lead. From the above relation the resistance A can be measured at any stage, and therefore the change in A and the corresponding strain can be determined.

To obtain direct or axial strain averaged from two sides of a tension or compression member and at the same time cancel out unwanted bending strains, two active gages A_1 and A_2 should be mounted back to back on opposite sides of the specimen and connected with two dummy gages, as shown in Fig. 7.16. In this case the strain in one gage, say A_1, will be greater than if it were sensing only the axial load and the strain in the other gage will be less by exactly the same amount. Thus the circuit adds their results and divides by two, which gives the true axial strain irrespective of any bending.

To measure bending only and to cancel out axial strains, gages A_1 and A_2 should be mounted on opposite sides of the specimen as before but connected in adjacent arms of the bridge, eliminating the dummy gages, as shown in Fig. 7.17. This circuit not only compensates for temperature but doubles the sensitivity as well because with A_1 in tension and A_2 in compression, located in adjacent arms, the resistance changes of opposite signs in effect add together. Whenever possible, in measuring bending strains, gages should be used in this manner, as this hookup provides the most efficient circuit with the least number of gages.

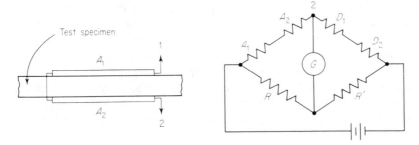

Figure 7.16 Bridge arrangement for measuring axial strain.

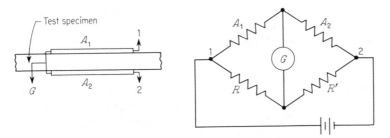

Figure 7.17 Bridge arrangement for measuring bending strain.

The strain indicator The instrument used in conjunction with strain gages is called a *strain indicator* (an adaptation of the Wheatstone bridge), a simplified form of which is shown schematically in Fig. 7.18. It is a battery-powered dc unit that gives optimum accuracy and sensitivity. In this instrument the active gage A and the dummy gage D are external to the indicator. G represents the galvanometer; the decade resistance is provided for a coarse balance of the bridge. The variable resistances R and R' are gages mounted back to back on a slender cantilever beam; a micrometer screw is used to deflect this beam and change the resistances R and R' to accomplish the fine balance. The micrometer also serves to measure the unbalance in the circuit when the bridge is used as a null indicator.

By the null method the galvanometer is balanced back to zero by turning the micrometer screw, which is graduated so that changes in reading give strains directly

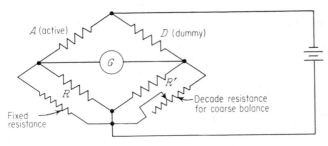

Figure 7.18 Strain indicator without amplification.

in micrometers per meter. The null method is slow, but it is a very accurate way of measuring the strain. A quicker method is to note the amount of deflection of the galvanometer from the zero position when provision is made to impress a definite voltage on the indicator, because the deflection is proportional to voltage. The null method is independent of the voltage used. Some indicators feature direct digital readout.

The indicating device used with a dc circuit is often a galvanometer, because it is one of the few instruments that will operate over a significant range without amplification, although some oscillographs with highly sensitive elements require no amplification.

The strain-gage signal is actually very low. For example, with a gage factor of 2.0 a stress change of 200 MPa in steel will change the resistance of a 120-Ω gage by only 0.7 Ω, and the bridge output for one active gage is only about 10 mV for a strain of 0.001, even though a mechanical advantage of 2 is provided by the gage factor. Therefore, the actual strain indicator includes electronic amplification to obtain a stronger signal.

Dynamic testing For rapidly fluctuating strains it is essential to use pen-and-ink recorders, oscillographs, or cathode-ray oscilloscopes for determining the strains produced, the selection of the instrument best adapted to a given problem depending on the rate of fluctuation of the strains. It is necessary to amplify strongly the weak but precise signal produced by the unbalance of the strain-gage bridge to develop enough power to operate any of these instruments. For these applications ac amplifiers are essential, but when using alternating current, it is necessary to balance the capacitance and the inductance as well as the resistance of the circuit. Alternating-current circuits permit amplification very simply and eliminate the contact potentials or thermal emf's in the gage because of the constantly changing polarity.

The type of record produced on an oscillograph film or on a pen-and-ink recorder chart by the amplified output of a 60-Hz ac bridge is shown in Figs. 7.19a and 7.19b. The strain variations picked up by the gage appear as a modulation of the 60-Hz output of the bridge produced when the gage is under no strain. The result is an envelope

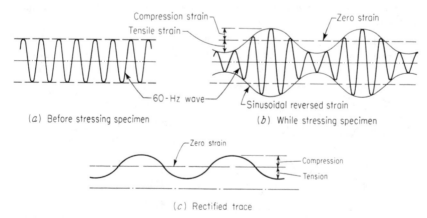

(a) Before stressing specimen   (b) While stressing specimen

(c) Rectified trace

Figure 7.19 Record of dynamic test.

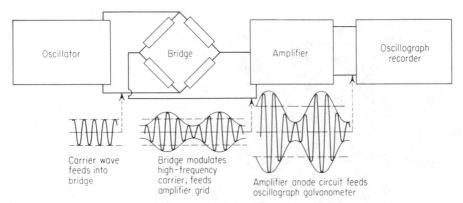

Figure 7.20 Hookup for high-frequency dynamic testing.

or unrectified record, and the pattern above the center line duplicates that below. Tension increases the gage resistance and correspondingly reduces the galvanometer swing; conversely, compression decreases the resistance and increases the swing. This serves to distinguish between tension and compression on the record. If the output is rectified by a copper oxide rectifier before recording, a single-line trace of the envelope of Fig. 7.19*b* is obtained as shown in Fig. 7.19*c*. This curve is easier to interpret.

The oscilloscope A 60-Hz ac source satisfactorily powers the bridge as long as the frequency of fluctuation of the strain picked up by the gage does not exceed about 6 Hz or one-tenth of the frequency of the impressed wave, commonly called the carrier wave. As the strain frequency approaches the carrier-wave frequency, the strain-wave pattern becomes less well defined. For high-frequency dynamic strain recordings it is desirable to have a carrier frequency at least 10 times the frequency of the gage pickup. This can be done by electronic oscillators, which are generators of high-frequency currents. A typical hookup for high-frequency dynamic testing is shown in Fig. 7.20.

An oscilloscope, such as the one shown in Fig. 7.21, provides a plot on a cartesian coordinate system of some function vs. time. It can do this at a much greater speed than any pen-and-ink recorder. Instead of paper it uses a cathode-ray tube (CRT) and in place of a pen an electron beam. The coordinate system, or graticule, is etched on the CRT.

The horizontal axis represents time (perhaps in μs), and the vertical axis represents any phenomenon that a transducer can translate into voltage. This includes mechanical quantities such as strain if they are sensed electrically by strain gages. Thus we might display the strain in some direction at some point of a vibrating beam. Both scales can be adjusted, and the display can be moved both horizontally and vertically. This is accomplished with four corresponding control knobs on the panel.

Even a cyclic signal would result in a jumbled display if it were not triggered properly. The "scope" is therefore instructed to start the display at a particular trigger point and continue it for a certain time window. Figure 7.22 demonstrates the principle. The trigger point is identified by setting both the level and the sign of the slope of the signal source, which is usually the displayed signal itself but may be a related

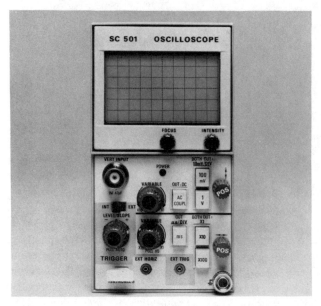

Figure 2.21 Oscilloscope. (*Courtesy of Tektronix, Inc.*)

one. The time window is determined by the width of the CRT and the horizontal scale (time base).

Thus for a vibrating beam, the strain variation is displayed as a stable picture. As the test proceeds, the amplitude and the period changes gradually alter the display in a manner discernable to the eye.

More sophisticated oscilloscopes have additional features, but they basically work the same way.

7.6 BRITTLE COATINGS

The methods of determining strain discussed so far, powerful and flexible as they are, have one major drawback: strains can be obtained only at predetermined locations on

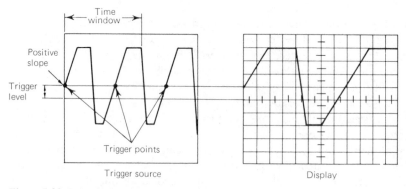

Figure 7.22 Triggering. (*Courtesy of Tektronix, Inc.*)

the test piece, usually only in predetermined directions. The method discussed in this article overcomes the disadvantage, but at the expense of accuracy, convenience, and versatility.

It has long been known that fine cracks develop in the mill scale of hot-rolled steel at stresses slightly beyond the yield point. This same effect can be observed, even at strains well below the yield point, when the surface is covered with a brittle coating. One type of commercially available brittle coating, called Stresscoat, is basically a limed-wood rosin K and dibutyl phthalate, with carbon disulfide as a solvent. Another product, Tens-Lac, consists of resins dissolved in methylene chloride. As load is applied to a coated specimen, the brittle coating will crack along lines normal to the direction of the maximum tensile stresses as soon as the tensile strains reach the level of strain sensitivity of the coating. For Stresscoat and Tens-Lac this strain is about 0.0005 to 0.0009, corresponding to a stress in steel of 100 to 180 MPa.[†] The actual level of strain sensitivity for each test is determined by calibration strips coated at the same time as the test specimen. By applying increments of load to the specimen and noting the load at which cracking of the coating begins at various locations, it is possible to determine the critically strained parts of the specimen and to note how other parts reach critical strains and corresponding stresses as the load is increased.

Calibration of the coating on each specimen is conducted using the apparatus shown in Fig. 7.23*a*. The calibration strips are of spring steel, perhaps 300 by 25 by 5 mm in size, coated on one side, and loaded as a cantilever beam using a known deflection. This deflection is such that the calibration strip will undergo a varying strain along its length, as shown on the scale accompanying the loading rig. After the strip is deflected, a mark can be made on it at the location where cracking begins. The strip is then placed in the special scale provided with the apparatus, from which the amount of strain causing the cracking is noted, as shown in Fig. 7.23*b*.

Factors affecting cracking The critical strain that causes the beginning of cracking of the coating varies with several factors, which must be carefully controlled to obtain reasonably accurate results.

The thickness of the coating should be controlled to within 75 to 200 μm. Within this range, cracking begins at a fairly repeatable strain, but for thinner and thicker coatings the initial cracking will occur at larger and less predictable strains.

The drying period is important: if insufficient time is allowed, the coating will not be brittle enough to crack at low strains, and furthermore the strain at initial cracking will be supersensitive to the thickness of the coating. In general, 15 to 24 h is the optimum drying time. After longer periods the coatings become erratic in their pattern formation.

Atmospheric conditions during the drying and the testing period have definite effects on the strain sensitivity of the coating. Gradual changes of temperature during the *drying* period are not so critical, but changes during the *test* period should not exceed ±0.5 K and the temperature of the calibration strip should not vary more than ±0.3 K from that of the specimen.

Sensitivity also varies with relative humidity, but ordinarily this does not change

[†]1 MPa = 145 psi.

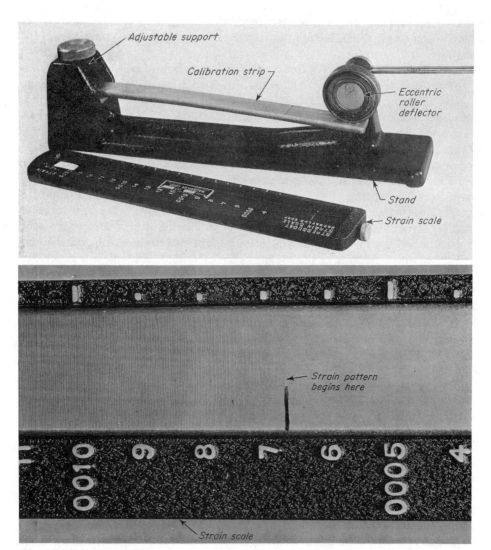

Figure 7.23 Calibration of brittle coating. (*Courtesy of Magnaflux Corp.*)

much during a test. In order to obtain the optimum coating for given atmospheric conditions, a series of different coatings that vary in brittleness are available.

Although the coatings are called *brittle*, they are actually somewhat plastic. Figure 7.24 shows how the strain necessary to crack a coating varies with the time taken to cause cracking. This influence of creep may be corrected by checking a calibration strip at the beginning of the test and then correcting all observed strains on the specimen to the values that would have been obtained if the load rate had been the same as that for the calibration strip.

If the temperatures are controlled properly, the brittle-coating method can produce results with an accuracy of about plus or minus 10 percent. Even if temperatures are

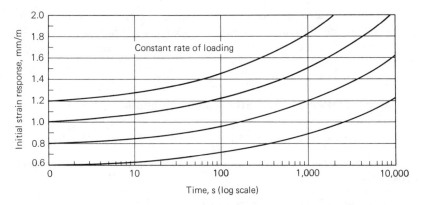

Figure 7.24 Variation of strain sensitivity with time of loading. (*From de Forest and Ellis, Ref. 24*)

not controlled, the method will indicate satisfactorily the location and the direction of the maximum stress developed. To avoid the difficulties involved in proper temperature control, some experimenters use a coating for qualitative determinations and then in later tests measure actual strains by use of strain gages applied at the critical points, as indicated by the tests with brittle coatings.

Test method The test procedure is generally as follows:

1. A coating appropriate for the temperature and humidity conditions in the laboratory is selected.
2. An aluminum undercoating is sprayed on both the test piece and the calibration strips after a thorough cleaning. This provides a bright background to facilitate observation of the cracks. The undercoating is dried at least 15 min.
3. The specimen and the strips are sprayed with the brittle coating and dried for 15 to 24 h.
4. The test piece is loaded in increments. At each load, note is made of any cracks that have just started, and the load is returned to zero. After the test piece is held at zero load for at least twice the time involved in the previous loading cycle, the next higher load is applied and any new cracking noted. If desired, the return to zero load may be omitted if proper corrections for creep are applied.
5. The coating is calibrated by loading a calibration strip in the calibrator in the same length of time as that used in loading the specimen. The point of initial cracking is marked, and the strain in the strain scale is read.
6. After completion of the test, the crack patterns may be more clearly shown for photographing by dye-etching (see Chap. 16).

This procedure applies to tensile strains only, but brittle coatings can also be used for analyzing compressive stresses. The procedure differs in that the specimen is coated while under a compressive load. After the coating dries, the load is gradually released. The crack pattern formed when the compression load is released is analogous to that produced when a coated specimen is subjected to tension. A similar condition is obtained by coating the nonloaded specimen, slowly loading in compression for an hour without cracking the coating, holding the load for at least another hour until the coating creeps sufficiently to become stress free, and then releasing the load in increments to cause cracking of the coating.

PROBLEM 7 Beam Deflection and Strain Measurements

Object: To determine the strain distribution and the deflection at one section of a simply supported beam.

Preparatory reading: ASTM E 529; chapters on stresses in and deflections of beams in a textbook on mechanics of materials.

Specimen: Wide-flange beam with strain gages attached.

Special apparatus: Strain-gage equipment and dial or LVDT deformeter.

Procedure:

1. Arrange the beam so that loads will be applied at two points symmetrical with the center of the span. Measure the span length, depth, flange width, and web thickness of the beam. Determine the location of each of the gage lines. Draw sketches showing locations of loads, reactions, gage lines, etc.

2. Arrange the deformeter so that it will measure the deflection at some point along the centerline on the top of the beam. Note the exact location on a sketch.

3. Compute the strains at the strain-gage locations and the deflection at the deformeter location as functions of the applied loads. Estimate the load that will produce initial yielding.

4. Examine the strain-gage instrumentation supplied and study the operating instructions. Connect the strain gages, including the compensating or dummy gage, and set the gage-factor dial for the gages used. Balance the circuit and zero the indicator for all gages, following the instructions in the manual.

5. Read or zero the deformeter, as appropriate.

6. Apply loads to the beam so that yielding is approached, but not reached. You might stop at a few intermediate points to take readings.

7. At the full load, read all strain gages and the deflectometer. Tabulate the results.

8. Remove the load gradually, and take another set of zero-load readings.

9. Compute the modulus of elasticity from the deflection obtained.

Report: Report all values obtained experimentally. Compute the stresses at the gage locations, sketch the stress distribution, and compute the resulting bending moment. Compare these results with the theoretical ones. Comment on special aspects as follows:

1. Does the plotted graph of strains vs. distances from the top of the beam substantiate the usual assumption that a plane section before bending remains a plane section after bending?

2. Explain discrepancies between theoretical and experimental results and discuss the possible sources of error in the test.

3. What advantages were derived from the symmetrical loading, considering shearing effects and the variations of stress along the gage lines?

DISCUSSION

1. If you had no large caliper, how could you determine the cross-sectional area of a concrete test cylinder of about 150 mm in diameter?

2. Suppose that you want to test an ordinary rubber band by hanging small weights on it. How could you determine the elongation?

3. What instrumentation could be used to make continuous observations of the strain in a hanger bar of a bridge?

4. Have you ever observed the kind of phenomenon on which brittle coatings are based? What happened?

5. Why are there both short and long strain gages?

6. Devise mechanical instruments that measure (*a*) the acceleration of a car, (*b*) wind speed, and (*c*) the roll (lean) of a sailboat. What are these devices called?

7. What differences must be considered in electric instrumentation for a structure's seismic response of a few hertz and an aircraft's vibration of several hundred hertz?

8. Consult a physics text and then explain the basic principle on which the LVDT is based.

9. If you want to determine the stress distribution in a linearly elastic eccentrically loaded compression block, how many strain gages do you need? By means of an example show a good arrangement and the equation of the stress plane. What if you were asked this question about a curved beam?

10. Can a resistance gage be used to measure quantities other than strain? Name some methods and their underlying principles.

READING

Perusal of the literature of instrumentation manufacturers is just as necessary as it is in the case of testing machines in order to keep abreast of developments. It will also be instructive to search the current periodical literature.

Basic Oscilloscope Operation, Tektronix, Beaverton, Oreg., 1978.

Doebelin, E. O.: *Measurement Systems*, rev. ed., McGraw-Hill, New York, 1975.

Perry, C. C., and H. R. Lissner: *The Strain Gage Primer*, 2d ed., McGraw-Hill, New York, 1962.

PART

TWO

CHARACTERISTICS OF TESTS

EIGHT

STATIC TENSION

8.1 TENSION VS. COMPRESSION

The terms *tension test* and *compression test* usually refer to tests in which a prepared specimen is subjected to a gradually increasing (i.e., "static") uniaxial load until failure occurs. In a simple tension test, the operation is accomplished by gripping opposite ends of the piece of material and pulling it apart. In a compression test, it is accomplished by subjecting a piece of material to end loading, which produces crushing action. In a tension test, the test specimen elongates in a direction parallel to the applied load; in a compression test, the piece shortens. Within the limits of practicability, the resultant of the load is made to coincide with the longitudinal axis of the specimen.

Except for certain arbitrarily formed test pieces, the specimens are cylindrical or prismatic in form and of approximately constant cross section over the length within which measurements are made. We try to obtain a uniform distribution of direct stress over critical cross sections normal to the direction of the load. The attainment of these ideal conditions is limited by the form and trueness to form of the test piece, by the effectiveness of the holding or bearing devices, and by the action of the testing machine.

The static tension and compression tests are the most commonly made and are among the simplest of all the mechanical tests. Since the beginning of scientific testing, tension tests, at least, have occupied a large share of attention, and great and probably well-merited confidence has been placed in the value and significance of both the tension and compression tests.

When properly conducted on suitable test specimens, these tests, of all tests, come closest to evaluating fundamental mechanical properties for use in design, although it should be observed that the tensile and compressive properties are not necessarily sufficient to enable the prediction of performance of materials under all loading conditions. When standard methods of test are employed, the results are acceptable criteria of quality of materials with which sufficient experience has accumulated to provide assurance that a given level of quality means satisfactory behavior in service. Such tests imply the standardization of specimens with regard to size, shape, and method of preparation and the standardization of the procedures of testing. As with any test, however, for newly developed materials, the tension and compression tests should be used with caution as quality-level indicators because the significance of these tests is limited by their correlation with performance.

Properly conducted tests on representative parts can be valuable in indicating directly the performance of such parts under loads in service. Suitable tests of specimens or fabricated parts subjected to specific treatments can be of use in evaluating quantitatively the effect of such treatments.

Although insofar as the sense or direction of stress is concerned compression is merely the opposite of tension, there are several factors that make the tension or the compression test the more desirable in a specific case. The most important of these factors are (1) suitability of the material to perform under a given type of loading, (2) differences in properties of a material under tensile and compressive loading, and (3) relative difficulties and complications induced by the gripping of or end bearing on the test pieces.

The use of the tension as opposed to the compression test is largely determined by the type of service to which a material is to be subjected. Metals, for example, generally exhibit relatively high tenacity and are therefore better suited to and are more efficient for resisting tensile loads than materials of relatively low tensile strength. The tension test is therefore very commonly employed with and is appropriate for general use with most cast, rolled, or forged ferrous and nonferrous metal alloys. The same is true for drawn and extruded plastics.

With brittle materials, such as mortar, concrete, brick, and ceramic products, whose tensile strengths are low compared with their compressive strengths and which are principally employed to resist compressive forces, the compression test is more significant and finds greater use.

The tensile strength of wood is relatively high, but wood cannot always be effectively utilized in structural members because of its low shear resistance, which causes failure at the end connections using bolts, split rings, or shear plate connectors before the full tensile resistance of a member can be developed. Thus, insofar as direct stresses are concerned, the compression test of wood is of greater practical significance than the tension test. Some materials, such as cast iron, although having a lower tensile than compressive strength, are used to resist either type of stress, and both types of test are sometimes made.

The use of tension and compression tests is not confined to the determination of the properties of the material in the form of prepared (shaped) specimens. Full-size tests of manufactured materials, fabricated parts, and structural members are commonly made. The variety of full-size fabricated parts and members to which tension or compression tests may be applied is large. In many cases, tests of this sort are essentially the same as tests on prepared specimens. For example, more or less standardized apparatus and test procedures are used to determine the properties of selected lengths of wire, rod, tubing, reinforcement bars, fibers, fabrics, cordage, and wire rope in tension, and brick, tile, masonry blocks, and certain types of metal castings in compression. The important feature of other full-size tests is the duplication, as nearly as possible, of the load conditions of service and the observation of the development of localized weaknesses as well as critical loads. A few of the full-size pieces on which tests are not uncommonly made may be mentioned: (in tension) eyebars, anchor chain, crane hooks, drawbars, and bolted and welded joints; (in compression) cast-iron and concrete pipe, built-up piers, pedestals, columns, and wall sections. Compression tests on columns, thin-walled fabricated parts and structures, and the like involve problems of elastic stability and usually require special procedures.

Particulars of tension tests are discussed in this chapter, and those of compression tests in the next.

8.2 SPECIMENS

Although certain fundamental requirements exist for all test specimens (see Art. 5.4) and certain shapes of specimens are customarily used for particular types of tests, prepared specimens for tension tests are made in a variety of forms. The cross section of the specimen is usually round, square, or rectangular. For metals, if a piece of sufficient thickness can be obtained so that it can be easily machined, a round specimen is commonly used; for sheet and plate stock, a flat specimen is generally employed. The central portion of the length is usually (but not always) of smaller cross section than the end portions in order to cause failure to occur at a section where the stresses are not affected by the gripping device. Typical nomenclature for tension specimens is indicated in Fig. 8.1. The gage length is the marked length over which elongation or extensometer measurements are made.

The shape of the ends should be suitable to the material and such as to fit properly the gripping device to be employed. The ends of round specimens may be plain,

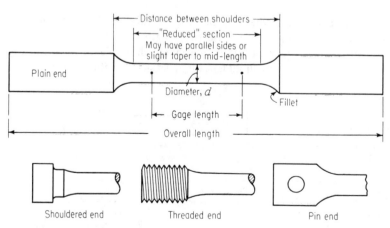

Figure 8.1 Typical tension specimens.

shouldered, or threaded. Plain ends should be long enough to accommodate some type of wedge grip. Rectangular specimens are generally made with plain ends, but sometimes they may be shouldered or contain a hole for a pin bearing.

The ratio of diameter or width of end to diameter or width of reduced section is determined largely by custom, although for brittle materials it is important to have the ends sufficiently large to avoid failure due to the compounding of the axial stress and the stresses from action of the grips. If a specimen is machined from larger stock, the reduction should be at least enough to remove all surface irregularities. The transition from end to reduced section should be made by an adequate fillet in order to reduce the stress concentration caused by the abrupt change in section; for brittle materials, this is particularly important. The effect of change of section on stress distribution is practically inappreciable at distances greater than about one or two diameters from the change.

To obtain uniform stress distribution across critical sections, the reduced portion of the piece is often made with parallel sides for its entire length, although many types of specimens are made with a slight taper from both ends of the reduced section to its midlength. Specimens of some materials (such as mortar briquets) are curved throughout the central portion of their length in order to prevent breakage at or near the grips; in such specimens, the stress is not uniform on the critical cross section, and all dimensions of the specimen must be standardized to obtain comparable results.

A specimen should be symmetrical with respect to a longitudinal axis throughout its length in order to avoid bending during application of load. Figure 8.2 illustrates various faults in the preparation of flat specimens.

The length of the reduced section depends on the kind of material to be tested and the measurements to be made. With ductile metals, for which elongation or reduction of area is to be determined, the length must be sufficient to permit a normal break; that is, the drawing out or necking down should not be inhibited by the mass of the ends. With brittle materials for which the elongation is very small and is not measured and for which fracture is plane, the length of the reduced section may be relatively short.

The gage length is always somewhat less than the distance between shoulders, but practice with regard to the ratio between these two lengths is not uniform. If extensometer measurements are to be made, it is considered desirable to have the gage length short of the distance between shoulders by at least twice the diameter of the

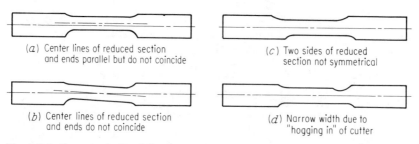

(a) Center lines of reduced section and ends parallel but do not coincide

(c) Two sides of reduced section not symmetrical

(b) Center lines of reduced section and ends do not coincide

(d) Narrow width due to "hogging in" of cutter

Figure 8.2 Common faults of flat test specimens.

test piece. The gage points should be equidistant from the center of length of the reduced section.

The percentage elongation (or strain) of a ductile metal specimen of given diameter depends on the gage length over which the measurements are made. It has been established by many tests that the elongation strain is practically constant for pieces of various sizes if the pieces are geometrically similar. For small cylindrical specimens of ductile metals, ASTM (ASTM E 8) calls for a gage length of four times the diameter. For larger specimens of ferrous metal, various ASTM specifications (ASTM A 36, A 615) use some specified gage length and thickness or diameter as a base, and the effect of different thickness or diameters is provided for by deductions from the permissible elongation in accordance with a stated rule.

Figures 8.3 through 8.7 show some representative standard specimens for tension tests of various materials. These examples illustrate the variety of tension specimens in general use.

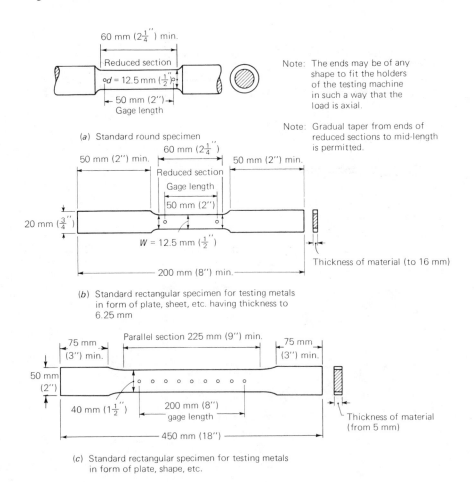

(a) Standard round specimen

(b) Standard rectangular specimen for testing metals in form of plate, sheet, etc. having thickness to 6.25 mm

(c) Standard rectangular specimen for testing metals in form of plate, shape, etc.

Figure 8.3 Standard (ductile) metal tension specimen (ASTM E 8).

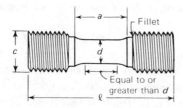

	Dimensions, mm			Dimensions, inches		
Specimen size:	Small	Medium	Large	Small	Medium	Large
a, min.	32	38	60	$1\frac{1}{4}$	$1\frac{1}{2}$	$2\frac{1}{4}$
c, appr.	20	30	48	$\frac{3}{4}$	$1\frac{1}{8}$	$1\frac{7}{8}$
d (±tol.)	12.5	20.0	30.0	0.50	0.75	1.25
ℓ, min.	95	100	160	$3\frac{3}{4}$	4	$6\frac{3}{8}$

Figure 8.4 Standard round tension specimen for cast iron (ASTM E 8, A 48).

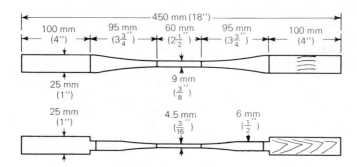

Specimen for test parallel to grain

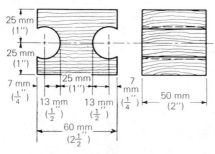

Specimen for test perpendicular to grain

Figure 8.5 Standard tension specimens for wood (ASTM D 143).

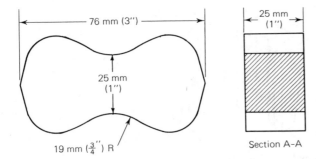

Section A-A

Figure 8.6 Standard tension specimen for portland cement mortar (ASTM 190).

8.3 APPARATUS

To apply the load, any tension, universal, or dynamic machine can be used, provided the capacity and sensitivity are appropriate for the specimen to be tested. The choice of a machine will largely depend on availability. This also applies to instrumentation, although an LVDT-type extensometer,[†] connected to some type of recorder, is generally employed.

An essential task is the selection of a proper gripping device. The function of the gripping device or shackles is to transmit the load from the heads of the tesing machine to the test specimen. The essential requirement of the gripping device is that the load be transmitted axially to the specimen; this implies that the centers of action of the grips be in alignment at the beginning and during the progress of a test and that no bending or twisting be introduced by action or failure of action of the grips. In addi-.

[†]Linear variable differential transformer (see Chap. 7).

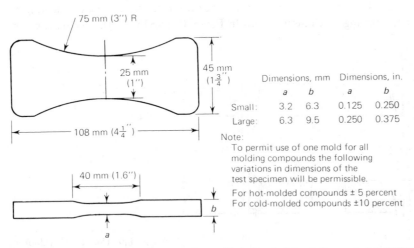

Dimensions, mm Dimensions, in.

	a	b	a	b
Small:	3.2	6.3	0.125	0.250
Large:	6.3	9.5	0.250	0.375

Note:
To permit use of one mold for all molding compounds the following variations in dimensions of the test specimen will be permissible.

For hot-molded compounds ± 5 percent
For cold-molded compounds ±10 percent

Figure 8.7 Standard tension specimen for molded electrical insulating materials (ASTM D 651).

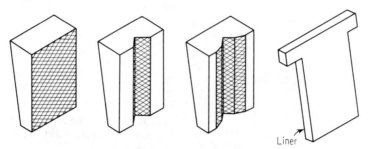

Liner

Figure 8.8 Wedge-grip units for tension tests of metals.

tion, of course, the device should be adequately designed to carry the loads and should not loosen during a test.

Wedge grips, illustrated in Fig. 8.8, are a common type of holding device. They are satisfactory for commercial tests of ductile metal specimens of adequate length, because slight bending or twisting does not appear to affect the strength and elongation of ductile materials. No adjustment to prevent bending can take place with grips of this sort. Wedge-type grips are usually unsatisfactory for use with brittle materials, because the crushing action of the wedges tends to cause failure at or near the grips. The faces of grips that contact the specimen are roughened or serrated to reduce slippage; for flat specimens the faces of the grips are flat, and for cylindrical specimens the grips have a V groove of suitable size. Adjustment is made by means of shims or liners so that the axis of the specimen coincides with the center line of the heads of the testing machine and so that the grips are properly located in the head. Correct and faulty settings of the grips are illustrated in Fig. 8.9.

Where assurance of more accurate alignment is necessary, which is highly important in tests of brittle materials, some type of universal joint is used in the holders at both ends; usually it is a spherically seated or pin-bearing arrangement (so-called self-aligning linkage). A schematic drawing of a device using spherically seated bearing at the heads of the testing machine is shown in Fig. 8.10 (ASTM E 8). The distance be-

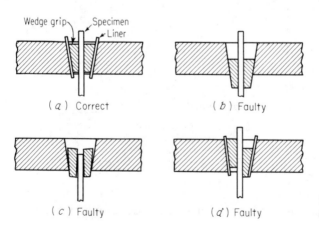

Figure 8.9 Correct and faulty setting of wedge grips.

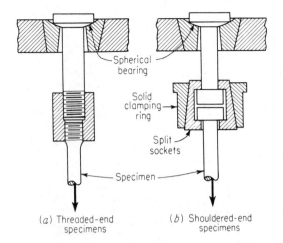

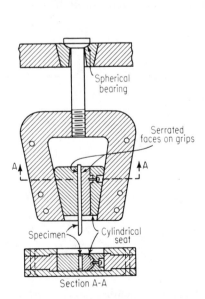

Spherical
bearing

Solid
clamping
ring

Split
sockets

Specimen

(*a*) Threaded-end
specimens

(*b*) Shouldered-end
specimens

Figure 8.10 Spherically seated holders (ASTM E 8).

tween spherical bearings should be as great as is feasible. Such devices are not always entirely effective; obviously, spherical seats do not adjust themselves easily if they are not properly lubricated and they may "freeze" at high loads, regardless of lubrication. More sophisticated versions exist, however.

A device for adequately gripping thin sheet-metal specimens and wire (Templin grips) is illustrated in Fig. 8.11. A device for testing wire is shown in Fig. 8.12.

The type of grip used for tests of mortar briquets is shown in Fig. 8.13. A common type of wire-rope socket is shown in Fig. 8.14.

For testing prismatic specimens of concrete, rigid steel plates are bonded to the

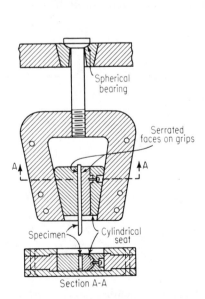

Spherical
bearing

Serrated
faces on grips

A

A

Specimen

Cylindrical
seat

Section A-A

Figure 8.11 Templin grips.

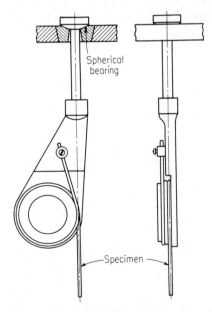

Spherical
bearing

Specimen

Figure 8.12 Snubbing device for testing wire.

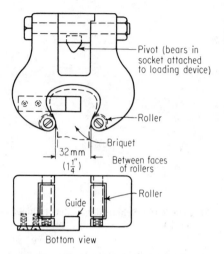

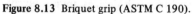

Figure 8.13 Briquet grip (ASTM C 190).

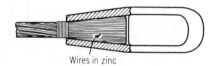

Figure 8.14 Wire-rope socket.

ends using an epoxy cement. Tensile loads are then applied to spherically seated axial steel rods connected to the end plates. Since the epoxy cement is stronger than the concrete, failure always occurs in the concrete.

Another type of test for determining the tensile strength of concrete is the splitting tension test, which is covered by ASTM C 496. As shown in Fig. 8.15, the splitting tension test uses a standard 150 by 300 mm† cylinder, which is loaded in compression along two axial lines 180° apart. Narrow strips of plywood are used as a cushioning material along these load lines. The splitting tensile strength is computed from

$$\sigma = 2P/\pi l d$$

†Or 6 by 12 in.

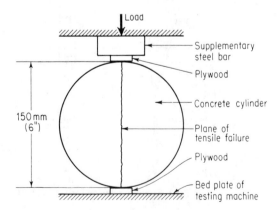

Figure 8.15 Splitting tension test of concrete.

where σ = splitting tensile strength
  P = maximum applied load
  l = length
  d = diameter

This type of test is simpler than any axial tension test, yet the test results agree reasonably well (about 15 percent higher) with those from the more conventional type of test [65]. A similar test is used for asphalt mixes.

8.4 PROCEDURE

In the commercial tension test of metals the properties usually determined are yield strength (yield point of ductile metals), tensile strength, ductility (elongation and reduction of area), and type of fracture. For brittle material, only the tensile strength and character of fracture are commonly determined. In more complete tests, as in much investigational work, determinations of stress-strain relations, modulus of elasticity, and other mechanical properties are included (see Chap. 2 for the basic parameters involved). The plan of the work and schedule of operations will obviously be adapted to the needs of the test.

Prior to applying load to a specimen, its dimensions are measured. Occasionally density values may be called for, requiring mass and volume determinations. Linear measurements are made with scales, calipers, or micrometers, depending on the dimension to be determined and the precision to be attained. In the simplest case, only the diameter or width and the thickness of the critical section are measured. Cross-sectional dimensions of metal specimens should ordinarily be read to a precision of about 0.5 percent. On cylindrical specimens, measurements should be made on at least two mutually perpendicular diameters.

If elongation measurements are to be made, the gage length is scribed or marked off. On ductile metal specimens of ordinary size, this is done with a center punch, but on thin sheets or brittle material, fine scratches should be used. In any case, the marks should be very light so as not to damage the metal and thus influence the break. Where much work is to be done, a double point or multipoint punch is sometimes used. It is convenient to lay round specimens in a V-shaped block while marking gage points. When a 200-mm gage length is used on steel specimens, marks are placed 25 mm apart.[†]

Before using a testing machine for the first time, the operator should become familiar with the machine: its controls, speeds, the action of the weighing mechanism, and the value of the graduations on the load indicator. Before a specimen is put into a machine, the weighing device of the machine should be checked for zero load indication and adjustments made if necessary.

When a specimen is put into a machine, the gripping device should be checked to see that it functions properly. If stops or guards are used to prevent the grips from flying out of their sockets when sudden failure occurs, the stops should be fastened in

---

[†]On an 8-in specimen, marks are 1 in apart.

place. The specimen should be so situated that it is convenient to make observations on the gage lines.

If an extensometer is to be used, the value of the divisions on the graph or indicator and the multiplication ratio should be determined before the extensometer is placed on the specimen. It should be placed centrally on the specimen and properly aligned. When collar-type extensometers are used, the axis of the specimen and the axis of the extensometer should be made to coincide. After it is clamped in place, the spacer bar (if any) is removed and the adjustments are checked. Often a small initial load is placed on the specimen before the extensometer is set at zero reading.

The speed of testing should not be greater than the rate at which load and other readings can be made with the desired degree of accuracy. If the testing speed has an appreciable influence on the properties of the material, the rate of straining of the test piece should lie within definite limits, although studies have indicated they may be reasonably broad.

Either the deformation (or strain) rate may be specified, which has been customary with screw-type machines,[†] or the load (or stress) rate, which has been usual in the case of hydraulic machines. As described in Chap. 6, modern servo-controlled machines allow either command to be implemented.

Various ASTM requirements have been established as maximum testing speeds; slower speeds may be and often are used. For tests involving yield-point determinations of metals, ASTM specifies a maximum loading rate of about 700 MPa/min (ASTM E 8). Sometimes the load is applied rapidly up to one-half the specified yield strength or yield point or up to one-quarter of the specified tensile strength, whichever is smaller, and then the rate is reduced to the value specified.

Above the yield point of ductile metals, higher speeds are permitted because variation in speed does not appear to have so much effect on the ultimate strength as on the yield strength; the elongation, however, is sensitive to variation in speed at high rates of loading.

For tests involving visual extensometer measurements, either the load is applied in increments and the load and deformation are read at the end of each increment, or the load is applied continuously at a slow rate and the load and deformation are observed simultaneously. The latter method is preferable.

After the test specimen has failed, it is removed from the testing machine, and if elongation values are called for, the broken ends of a specimen are fitted together and the distance between gage points is measured with a scale or dividers and scale to the nearest 0.2 mm. The diameter of the smallest section is calipered, preferably with a micrometer caliper equipped with a pointed spindle and anvil, for determining reduction in area. The same degree of precision should be employed as was used in measuring the original diameter. For symmetrical "cup-cone" fractures of rectangular specimens, Johnson recommended measurements for determining reduction of area as shown in Fig. 8.16 [104]. For irregular fractures, several measurements must be made, depending on the nature of the break.

[†]Actually, the "idling" or "no-load" speed is often specified for convenience, but this is not a good measure of actual test speed.

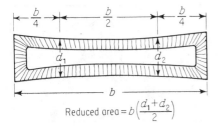

Figure 8.16 Determination of reduced section of rectangular specimens. (*From Withey and Aston, Ref. 104.*)

8.5 OBSERVATIONS

Observations made during a test are recorded on an appropriate form, prepared before starting the test. Identification marks and similar pertinent information are noted. Original and final dimensions and critical loads are recorded as they are observed. If extensometer measurements are made manually, a log of the load and the corresponding deformations is made. Some testing machines are equipped with an automatic attachment for drawing the load/stress vs. deformation/strain diagram (see Fig. 8.17).

The character of the fracture and the presence of any defects are noted. The test conditions should also be recorded, particularly type of equipment used and speed of testing. Stresses, strains, strength, and the ratios of elongation and reduction of area

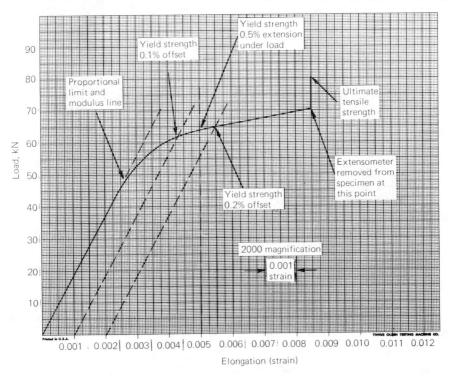

Figure 8.17 Load-strain diagram from X-Y plotter. (*Courtesy of Tinius Olsen Testing Machine Co.*)

TENSION TEST OF METAL — LOG SHEET

APPARATUS:
  Machine: 250-kN Olsen UTM (No. 12)
  Dial extensometer: Federal (No. 61),
    with multiplier of 2

TEST SPEED:
  To yield strength: 1.25 mm/min
  After yielding: 5.00 mm/min

SPECIMEN: Mild steel
  Mark or ID: 618
  Total length: 470 mm
  Length between shoulders: 285 mm
  Gage length: 200 mm
  Number of divisions: 8
  Diameter of ends: 32 mm
  Diameter of reduced section: 25.23 mm

OBSERVATIONS:
  Elongation in gage length: 62.5 mm
    per div.: 5.0, 5.5, 6.2, 8.8, 19.5, 6.7,
    5.8, 5.0 mm
  Percent elongation: 100(62.5/200) = 31.25%
  Percent reduction in area: 100(500-187.5)/500
    = 62.50%
  Type of fracture: Three-quarter cup cone,
    fine-grained in center, silky at edge
  Modulus of elasticity: 202 E6/1.02 E-3 = 198 GPa

Operator: H.D.
Recorder: G.H.
Date: 80-03-21
Temperature: 22°C

      * Data used for E
    ** Extensometer removed
  *** Specimen ruptured

Load kN	Deform. mm	Stress MPa	Strain m/m
14.5	0.050	29	0.000125
27.5	0.100	55	0.000250
39.0	0.150	78	0.000375
52.5	0.200	105	0.000500
63.0	0.250	126	0.000625
75.5	0.300	151	0.000750
89.0	0.350	178	0.000875
101.0	0.400	202*	0.001000
112.5	0.450	225	0.001125
124.0	0.500	248	0.001250
134.0	0.550	268	0.001375
134.5	0.576	269	0.001440
133.5	0.600	267	0.001500
133.0	0.752	266	0.001880
131.5	1.000	263	0.002500
134.0	1.248	268	0.003120
134.0	1.876	268	0.004690
134.5	2.500**	269	0.006250
134.5	2.5	269	0.0125
157.5	5.0	315	0.0250
175.5	7.5	351	0.0375
200.0	12.5	400	0.0625
212.5	17.5	425	0.0875
221.0	22.5	442	0.1125
225.0	27.5	450	0.1375
226.5	32.5	453	0.1625
226.5	37.5	453	0.1875
225.5	42.5	451	0.2125
224.5	47.5	449	0.2375
220.5	52.5	441	0.2625
164.5***	62.5	329	0.3125

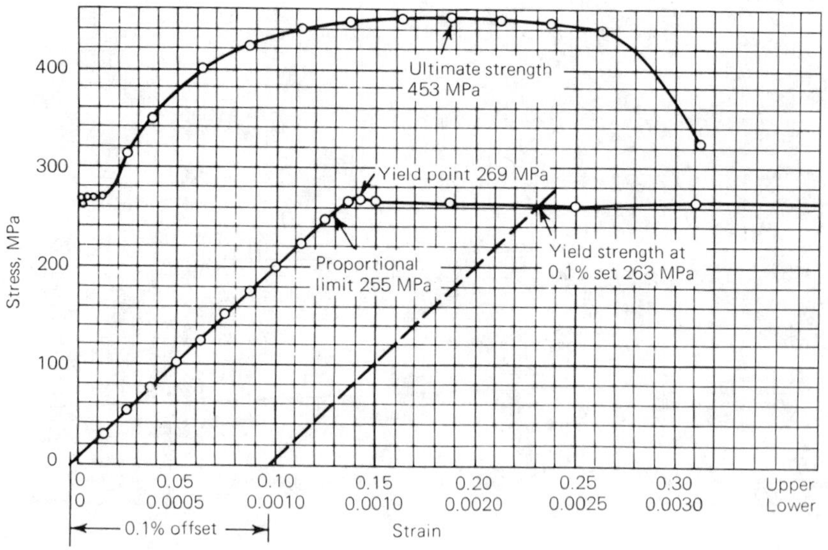

Figure 8.18 Log sheet with stress-strain diagram.

are computed on the basis of the original dimensions. A log sheet and a stress-strain diagram prepared from it are shown in Fig. 8.18, as an example of a record manually obtained.

When the test is performed on a highly ductile material such as mild steel, it is useful to draw two stress-strain curves:

1. A primary diagram showing the entire stress-strain curve up to the point of rupture. The elastic portion may well become an almost vertical line, since the strain at rupture is perhaps 200 times larger than that at yielding.
2. A secondary diagram, with the strain plotted to a scale 100 times larger than in the primary diagram. The elastic portion should now appear as a sloped straight line. The slope (Young's modulus in graphical form) should be about 40° to 60°. This can be accomplished by choosing an appropriate scale to represent the stress.

The two diagrams should have the same vertical scale and may conveniently be drawn on the same graph, which will show the two different horizontal scales.

A review of Chap. 2 may be profitable at this point.

The elongation is the increase in the gage length, expressed as a ratio of the original gage length. Both the percentage increase and the original gage length are reported. In ductile metals, if the break occurs near one end of the gage length, some of the effects of the localized drawing out or necking down will extend beyond the gage length. Hence, when the break occurs outside the middle third, specifications often call for a retest, although an approximate method for obtaining the elongation may be used as shown in Fig. 8.19.

The reduction of area is the difference between the area of the smallest cross section (at the break) and the original cross-sectional area, expressed as a ratio of the original cross-sectional area.

Tensile fractures may be classified as to form, texture, and color. Types of fracture as regards form are symmetrical: cup-cone, flat, and irregular or ragged; or asymmetrical: partial cup-cone, flat, and irregular or ragged. Various descriptions of texture are silky, fine grain, coarse grain or granular, fibrous or splintery, crystalline, glassy, and dull.

Certain materials are effectively identified by their fractures. Mild steel in the form of the standard cylindrical specimen usually has a cup-cone type of fracture of silky texture. Wrought iron has a ragged fibrous fracture, whereas the typical fracture of cast iron is gray, flat, and granular. An examination of the fracture may give a pos-

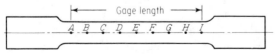

For breaks between C and G; Elongation = Final AI − Original AI

For breaks between A and C but nearer B than A;
Elongation = Final $(AC + 2CF)$ − Original AI

For breaks within one half division of A; Elongation = Final $2AE$ − Original AI

Figure 8.19 Determination of approximate elongation.

sible clue to low values of strength or ductility of the specimen. Nonaxial loading will cause asymmetrical types. Lack of symmetry may also be caused by nonhomogeneity of the material or a defect or flaw of some sort, such as segregation, a blowhole, or an inclusion of foreign matter, such as slag. On the fractured surface of material that has been cold-worked or has an internal stress condition due to certain heat-treatments, streaks or ridges often radiate outward from some point near the center of the section; this is sometimes referred to as a "star fracture." A description of the fracture should be included in every test report. Illustrations of a number of typical fractures are shown in Fig. 8.20.

8.6 EFFECTS OF SPECIMEN VARIABLES

Although the object of a tension test is to determine various material "properties," the results strictly pertain, as noted already, only to the specimen tested. To different degrees the size, shape, and history of a specimen affect its mechanical properties.

In general, with metallic materials, if the metal is of *uniform quality*, the size of geometrically similar specimens does not appreciably affect the results of the tension test. Various investigations on structural steels have borne this out [54, 66]. However, it is important to remember that, in the course of making or processing parts or shapes, the quality of the metal often varies with the size of the piece being produced. Thus test results for specimens of different sizes may reflect the effect of massiveness on properties. In the case of hot-rolled steel, the ductility is affected to some extent by the work of rolling, although the yield and tensile strengths are but little affected. The strength of cold-drawn wire is markedly influenced by the drawing process. Because of the strain-hardening effect, there is a considerable increase in yield point and ultimate strength of cold-worked metals, but these changes are accompanied by a marked decrease in ductility. In the case of cast metals, tensile strength decreases with increasing specimen diameter, but the differences represent largely actual differences in the properties of the specimens, as cast, rather than a real effect of size.

The total elongation of a ductile metal at the point of rupture is due to plastic elongation. This is more or less uniformly distributed over the gage length, on which is superimposed a localized drawing out or extension of the necked section, which occurs

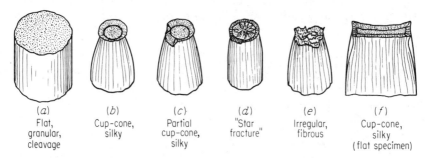

(a)	(b)	(c)	(d)	(e)	(f)
Flat, granular, cleavage	Cup-cone, silky	Partial cup-cone, silky	"Star fracture"	Irregular, fibrous	Cup-cone, silky (flat specimen)

Figure 8.20 Typical tensile fractures of metals.

just before rupture. The plastic elongation is practically independent of the shape of the cross section and gage length and is small compared with the final localized drawing out, which is affected by the shape of the piece. The elongation in each of eight segments of a typical mild-steel tension specimen is illustrated in Fig. 8.21. The corresponding elongations over various gage lengths symmetrical about the break are also shown. The length affected by the final localized drawing out is of the order of two or three times the diameter of the specimen. It is apparent, then, why the diameter of piece and gage length (or ratio of diameter to gage length) must be fixed if comparable elongations are to be obtained and why specifications call for rejection of a test if the break is too near the ends.

The requirement of geometrical similarity of test pieces for comparable elongations was first stated by J. Barba in 1880 and is often referred to as Barba's law [5]. Numerous investigations since then have confirmed this general finding that when the

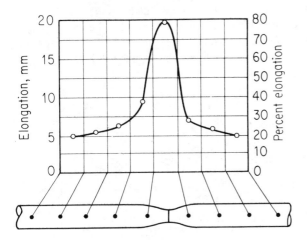

(a) Elongation in each division
(see Fig. 8.18 for values)

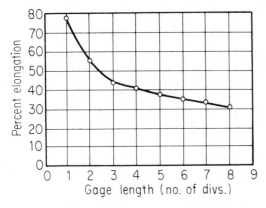

(b) Percent elongation vs. gage length

Figure 8.21 Effect of gage length on percentage of elongation.

gage length $L = k\sqrt{A}$, where A is the cross-sectional area and k is a constant for the type of specimen, the elongation is practically constant [66].

If the shoulders of a test bar are too close together, if the piece is notched or grooved transversely or contains holes, or if the sides of the specimen are curved, the strength and ductility of the piece may be appreciably affected. The severity of this effect depends on the abruptness and relative magnitude of the change in section and the ductility of the material.

For a series of specimens of ductile metal that enlarge abruptly at the ends of the gage length, the effect of the length-diameter ratio (L/d) on both the elongation and reduction of area is shown in Fig. 8.22a [97]. For values of L/d greater than about 2, the reduction of area is independent of L/d, but for lower values it is reduced because the enlarged ends provide lateral restraint against the reduction of area. In the extreme case it is reduced to zero. The elongation curve does not continue sharply upward to the left as shown in Fig. 8.21b and by the dashed line in Fig. 8.22a, but it also turns downward to zero because of this same restraint. Figure 8.22b shows the stress-strain diagrams for the same ductile specimens. When L/d exceeds about 1, the only effect is on the segment of the curve beyond the maximum strength at F. The shorter the length L, the longer the segment of the curve beyond point F. When L is shortened so

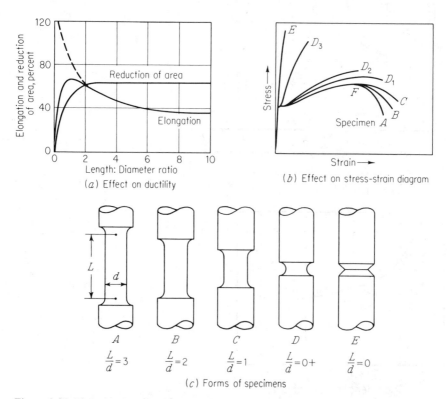

Figure 8.22 Variation in strength and ductility of steel with length of reduced section. (*From Upton, Ref. 97.*)

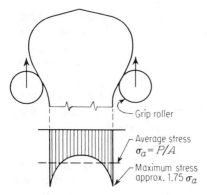

Figure 8.23 Stress distribution in mortar briquet.

that the ends give support to the reduced section, the curve is raised and shortened as shown by curves D_1, D_2, and D_3. Finally, the specimen E with zero length L gives a high strength with very little deformation.

The effect of notching is (1) to suppress the drawing out of the reduced section, owing to the support given by the mass of the adjacent larger sections, and (2) to cause stress concentration in the material at the base of the notch. The former tends to increase the apparent strength and reduce the apparent ductility of ductile materials. The latter tends to cause a reduction in apparent strength, and with brittle materials, which have little elongation up to rupture, this is very pronounced. The effect of notching in a ductile material is shown by comparing the strength of a standard specimen with that of a grooved specimen of the same material, where the diameter at the base of the groove is the same as that of the standard specimen. For a certain high-carbon steel, the ultimate strength of the standard specimen was 703 MPa, whereas that for a specimen with a 0.8-mm (wide) square groove was 1124 MPa, both computed in the conventional manner [95]. In the latter specimen the usual drawing out and reduction in area were prevented, thus causing an increase in strength greater than the decreasing effect of high stress concentrations at the edge of the groove. In a brittle material, where the elongation is of little consequence, the groove would have caused a decrease in strength.

Owing to the effects of curvature of the side of the piece and to the closeness of the grips, the stress distribution across the net section of the standard mortar tension briquet is not uniform but varies as shown in Fig. 8.23. The ratio of maximum to average stress on the critical cross section has been estimated from photoelastic studies to be about 1.75 [1]. Since mortar is a brittle material, these stress concentrations reduce the ultimate load.

8.7 EFFECTS OF TEST VARIABLES

As has been repeatedly pointed out, the test conditions and the condition of the material at the time of test have a very important influence on results. Reports of investigations to determine these effects comprise a vast literature extending over many years. One object of many such investigations is to evaluate the effects of test conditions with a view to selecting a standard procedure that will give results having a minimum of variability with reasonable fluctuation in test conditions; another object is to develop a basis for projecting the results of tests made under given conditions to probable behavior under some other conditions.

It is not possible here to discuss at length the effect of the numerous variables of testing. However, the general effect of a few of the more important variables will be pointed out in order to provide some appreciation of sources of error.

Eccentricity Eccentric loading produced by the gripping device causes nonuniform stress distribution in the test bar. An example of this is shown in Fig. 8.24, where the separate stress-strain curves for three gage lines 120° apart around a cylindrical test specimen are plotted. In the test for which the results are shown, the wedge grips at the ends of the specimen were out of alignment by 0.9 mm, a relatively small amount. This caused certain parts of the specimen to reach the proportional limit before other parts and thus resulted in a proportional limit, as determined from the average stress-strain diagram, somewhat lower than if the specimen had been stressed uniformly. The average stress-strain diagram gives no apparent indication of the effect of eccentric loading. In tests to determine elastic properties of materials, the effect of eccentric loadings is important. Under conditions of slightly eccentric loading, the averages of strain measurements taken on two opposite elements appear to give satisfactory values of the modulus of elasticity [92]. The strength of ductile materials does not appear to be greatly affected by slight eccentricities of load; the strength of brittle materials may be appreciably affected.

Test speed Over a wide range of speed, the rate of loading has an important effect on the tensile properties of materials. Strengths tend to increase and ductility to decrease

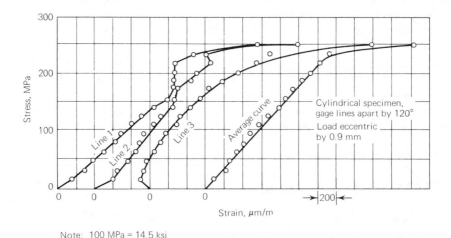

Note: 100 MPa = 14.5 ksi

Figure 8.24 Effect of eccentric load on strains.

with increased speeds. For example, certain tests have indicated that with a speed ratio of about 14 000:1 the yield point of mild steel was increased about 30 percent [15]. In general, the change in strength or elongation appears to vary approximately as the logarithm of the speed [40, 55]. The effect appears to be more pronounced for materials having low melting points, such as lead, zinc, and plastics, than for materials having high melting points, such as steel. With some materials, notably wood [104], but apparently also steel [70], the effect of very slowly applied loads (long-time tests) is to decrease the strength from that observed at normal testing speeds.

The term *viscoelasticity* is used to designate a time-dependent mechanical behavior that is a function of both elastic and viscous components. The strength-time data shown in Fig. 8.25 for rapidly loaded rectangular-shaped specimens having a 100-mm uniform length are typical of many plastic materials that exhibit this characteristic. The influence of temperature at time of test for such materials is also shown. In general, these materials exhibit a lowered modulus of elasticity at slower strain rates and also at higher temperatures.

Fortunately, investigations have shown that over the range of speed used with ordinary testing machines the effects of moderate variation in speed on the tensile properties of metals is fairly small, and rather wide tolerances can be permitted without introducing serious error in the results of tests for ductile metals [40]. For example, in tests of standard specimens of a structural steel it was found that an eightfold increase in rate of *strain* increased the yield point by about 4 percent, the tensile

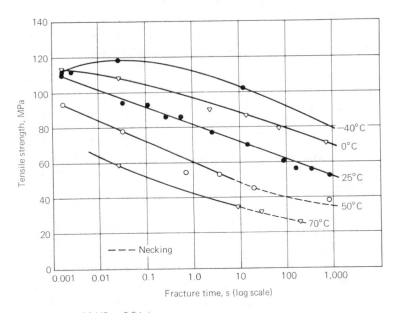

Note: 20 MPa = 2.9 ksi

Figure 8.25 Strength-time data for **PMMA** plastic. (*From Symposium on High Speed Testing, Ref. 88.*)

strength by about 2 percent, and decreased the elongation by about 5 percent. In the machine in which these tests were made, this change corresponded to a change in idling speed of the head from 1.25 to 10 mm/min. The effect of speed variations within the range of normal rates of loading on the strength of brittle materials such as cast iron appears to be small.

Fracture A number of factors affect the character of the fracture. As discussed in Chap. 2, the two fundamental types of tensile fractures are sliding (or shear) and separation (or cleavage) [27, 74, 80, 95]. If resistance to sliding is the greater, the material will fail by separation, overcoming the cohesive force with very little plastic elongation, and the material is said to be brittle. If the strength depends on its resistance to sliding, i.e., resistance to separation is the greater, there is considerable plastic elongation and reduction of area before fracture occurs; such behavior characterizes ductile materials. Both these properties are functions of the temperature and the rate at which the load is applied. Experiments show that the resistance to separation is affected less than the resistance to sliding. Some materials that show considerable ductility under slowly applied loads fail with little plastic elongation when the load is applied suddenly. Both the resistance to separation and the resistance to sliding increase with decreasing temperature, but the resistance to sliding increases much more markedly; thus it is possible to obtain separation fractures (brittle behavior) in plain unnotched steel bars cooled in liquid air. For an illuminating discussion of this, see Ref. 26.

A state of triaxial stress may reduce ductility or cause a separation fracture to occur in ductile materials under tension. Since the maximum shear is a function of the difference between the maximum and minimum principal (direct) stresses, as the magnitudes of the three principal stresses approach each other, the maximum shear may become very small, even though the principal stresses are high. A state of triaxial stress may be induced by abrupt changes in section or the presence of irregularities in a piece of material subject to uniaxial load; thus it is possible to reduce ductility markedly and to cause separation fractures at moderately low temperatures in steel [74].

After a standard test specimen of ordinary steel has necked down (shearing or sliding action), the state of stress in the central portion of the necked section is no longer that of simple tension; radial as well as axial stresses act on the crystals that compose the material. The maximum principal stress may be several times the maximum shear stress, instead of having the 2 to 1 relation that existed between tension and shear before necking began. Hence, at moderately low temperatures, the type of failure may be mixed, i.e., a separation failure in the central portion of a cup-cone fracture and sliding along the cone at the edges. However, experimental evidence indicates that at room temperatures the mode of failure of mild steel at the bottom of the cup can be sliding, even though the bottom of the cup appears granular to the naked eye. This has been attributed to a higher shear stress near the center than near the edges, which also accounts for the observation that the first crack can begin at the *center* of the section [26, 74].

8.8 LOW- AND HIGH-TEMPERATURE TESTS

The tensile behavior of most materials at ordinary room temperatures does not indicate the characteristics at either very low or very high temperatures. Therefore tests are occasionally conducted in special environmental chambers.

Tests at low temperatures Many materials are used at low temperatures. Such applications are made in refrigeration equipment, many chemical operations, and in both machines and structures located in areas where cold weather occurs. Most metals show an increase in tensile strength at lower temperatures, but their yield strengths are not always affected to the same degree, as shown in Fig. 8.26. In general, ductility decreases with lower temperatures, so that their toughness is materially reduced under impact loadings.

Tests at high temperatures The loading of the components of rockets, guided missiles,

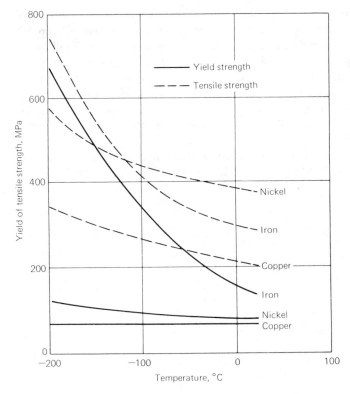

Note: 200 MPa = 29 ksi

Figure 8.26 Variation of yield and tensile strengths of metals with temperature. (*From Amer. Soc. Metals Symposium, Ref. 108.*)

and aircraft power plants may occur within a few seconds, and the temperatures developed in some of the parts are very high. Obtaining satisfactory information for design of such parts has required a special set of high-temperature-testing techniques, since the test temperatures are usually in ranges where the metallurgical structure of the material is unstable and the creep rate comparatively high. These test procedures involve the following factors in determining the strength of a material:

1. Loss of strength at the elevated temperature
2. Creep rate of the material during the test
3. Effect of strain rate on the resultant strength level
4. Effect of aging or other metallurgical changes during the test

The *high-heating-rate test* for the evaluation of materials for missiles or similar service is performed by loading a special test specimen with weights, either directly or by means of levers, and heating the specimen, acting as a resistance unit, with an electric current until it fractures. The temperature and elongation are recorded by automatic equipment. The temperature-elongation curve shown in Fig. 8.27 resembles an ordinary stress-strain curve; the increase in temperature causes yielding and rupture to develop, whereas in the common tensile test the increase in stress causes failure. In Fig. 8.27, *OA* represents the initial strain due to loading. During the early part of the test, as the temperature increases, the thermal expansions and decrease in the modulus of elasticity cause elastic elongations to occur. As for determinations of yield strength in an ordinary test, a line at 0.2 percent offset is drawn and extended to cross the plastic portion of the curve at *B*. This point determines the *yield temperature* for the given material, stress, and heating rate.

The results from a series of tests at a given heating rate but at different stresses are shown by the solid line in Fig. 8.28. The dashed line shows the results from con-

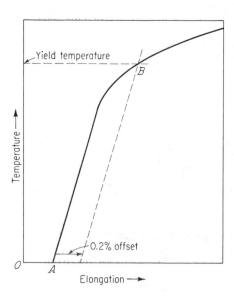

Figure 8.27 Determination of yield temperature for a given stress and heating rate.

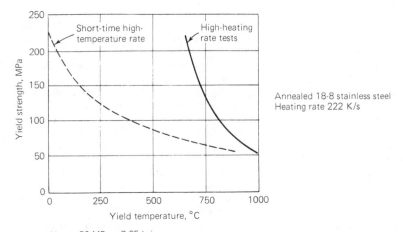

Note: 50 MPa = 7.25 ksi

Figure 8.28 Comparison of short-time high-temperature and high-heating-rate yield strengths. (*From Amer. Soc. Metals, Ref. 109.*)

ventional short-time high-temperature tests in which the material is held for a while (or "soaked") at a given temperature without load and then is loaded to rupture over a period of time, usually several minutes in duration. The latter results show yielding at a lower temperature due to the effects of creep, the slower strain rate, and the metallurgical changes that occur during the test.

It should be noted that the data shown here are for relatively short-time tests. More commonly materials are subjected to elevated temperatures for extended periods, in which case creep takes place. Creep is the subject of Chap. 15.

PROBLEM 8 Tension Test of Steel

Object: To determine the strength and several properties of a ductile steel, to observe the behavior of the material under load, and to study the fracture. The specific items to be determined are
1. Elastic strength in tension
    a. Proportional limit
    b. Upper yield point
    c. Lower yield point
    d. Yield strength for an offset of 0.1 percent
2. Tensile strength
3. Ductility
    a. Percentage elongation
    b. Percentage reduction of area
4. Modulus of elasticity
5. Modulus of resilience
6. Modulus of toughness
7. Type and character of fracture
Preparatory reading: ASTM E 8.
Specimen: Low-carbon steel.
Special apparatus: Extensometer (dial- or LVDT-type).

Procedure:
1. Determine the average cross-sectional dimensions of the specimen with a micrometer caliper. Scribe a line along the bar, and with a center punch lightly mark an appropriate length symmetrical with the length of the bar.
2. Firmly grip the upper end of the specimen in the fixed head of the testing machine. Place the specimen so that the punch marks face the front of the machine.
3. Measure the gage length and determine the multiplication ratio, if any, of the extensometer. Determine the value of the divisions on the dial indicator or readout. Firmly attach the extensometer to the specimen so that its axis coincides with that of the specimen, and remove the spacer bar (if present). Adjust the testing machine and the extensometer to read zero, setting the latter so that most of its range will be available. Grip the lower end of the specimen, taking care not to jar the extensometer.
4. Select suitable increments of strain to secure at least 15 readings below the probable proportional limit. Apply load at a slow speed, and make simultaneous observations of load and strain without stopping the machine. Continue loading until the yield point is passed. Stop the machine (but hold the load) to remove the extensometer (unless it is of the "breakaway" type).
5. Again apply the load continuously. When the strain has increased to 0.0125, as measured with dividers and scale, observe the load. Thereafter, for each 0.025 increase in strain, observe the load. For this part of the test the speed of the machine may be increased to about 5 mm/min. Record the maximum and breaking loads.
6. Remove the broken specimen from the machine. If the specimen is jammed in the grips, use a hammer to strike the sides of the specimen. Do not strike the grips or grip handles.
7. Observe the location and character of the fracture, and measure the dimensions of the smallest section. Fit the broken parts together and measure the gage length.

Report: Plot a stress-strain diagram for the test in accordance with general instructions and compute all properties called for. Comment on the following:
1. Did the test specimen used conform with ASTM standards?
2. Was the test carried out in accordance with ASTM standards?
3. State whether or not you would accept the material as satisfactory and in conformity with the appropriate ASTM specification. Indicate how it may have failed to meet requirements.

DISCUSSION

1. What is the relation between nominal and actual stress for loads beyond the yield point?
2. Why is it necessary to state the gage length when reporting the percentage of elongation?
3. Discuss the variation in the percentage elongation with the size and shape of the bar.
4. How can the work required to rupture the specimen be determined?
5. What errors may be introduced if the axis of the extensometer and that of the specimen do not coincide?
6. Are wedge grips suitable for tests of brittle material? Explain.
7. Distinguish clearly between proportional limit and elastic limit.
8. Distinguish between yield point and yield strength.
9. What are the advantages of a stress-strain diagram over a load-elongation diagram for showing the results of the test?
10. How seriously are the results of a tension test of mild steel affected if the specimen dimensions vary a little from the standard?

READING

Gilliam, E.: *Materials Under Stress*, Butterworth, London, 1969.
Tension Testing of Non-Metallic Materials, ASTM, Philadelphia, 1956.

STATIC COMPRESSION

9.1 COMPRESSION

It was already pointed out in Art. 8.1 that, in theory at least, the compression test is merely the opposite of the tension test with respect to direction or sense of the applied stress. General reasons for choice of one or the other type of test were stated. Also, a number of general principles and concepts were developed throughout the chapter on tension testing that apply equally well to compression testing. There are, however, several special limitations to the compression test that deserve attention:

1. The difficulty of applying a truly concentric or axial load.
2. The relatively unstable character of this type of loading contrasted with tensile loading. There is always a tendency for bending stresses to be set up and for the effect of accidental irregularities in alignment within the specimen to be accentuated as loading proceeds.
3. Friction between the heads of the testing machine or bearing plates and the end surfaces of the specimen due to lateral expansion of the specimen. This may alter considerably the results that would be obtained if such a condition of test were not present.
4. The relatively large cross-sectional areas of the compression-test specimen, required to obtain a proper degree of stability of the piece. This results in the necessity for a relatively large-capacity testing machine or specimens so small and therefore so short that it is difficult to obtain from them strain measurements of suitable precision.

It is presumed that the simple compression characteristics of materials are desired and not the column action of structural members, so that the discussion here is confined to the short compression block. Still, for some materials the failure modes in compression may well include microscopic buckling, as is the case, for example, with wood loaded parallel to its longitudinal fibers.

9.2 SPECIMENS

For uniform stressing of the compression specimen, a circular section is to be preferred over other shapes. The square or rectangular section is often used, however, and for manufactured pieces such as tile, it is not ordinarily feasible to cut out specimens to conform to any particular shape.

The selection of the ratio of length to diameter of a compression specimen appears to be more or less of a compromise between several undesirable conditions. As the length of the specimen is increased, there is an increasing tendency toward bending of the piece, with consequent nonuniform distribution of stress over a right section. A height-to-diameter ratio of 10 is suggested as a practical upper limit. As the length of the specimen is decreased, the effect of the frictional restraint at the ends becomes relatively important; also, for lengths less than about 1.5 times the diameter, the diagonal planes along which failure would take place in a longer specimen intersect the base, with the result that the apparent strength is increased. A ratio of length to diameter of 2 or more is commonly employed, although the height-to-diameter ratio used varies for different materials. To accommodate a compressometer of desired precision, it is sometimes necessary to use a relatively long specimen.

The actual size of the specimen depends on the type of material, the type of measurements to be made, and the testing apparatus available. For homogeneous materials for which only ultimate strength is required, relatively small specimens may be used. The size of specimens of nonhomogeneous materials must be adjusted to the size of the component particles or aggregates.

The ends to which load is applied should be flat and perpendicular to the axis of the specimen or made so in effect by the use of caps and adjustable bearing devices.

Gage lengths for strain measurements preferably should be shorter than the specimen length by at least the diameter of the specimen.

Specimens for compression tests of metallic materials recommended by ASTM (ASTM E 9) are shown in Fig. 9.1. The short specimens are intended for use with bearing metals, the medium-length specimens for general use, and the long specimens for tests to determine the modulus of elasticity. Specimens for compression tests of sheet metal must be loaded in a jig that provides lateral support against buckling without interfering with the axial deformations of the specimen. The details of such jigs and the corresponding specimens are covered by ASTM E 9.

For concrete, the standard specimens in the United States are cylinders twice the diameter in height. For concrete with aggregates of maximum size not greater than 50 mm, the standard size cylinder is 150 by 300 mm;[†] for concretes containing aggregates of maximum size up to 65 mm, a 200-by-400-mm cylinder is used (ASTM C 31). It is common practice in many laboratories to use 75-by-150-mm cylinders

[†]Or 6 by 12 in.

Suggested dimensions for specimens

Type	Dia. d, mm	Hgt. h, mm	Dia. d, in.	Hgt. h, in.
Short	30	27	$1\frac{1}{8}$	1
Medium-length	13 to 30	39 to 90	$\frac{1}{2}$ to $1\frac{1}{8}$	$1\frac{1}{2}$ to $3\frac{3}{8}$
Long	20 to 32	160 to 320	0.798 to $1\frac{1}{4}$	$6\frac{3}{8}$ to $12\frac{1}{2}$

$h = 8d$ to $10d$

$h = 3d$

$h = 0.9\,d$

Short specimen

Medium-length specimen

Long specimen

Figure 9.1 Compression specimens for metallic materials in other than sheet form (ASTM E9).

for concrete with aggregates up to 20 mm, and for tests of mass concrete with aggregates up to 150 mm, 450-by-900-mm cylinders are used. Cubes are in use in Europe, 200 mm being a normal size.

For mortars the 50-by-100-mm cylinder is often used, although ASTM specifies a 50-mm cube (ASTM C 109).

Specimens for compression tests of small, clear pieces of wood parallel to the grain are 50-by-50-by-200-mm† rectangular prisms. Compression tests perpendicular to the grain are made on nominal 50-by-50-by-150-mm specimens, as shown in Fig. 9.2. The load is applied through a metal bearing plate 50 mm in width placed across the upper surface at equal distances from the ends and at right angles to the width (ASTM D 143).

†Or 2-by-2-by-8 in.

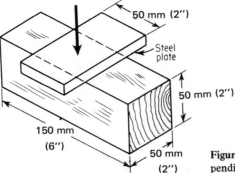

50 mm (2'')

Steel plate

50 mm (2'')

150 mm (6'')

50 mm (2'')

Figure 9.2 Method for compression test of wood perpendicular to the grain (ASTM D 143).

The compressive strength of building brick is determined on a half brick with approximately plane and parallel surfaces, tested flatwise (ASTM C 67).

Refer to ASTM specifications for types of specimens for compression tests of other materials such as drain tile (ASTM C 4), structural clay tile (ASTM C 67), sewer pipe (ASTM C 14), refractory brick (ASTM C 133), vulcanized rubber (ASTM D 395, D 575), timber in structural sizes (ASTM D 198), and building stone (ASTM C 170).

9.3 APPARATUS

While universal testing machines (UTMs) can be and often are used for compression testing, there are machines designed specifically for the purpose. Many are used primarily for routine compression tests of concrete specimens. Because these specimens are rather large, testers of relatively high capacity are needed. Yet if tensile capability is eliminated, the machines can not only be made less expensive than UTMs but also light enough to be called portable, meaning suitable for mobile quality control laboratories at a construction site. The machine shown in Fig. 9.3 has a capacity of more than 1 MN, but a mass of less than 300 kg. Standard cylinders of 60-MPa concrete can be tested, yet only a few workers are needed to lift the tester.

The way the load is applied to the specimens requires attention and care. The ends of compression specimens should be plane, or flat so as not to cause stress concentrations and should be perpendicular to the axis of the piece so as not to cause bending due to eccentric loading.

Figure 9.3 Portable hydraulic compression tester. (*Courtesy of Forney's, Inc.*)

The end surfaces of metal specimens can be machined flat and at right angles to the axis. Wood test pieces can usually be trimmed so that these conditions are satisfied. For materials such as concrete, stone, and brick, however, a bedment, with or without the use of accompanying capping plates, is usually necessary. Materials commonly used for bedments are plaster of paris, Hydrostone (a high-strength gypsum compound), quicksetting cements, and sulfur compounds. In setting capping plates, precaution should be taken to ensure perpendicularity between the bearing surface and the axis of the specimen. A jig is usually employed for the purpose. It is desirable for the capping material to have a modulus of elasticity and strength at least equal to that of the material of the specimen. The cap should be as thin as is practicable. If a capping compound contains water, it may affect the strength of absorbent materials such as brick, so a coat of shellac or a sheet of waxed paper is placed on the ends of the specimen before capping. Loose materials such as sand or small steel balls have not proved successful for end bedments. Soft bedments such as rubber sheets and fiber boards should be avoided, since they tend to flow laterally under load and cause the specimen to split.

Plain bearing plates or capping plates should have machined, flat, parallel surfaces. The material of the bearing plate should be strong and hard relative to the test specimen. See ASTM C 39 and C 192 for typical detailed requirements.

Usually one end of the specimen should bear on a spherically seated block. Figure 9.4 shows satisfactory arrangements of the specimen and block. The purpose of the block is to overcome the effect of a small lack of parallelism between the head of the machine and the end face of the specimen, giving the specimen as even a distribution of initial load as possible. It is desirable that the spherically seated bearing block be at the upper end of the test specimen. In order that the resultant of the forces applied to the end of the specimen not be eccentric with respect to the axis of the specimen, it is important for the center of the spherical surface of this block to lie in the flat face that bears on the specimen and for the specimen itself to be carefully centered with respect to the center of the spherical surface. Owing to increased frictional resistance as the load builds up, the spherically seated bearing cannot be relied on to adjust itself to bending action that may occur during the test. Under some conditions of test, the spherically seated bearing block may be omitted, and under

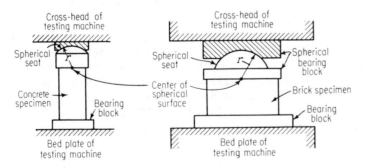

Figure 9.4 Spherical bearing blocks for compression tests.

others two such blocks may be required. The block should be of the same or slightly larger diameter than the specimen.

9.4 PROCEDURE

In commercial tests, the only property ordinarily determined is the compressive strength. For brittle materials in which fracture occurs, the ultimate strength is definitely and easily determined. For materials in which there is no unique phenomenon to mark ultimate strength, arbitrary limits of deformation are taken as the criteria of strength (see, for example, ASTM B 22 and D 575).

In tests to determine the yield strength of metals in compression, the usual criteria (described in Chap 2) may be used.

Dimensions should be determined with appropriate precision. Recommended precisions for cross-sectional measurements for ordinary work are as follows: metals, to the nearest 0.02 mm; concrete and wood, to the nearest 0.2 mm. On cylindrical specimens, measurements should be made on at least two mutually perpendicular diameters. If densities are required, specimens should ordinarily be weighed with a precision of about 0.5 percent.

Great care must be exercised to obtain accurate centering and alignment of the specimen and bearing blocks in the testing machine. For careful work, an effort should be made to have the axes of the specimen and bearing blocks coincide with an axis through the centers of the head and base plate of the machines within 0.2 mm. While the head of the machine is being lowered to contact with the spherical bearing block, it is desirable to rotate slightly by hand the upper part of the block in a horizontal plane to facilitate the seating of the block.

In testing metals the ends of the specimen and the faces of the bearing blocks should be cleaned with acetone or another suitable solvent immediately before testing to remove grease or oil, which would influence the frictional restraint at the end surfaces (ASTM E 9).

In a compression test, absolutely uniform stress distribution is practically never attained. In making precise stress-strain determinations with a view to finding the proportional limit, it is therefore desirable to measure strains along at least three gage lines 120° apart around a cylindrical piece. For ordinary determinations of the modulus of elasticity, an averaging type of compressometer is usually sufficient.

For routine strength testing in compression, maximum test speeds are generally specified. The trend is to specify a stress rate, which for concrete might range from 8 to 20 MPa/min.[†] For a cyclinder 300-mm in length made of concrete with a modulus of 20 GPa, this would result in a deformation rate of about 0.12 to 0.30 mm/min[‡] in elastic range. For convenience in the use of screw-gear machines, idling crosshead speeds are still sometimes specified. Thus ATMS C 39 gives 1.25 mm/min as the maximum value for concrete; the speeds for other materials tend to be similar or

[†]1 MPa = 145 psi.
[‡]1 mm = 0.039 in.

slower. In many cases any convenient speed is permitted up to one-half of the expected maximum load, after which the rate is specified. There are also specifications for the minimum time to apply the last half of or the entire load.

In materials investigations intended to determine stress-strain behavior, slower speeds than those given by specifications are advisable unless fully automatic recording instrumentation is used. For a normal visually observed and manually recorded stress-strain test of concrete, for example, a stress rate of about 4 to 7 MPa/min and a strain rate of perhaps 0.000 4/min seem appropriate. This will allow several minutes for the maximum load to be reached, time enough to take a reasonable number of careful readings.

9.5 OBSERVATIONS

Identification, dimensions, critical loads, compressometer readings (if taken), type of failure, including sketches, and other data are recorded on a form appropriate to the type of test and the extent of the data to be taken.

Brittle materials commonly rupture either along a diagonal plane, or with a cone- (cylindrical specimens) or a pyramidal- (square specimens) shaped fracture, sometimes called an *hourglass* fracture (see Fig. 9.5). Cast iron usually fails along an inclined plane, and concrete exhibits the cone type of fracture. Such fractures are essentially shear failures.

For a material that resists deformation and failure by internal friction as well as by cohesion and that behaves in accordance with the Mohr theory of rupture, the angle of rupture is not $45°$ (the plane of maximum shear stress) but is a function of the angle of internal friction ϕ. In Fig. 9.6 is shown, by means of the Mohr stress circle, the state of stress at failure in an element subjected to uniform principal stress in one direction only. From the representation of the angles of break on the Mohr circle diagram, it may be seen that $\alpha = 45° - \phi/2$ or $\theta = 45° + \phi/2$. (For an exposition of the Mohr circle and its use, see textbooks on strength of materials.)

The behavior of such materials as cast iron, concrete, or ceramics does not conform exactly to that predicted by the Mohr theory of rupture, in part because their nonhomogeneous composition causes irregularities in the stress pattern. Further, the angle of rupture may deviate somewhat from the theoretical value owing to the

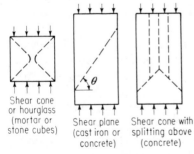

Figure 9.5 Types of failure of brittle materials under compressive loading.

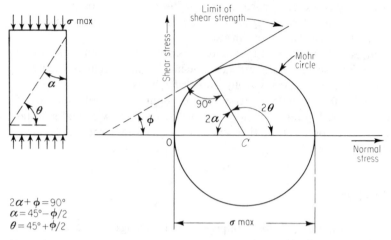

Figure 9.6 Relation between angle of rupture α and angle of friction ϕ.

complex stress condition induced in the end portions of compression specimens by restraint to lateral expansion under load caused by friction of the bearing plates on the end surfaces. This effect of lateral restraint of the ends becomes more pronounced in short specimens.

The observed values of θ for a number of materials including cast iron, sandstone, brick, and concrete vary roughly between $50°$ and $60°$ for specimens long enough for normal failure surfaces to develop [104], as shown in Fig. 9.5.

If the specimen is so short that a normal failure plane cannot develop within its length, then the strength is appreciably increased, and other types of failure, such as crushing, may occur. With brittle materials in short specimens, when there is a combination of high compressive strength and unrestrained lateral expansion at the ends, the pieces often fail by separation into columnar fragments, giving what is known as a splitting failure or columnar fracture. Lateral flow of a bedment tends to produce a splitting failure.

Wood under compressive loading exhibits a behavior peculiar unto itself. It is anything but an isotropic material, being composed of cells formed by organic growth that align themselves to form a series of tubes or columns in the direction of the grain. As a result of this structure, the elastic limit is relatively low, there is no definite yield point, and considerable set takes place before failure. These properties vary with the orientation of the load with respect to the direction of the grain. For loads normal to the grain, the load that causes lateral collapse of the tubes or fibers (crushing) is the significant load. For loads parallel to the grain, not only is the "elastic" strength important but so is the strength at rupture. Rupture often occurs because of collapse of the tubular fibers as columns. Various types of failure of wood loaded parallel with the grain are described by the sketches in Fig. 9.7.

Ductile and plastic materials with some tenacity bulge laterally and take on a barrel shape as they are compressed, provided, of course, the specimen does not bend or buckle. Material of relatively low ductility and case-hardened pieces develop surface cracks parallel with the loading axis as failure becomes pronounced.

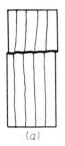    

(a)　　　　(b)　　　　(c)　　　　(d)　　　　(e)

(*a*) Crushing (plane of rupture approximately horizontal).

(*b*) Wedge split (note direction of split: radial or tangential).

(*c*) Shearing (plane of rupture at acute angle with horizontal).

(*d*) Splitting.

(*e*) Shearing and splitting parallel to grain. (Usually occurs in cross-grained pieces).

Figure 9.7 Types of failure of wood under compression parallel to the grain (ASTM D 143).

9.6 EFFECTS OF VARIABLES

In compression, as in tension, variations in specimen configuration and test conditions may affect the results significantly.

Size and shape Almost all compression specimens are right regular prisms or cylinders, so that one important shape variable is the length-to-width (or height-to-diameter) ratio. Its effect on the compressive strength of brittle materials is illustrated in Fig. 9.8,

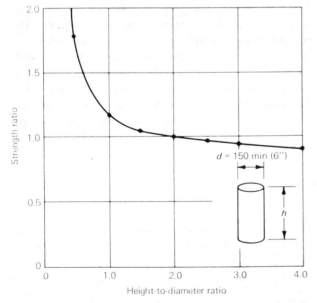

Figure 9.8 Effect of height of concrete cylinders on strength. *(Based on H. F. Gonnerman, "Effect of Size and Shape of Test Specimen on Compressive Strength of Concrete," Proc. ASTM, vol. 25, pt. II, 1925.)*

Table 9.1 Correction factors for concrete cylinders not of standard length[†]

Height-to-diameter ratio	2.00	1.75	1.50	1.25	1.10	1.00	0.75	0.50
Multiplying factor	1.00	0.89	0.96	0.94	0.90	0.85	0.73	0.60

[†]Based on ASTM C 42.

which summarizes the results of tests on concrete cylinders of standard diameter and varying height. The strengths are related to that of a standard cylinder with a height equal to twice the diameter. ASTM C 42 specifies multiplying factors, shown in Table 9.1, which are to be applied to concrete cylinders of other than 2 : 1 proportions.

Table 9.2 is based on the same test program as Fig. 9.8. It shows the relative strengths of cylindrical and square prismatic specimens with height-to-diameter ratios of 1 : 1 and 2 : 1.

The relative compressive strengths of concrete cylinders of various sizes, all with normal 2 : 1 shape ratios, are illustrated in Fig. 9.9. The data are adapted from tests made at the former U.S. Bureau of Reclamation[†] [8].

End conditions The end conditions at the time of test, the capping method, and the end conditions before capping may have a pronounced effect on the compressive strength of concrete test cylinders [30, 96]. Cylinders molded with machined plates so as to produce convex ends and tests without capping give pronounced reductions in strength even for a small amount of convexity. For a convexity of only 0.25 mm in a cylinder 150-mm in diameter, tests of 1 : 2 and 1 : 5 mixes have shown reductions in strength of about 35 and 20 percent respectively [30]. This shows the importance of having plane ends on test specimens. Tests have also shown that the higher the compressive strength of the capping material, the higher the indicated strength of the concrete and the less the effect of irregular ends before capping on the indicated strength. With caps of plaster of paris or steel shot the indicated strength of normal concrete may be reduced as much as 10 percent even for flat-ended cylinders, but for irregular ends before capping, the strengths may be reduced as much as 25 percent. The results of tests showing the relative strengths obtained with several types of caps are summarized in Table 9.3.

[†]Now the U.S. Water and Power Resources Service.

Table 9.2 Relative strengths of concrete specimen shapes[†]

Width by height[‡]	150 × 150 mm	150 × 300 mm	200 × 200 mm	200 × 400 mm
□ Square prism	113	93	115	91
○ Cylinder	115	100		96

[†]Based on H. F. Gonnerman, "Effect of Size and Shape of Test Specimen on Compressive Strength of Concrete," *Proc. ASTM,* vol. 25, pt. II, 1925.
[‡]50 mm = 2 in.

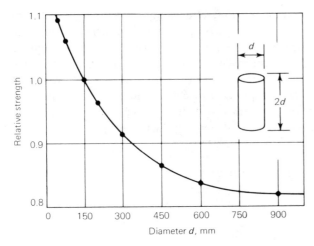

Note: 150 mm = 6"

Figure 9.9 Effect of size on concrete cylinders on strength. *(Based on R. F. Blanks and C. C. McNamara, "Mass Concrete Tests in Large Cylinders," Proc. ACI, vol. 31, 1935, and discussion in vol. 32, 1936.)*

Table 9.3 Effect of capping materials and end conditions before capping on compressive strength of concrete cylinders[a]

Type of capping material	20-MPa (3-ksi) concrete				55-MPa (8-ksi) concrete			
	Plane ends	Beveled ends[f]	Con-vex ends[g]	Con-cave ends[h]	Plane ends	Beveled ends[f]	Con-vex ends[g]	Con-cave ends[h]
Hydrostone[b].	100	99	102	99	100	94	97	93
Castite[c].	97	98	94	101	102	101	96	89
Shot[d].	88	90	93	75	89	90	90	56
Plaster of paris[e] . . .	97	85	88	88	87	66	54	34

Note: Cylinders with plane normal ends and Hydrostone caps taken as relative strength of 100.
  [a]G. E. Troxell, "The Effect of Capping Methods and End Conditions before Capping upon the Compressive Strength of Concrete Test Cylinders," *Proc. ASTM,* vol. 41, 1941. [96].
  [b]A gypsum compound; 1-h strength, 34 MPa; modulus, 11 GPa.
  [c]A sulfur-silica mixture; 1-d strength, 59 MPa; modulus, 15 GPa.
  [d]1.6-mm steel shot. Oiled. Results with dry shot practically the same.
  [e]A gypsum compound; 1-h strength, 12 MPa; modulus, 345 GPa.
  [f]Plane but not perpendicular to axis. Slope of 1.6 mm in 76-mm diameter.
  [g]Spherical bulge of 1.6 mm.
  [h]Spherical depression of 1.6 mm.

Test speed The speed of testing has a definite effect on compressive strength, although the effect is usually fairly small over the ranges of speed used in ordinary testing. The results of tests on concrete indicate that the relation between strength and rate of loading is approximately logarithmic—the more rapid the rate, the higher the indicated strength [22, 39]. The strength of a specimen loaded, say, at 40 MPa/min would be almost 15 percent greater than the strength of a specimen loaded at 1 MPa/min. The modulus of elasticity also appears to increase with the loading rate, although most observers have attributed this effect to reduced creep during the test period.

PROBLEM 9 Compression Test of Concrete

Object: To study the behavior of concrete under compressive loading, and to determine the following physical and mechanical properties:
1. Proportional limit
2. Yield strength at 0.01 percent offset
3. Compressive strength
4. Initial tangent modulus of elasticity
5. Secant moduli of elasticity at stresses of 5, 10, and 15 MPa
6. Density

Preparatory reading: ASTM C 39.

Specimen: Concrete cylinder or cube specimens.

Special apparatus: Compressometer.

Procedure:

1. From the instructor (or your Report 18) obtain data regarding the kinds and proportions of the constituent materials, the water-cement ratio and the consistency of the mix, the curing and storage conditions, and the age of the specimen. From these data predict the ultimate strength by consulting Figs. 18.1 and 18.2.

2. Determine the mean diameter of the cylinder at its midsection and its average length, making measurements to 0.1 mm. Weigh the specimen to 0.01 kg.

3. Cap each end of the specimen with the material provided, using a machined metal capping device.

4. Study the action of the compressometer, note its gage length and multiplication ratio, and determine the strain corresponding to the least reading of the dial. Attach the compressometer to the central portion of the specimen and remove the spacer bars.

5. After the caps have hardened, center the specimen in the testing machine and center the spherical bearing block on top of the specimen. Centering operations should be carried out by actual measurements.

6. Adjust the compressometer dial to read zero and make sure that most of its range is available.

7. Apply load continuously at a speed of, say, 0.1 mm/min (or 100 kPa/s), and read the compressometer after each load increment of about one-twentieth of the estimated ultimate load. If the strain approaches the range of the dial, hold the load and set the hand back a full revolution by means of the adjusting screw. At a load of three-quarters of the estimated ultimate load remove the compressometer. Thereafter apply the load continuously. Record the maximum load.

8. Draw a sketch to show the type of failure.

Report:

1. Plot a stress-strain diagram. Draw a smooth curve through the plotted points. Note that the curve may not pass through the origin of the graph. Mark the proportional limit and yield strength (0.01 percent offset) on the diagram. Determine the several moduli of elasticity specified.

2. Compute the compressive strength and the density of the concrete.

3. Tabulate the test results in a suitable form, and exchange the information with other groups, if possible.

4. Using all test data, correlate the results, with particular emphasis on the relation between the water/cement ratio and strength.

DISCUSSION

1. Discuss why cross-sectional areas of compression specimens must be larger than those of tension specimens.
2. Inspect Table 9.2 and estimate the relative strength of a 200-by-200-mm concrete test cylinder.
3. If a compression machine is equipped with an LVDT, can smaller specimens be tested than otherwise?
4. Devise a method to test the compressive strength of golf balls, as if you were writing a standard.
5. Column curves usually start with a "short compression block" strength value corresponding to a slenderness ratio of zero. What is the meaning of that value?
6. Why is the compression test the one most frequently made for concrete?
7. What is the purpose of a spherical bearing block in a compression test? List the various precautions that should be taken in positioning it.
8. Are the strength correction factors given in Table 9.1 for concrete specimens having height-to-diameter ratios below 2 rational or empirical values?
9. Can you think of any instances where materials are *intended* to fail in compression?
10. List some uses to which a compression machine can be put besides compression testing.

READING

Test Methods for Compression Members, ASTM, Philadelphia, 1966.

STATIC SHEAR

10.1 SHEAR

A shearing stress acts parallel to a plane, as distinguished from tensile and compressive stresses, which act normal to a plane. Direct shear and torsional shear are the loadings causing shear conditions that are of principal interest in materials testing.[†]

Direct shear If the resultants of parallel but opposed forces act through the centroids of sections that are spaced infinitesimal distances apart, it is conceivable that the shearing stresses over the sections would be uniform and a state of pure direct shear would exist. This condition may be approached but is never realized practically. An approximation of pure direct shear is the case of a rivet in shear as shown in Fig. 10.1a; here, for practical purposes, direct shear may be considered to exist within the rivet on planes xx and yy. Because two planes are involved simultaneously, the condition is called "double shear."

In some tests only one surface is subjected to shear, as in the case of the wood specimen depicted in Fig. 10.1b. This situation is referred to as "single shear."

One may think of the two specimens in Fig. 10.1 as being very short beams that resemble a simply supported beam and a cantilever beam respectively. This has two pertinent implications: bending stresses are not totally absent, and the shear stresses are greatest near the middle of the specimen and zero at the top and bottom.

Torsion The applied forces are parallel and opposite but do not lie in a plane containing the longitudinal axis of the body; thus a couple is set up that produces a twist about a longitudinal axis. The twisting action of one section of a body with respect to a contiguous section is termed *torsion*. Figure 10.2 represents a piece of shafting sub-

---

[†]For granular materials, such as soils, the triaxial compression test is a more useful and reliable indicator of shear strength than the direct shear test.

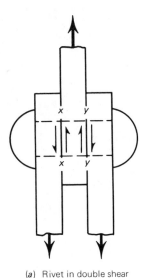

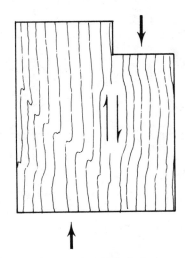

(a) Rivet in double shear

(b) Wood block in single shear

Figure 10.1 Direct shear loadings.

jected to a torque. Torsional shearing stresses on circular cross sections vary from zero at the axis of twist to a maximum at the extreme fibers. In the absence of other loads, pure shear exists at each point.

Pure shear stress At any point in a stressed body, the shearing stresses τ in any two mutually perpendicular directions are equal in magnitude. If at some point and on some pair of planes only shear stresses act, the material at that point is said to be in pure shear. The pure shears are greater than those on any plane through the point. The pure-shear condition is illustrated by Fig. 10.3a, which represents an elemental block on which the stresses are uniformly distributed. On all planes inclined to the planes of maximum shear, normal tensile or compressive stresses act; on mutually perpendicular planes at a 45° angle to the planes of maximum shear, the tensile and compressive stresses are a maximum and the shear stress is zero. The maximum normal stresses are equal in magnitude to the maximum shearing stresses. Conversely, pure shear is induced by equal and opposite normal stresses, as shown in Fig. 10.3b. The Mohr circle representation of the state of stress induced by pure shear is illustrated in Fig. 10.3c.

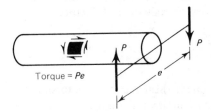

Torque = Pe

Figure 10.2 Torsional loading.

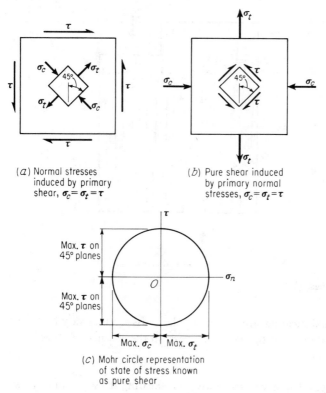

(*a*) Normal stresses induced by primary shear, $\sigma_c = \sigma_t = \tau$

(*b*) Pure shear induced by primary normal stresses, $\sigma_c = \sigma_t = \tau$

(*c*) Mohr circle representation of state of stress known as pure shear

Figure 10.3 Relation between pure shear and normal stresses.

Shearing strain The strain that accompanies shear may be thought of as arising from the effort of thin parallel slices of a body to slide one over the other. Shearing strain or "detrusion" is a function of the change in angle between adjacent sides of an elementary block as it is distorted under shearing stresses, which is illustrated in Fig. 10.4*a*. The total change in angle is more conveniently represented by a diagram such as Fig. 10.4*b*, where it may be seen that the shearing strain is the tangent of the angular distortion. However, within the range of elastic strength of materials used for construction, the shearing strains are small, and the angle is commonly expressed in radians.

In some materials (in particular, materials composed of granular elements, such as concrete and soil) the resistance to rupture by shear is a function not only of the shearing strength of the material but also of the frictional resistance to sliding on the surface of rupture. For these materials it is necessary to evaluate both factors.

10.2 SCOPE AND APPLICABILITY

The types of shear tests in common use are the direct shear test and the torsion test. In certain instances, shear properties are evaluated by indirect methods.

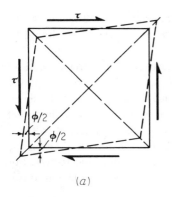

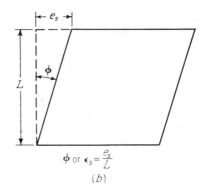

$$\phi \text{ or } \epsilon_s = \frac{e_s}{L}$$

(a)   (b)

Figure 10.4 Shearing strain.

In the direct shear test, sometimes called the "transverse" shear test, it is usually the procedure to clamp or support a prism of the material so that bending stresses are minimized across the plane along which the shearing load is applied. Although the method suffices for an indication of what shearing resistance may be expected in rivets, crankpins, wooden blocks, and so forth, nevertheless, owing to bending or to friction between parts of the tool or to both, it gives but an approximation to the correct values of shearing strength. The results of such a test depend to a considerable degree on the hardness and the sharpness of the edges of the hardened plates that bear on the specimen. The transverse shear test has the further limitation of being useless for the determination of the elastic strength or of the modulus of rigidity, because of the impossibility of measuring strains.

The punching shear test is also a form of the direct shear test; its use is restricted to tests of flat stock, principally metal. When a metal plate is punched, the punched area is removed by a slicing motion within a narrow ring of material adjacent to the cutting edge of the punch. The greater the clearance between the punch and the die, the greater the bending stresses accompanying the development of shear. Punching shear tests of brittle materials, such as concrete and cast iron, usually give higher nominal strengths than do simple transverse shear tests of the same materials, probably because of the greater friction and the smaller bending developed in the punching test. However, the results of punching shear tests are unsatisfactory as measures of shear strength and should be considered as giving simply a representation of the overall load needed to cause punching.

For a more precise determination of shearing properties, the torsion test is made, employing either solid or hollow specimens of circular section. In a torsion test, the specimen can be of such a length that a strainometer (called a troptometer in this type of test) can be attached to assist in determinations of proportional limit and yield strength in shear, shearing resilience, and stiffness (modulus of rigidity or modulus of elasticity in shear), the latter being obtained from the angle of twist and the applied torque. The ultimate shearing strength or shearing modulus of rupture is usually obtained. The ductility of the material is determined from the amount of twist up to rupture, toughness is represented by the amount of twist and the strength, and uniformity is indicated by the spacing, distribution, and appearance of the lines of twist. For accurate determination of the elastic strength, a tubular test specimen should be used.

The torsion test may be of especial use in the investigation of noncircular sections or of circular sections having various surface irregularities such as keyways and splines. In such cases, the object is to find the torsional resistance of the part, not the shearing strength of the material. Torsion tests can be used in investigations of the effect of various heat-treatment operations as well, particularly for parts subjected to treatments that tend to have a greater effect on the metal near the surface than on the interior of the member. For such studies the full-size part is subjected to test. For example, complete automobile- and truck-axle assemblies have been subjected to torsion tests.

The torsion method is inapplicable to determinations of the *shearing* strength of brittle materials, such as cast iron, since a brittle specimen fails in diagonal tension before the shearing strength is reached. However, the torsion test has been applied to cast iron [110, 111] and concrete [2] to determine other shearing properties or nominal overall torsional resistance.

It is worthy of note that the pieces subject to torque in service are usually machine parts and have to withstand impact loading and reversals of stress. In the interpretation of experimental investigations on such parts, these conditions should receive consideration.

Helical springs tested in axial compression or tension constitute one type of shear test, since the stresses developed are largely those of torsional and direct shear but principally the former. The modulus of rigidity of the material can be determined from the applied loads and the deflections, whereas the shearing stresses developed are a direct function of the loads.

The torsional rigidity of small rods and wire is sometimes determined from observations of the period of the torsional vibrations when a specimen is hung vertically with a known mass at the lower end. The modulus of rigidity may also be determined by applying a known torque to the lower end and measuring the angle of twist; the torque may simply be applied through a drum on the lower end of the rod by means of a system of strings, pulleys, and weights.

There are only a few standardized torsion tests. These include a stiffness test for plastics at various temperatures (ASTM D 1043) and a test of wire (ASTM E 558).

10.3 THE DIRECT SHEAR TEST

For the direct shear test of metals, a bar is usually sheared in some device that clamps a portion of the specimen while the remaining portion is subjected to load by means of suitable dies. In the Johnson type of shear tool, a bar of rectangular section, about 25 by 50 mm, or a cylindrical rod of about 25-mm diameter is used. As shown in Fig. 10.5a, the specimen A is clamped to a base C. Force applied to the loading tool E ruptures the specimen in single shear. If the specimen is extended to B and bridges the gap between the dies D, it is subjected to double shear. The dies and the loading tool

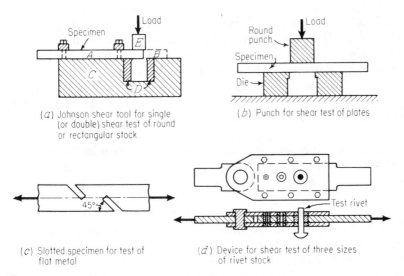

(*a*) Johnson shear tool for single (or double) shear test of round or rectangular stock

(*b*) Punch for shear test of plates

(*c*) Slotted specimen for test of flat metal

(*d*) Device for shear test of three sizes of rivet stock

Figure 10.5 Methods of testing metals in direct shear.

are made of tempered tool steel ground to an edge. For metal plates, a round punching device is sometimes used, as illustrated schematically in Fig. 10.5*b*. In some tests of steels, a slotted specimen is used, as illustrated in Fig. 10.5*c*. For tests of rivet stock, a device illustrated in Fig. 10.5*d* has been employed [101]. Direct shear tests are ordinarily made in compression- or tension-testing machines.

For direct shear tests of wood, a special tool and specimen developed by the Forest Products Laboratory are used as shown in Fig. 10.6 (ASTM D 143). Failure tends to occur along the shear plane marked. For shear tests of glued joints, a specimen somewhat similar to that depicted in Fig. 10.6 is glued along the dashed line and then loaded as indicated. Special tests are used for plywood (see Chap. 20).

In the direct shear test, the testing device should hold the specimen firmly and

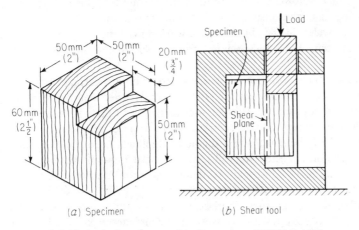

(*a*) Specimen

(*b*) Shear tool

Figure 10.6 Method of testing wood in direct shear (ASTM D 143).

preserve good alignment, and the load should be applied evenly at right angles to the axis of the piece as shown in Fig. 10.5a. In the single shear test, when equipment similar to that shown in Fig. 10.5a is used, the specimen must extend sufficiently beneath the loading tool E to avoid high bearing stresses. Likewise, in the double shear test, the specimen B must overlap the second die D sufficiently to avoid high bearing stresses. The speed of the crosshead for applying the load should not exceed 1.25 mm/ min for metals, stone, and concrete; for wood, the standard crosshead speed is 0.6 mm/min.

The only critical value that can be observed in the direct shear test is the maximum load P. If A is the area subjected to the force, then the average shearing strength is taken simply as P/A. The shape and texture of the fractured surface should be reported.

Several series of tests of rivets and riveted connections show that the unit single shear strength of steel rivets is usually *greater* than the unit double shear strength. The difference may be as much as 20 percent, depending on the material of the rivets and of the connected plates and on the arrangement of the rivets [101].

Care should be taken to distinguish between pure shear failures and failures that may occur as a result of bending stresses or of diagonal tensile stresses.

10.4 TORSION SPECIMENS

The principal criteria in the selection of the torsion-test specimen appear to be that (1) the specimen should be of such a size as to permit the desired strain measurements to be made with suitable accuracy, and (2) it should be of such proportions as to eliminate from that portion of the specimen on which measurements are made the effect of stresses due to gripping the ends. The ends should be such that they can be securely gripped without developing stresses sufficiently localized to cause failure in the grips. Ordinarily, the grips in the chucks of the machine are in the form of serrated blocks or cams, some types of which automatically tighten as the torque is applied. Care must be exercised in the gripping of the specimen so that bending is not introduced. Centering points are usually provided in the chucks of the torsion machine for insertion into small centering holes in each end of the specimen; thus the specimen can be accurately centered in the machine.

It is practically impossible to determine the proportional limit shearing strength of the extreme fibers of a solid torsion specimen. A thin tubular specimen is preferable for the determination of this property. Tubular specimens for ultimate shear-strength determinations should have short reduced sections with a ratio of length of reduced section to diameter (L/D) of about 0.5 and a diameter-thickness ratio (D/t) of about 10 or 12. For determinations of shearing yield strength and modulus of rigidity, a hollow specimen having a length of at least 10 diameters and a ratio of diameter to wall thickness of about 8 or 10, for its reduced section, is to be preferred [91, 64]. For larger ratios of diameter to thickness there is a tendency for failure to occur by buckling, owing to inclined compressive stresses; this would appreciably affect the value determined for the yield strength. The actual dimensions of the specimen used are commonly chosen to suit both the size and type of testing machine available, as well as the product to be tested.

In making a torsion test on tubing, the ends usually must be plugged so that pressure from the jaws of the machine will not collapse the tubing. These plugs should not be so long that they extend within the test section. Occasionally it becomes necessary to test tubing that will not fit into the chucks of the torsion machine. In such cases, steel plug adapters may be welded, riveted, or screwed to the ends of the tube and the projecting ends of the adapters reduced in diameter to fit the chucks of the machine.

For the torsion test of plastics (ASTM D 1043), specimens with square cross sections and featuring gripping holes near the ends are specified.

10.5 TORSION APPARATUS

Although some universal testing machines have torsional capability, the torsion test of metals is usually carried out in a special testing machine designed for the purpose. An illustration of one type of torsion testing machine is shown in Fig. 10.7. A suitable driving mechanism turns a chuck with hardened serrated jaws, and the applied torque is transmitted through the test specimen to a similar chuck at the weighing head, which actuates some type of torque indicator. By one method, a lever system actuates a beam over which a movable poise can be made to travel; the beam is graduated in torque units, say $N \cdot m/rad$. On some machines, the weighing system involves a pendu-

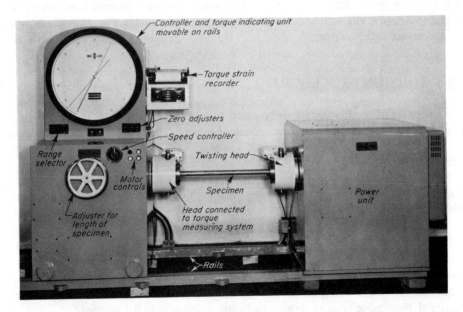

Figure 10.7 Torsion testing machine with electronic torque-bar indicator. (*Courtesy of Tinius Olsen Testing Machine Co.*)

lum connected to the pointer of a dial, which is graduated in N · m/rad; as torque is transmitted through the specimen, the pendulum rotates from its vertical equilibrium position until its static moment balances the applied torque. In other types of machines, a lever or arm attached to the chuck actuates (1) a hydraulic capsule connected to a load dial or (2) a series of levers connected to a small beam, the deflection of which is transmitted electronically to the load dial as in the machine shown in Fig. 10.7. To allow for longitudinal deformation of a specimen during a test, as well as to accommodate specimens of various lengths, the part of the machine that carries the load-indicating mechanism is adjustable in position, often being mounted on rollers. This arrangement avoids the superposition of stress due to any axial loading on the stress due to torsion. A capacity of 10 kN · m/rad[†] is probably satisfactory for most ordinary testing of bar stock, although machines having a capacity of over 200 kN · m/rad have been designed. Machines of low capacity, less than 1 kN · m/rad, are used for testing wire and plastic specimens.

Various devices are used to measure the strain or angular twist in a torsion specimen. These torsion indicators or *troptometers* ordinarily consist of two collars secured to the specimen a given distance or gage length apart, with some means of measuring the relative angular displacement of the collars. In one type, a vernier attached to one collar moves around a graduated circle attached to the other collar; in another type, mirrors are attached to the collars, and observations are made with telescopes and scales. A very simple torsion meter consists of long radial arms attached to the collars, the arms being arranged to move around graduated arcs as the specimen twists. The precision of the first and third types may be approximately ±0.0005 radian, but troptometers of the mirror type have been made to give a precision of ±0.000 05 radian (50 μrad). The simple mechanical troptometer shown in Fig. 10.8 is calibrated in increments of 0.2 degree.

10.6 TORSION PROCEDURE

The purposes of torsion tests usually parallel those for tension tests. The proportional limit and the ultimate strength in shear are of interest, and often the modulus of rigidity and stress-strain behavior in general. The specific procedures depend on the object of the test.

Before the test is started, the specimen is carefully measured. In the elastic range, the extreme fiber stress τ is related to the torque T by the torsion formula for circular shafts

$$\tau = \frac{Tr}{J}$$

where r is the outside radius and J is the polar moment of inertia of the cross section. Since the latter is a function of the fourth power of the radius or radii, the cross-sectional dimensions must be determined very accurately, to 0.1 percent.

[†]1 kN · m/rad = 8850 lbf-in.

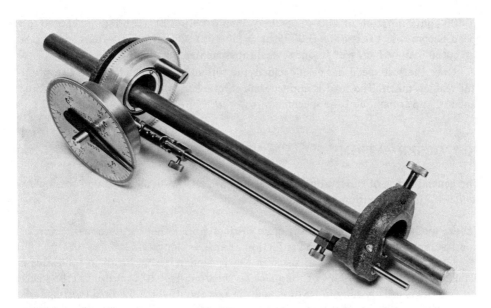

Figure 10.8 Troptometer. (*Courtesy of Tinius Olsen Testing Machine Co.*)

The gage length must also be noted and the troptometer carefully attached. In inserting the specimen into the machine, one must take care that it is properly centered in the heads and securely gripped. For the test speed, strain control normally applies, which for the elastic range is equal to the specimen's total twist at the proportional limit divided by the desired test duration to that point.

From Fig. 10.9a, which shows a twisted shaft, and from the definition of the modulus of rigidity $G = \tau/\phi$, the angle of twist can be expressed as

$$\theta = \frac{L}{r}\,\phi = \frac{L}{r}\,\frac{\tau}{G}$$

From the specimen L/r can be obtained, and G/τ, although unknown, can be estimated for the material. The duration of loading should suffice to take a reasonable number of readings at a comfortable pace. If, for example, a specimen has an L/r of 40 and a G/τ

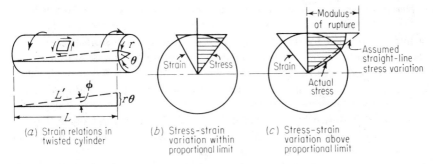

Figure 10.9 Strain and stress relationships in torsion.

of 400 and a duration of 400 s is desired, then $\theta = 40/400 = 0.1$ and the desired test speed becomes $\omega = 0.1/400 = 0.0025$ rad/s (about 0.024 rev/min). After yielding, the test speed may be increased to one several times greater.

Once the test speed has been decided, the torque is zeroed and an initial troptometer reading taken. The load is applied, and torque and twist readings are taken simultaneously up to failure of the specimen.

10.7 OBSERVATIONS

The general types of observations and records of tests in direct shear and torsion are similar to those of tension and compression tests.

Failure mode The shear fracture is quite distinct from either the tension or compression fracture: there is no localized reduction of area or elongation.

. For materials that break in shear in the torsion test, the break in solid rods is plane and normal to the axis of the piece, as shown in Fig. 10.10a. For ductile steels, the fracture is usually silky in texture, and the axis about which the final twisting took place may usually be observed. Since the surfaces of the break may not be quite smooth, the outer portions in moving past each other act like cams, pushing the piece apart in the direction of its length. The center portions not yet broken by shear are possibly broken in tension by this cam action. In brasses, bronzes, and wrought iron, the material often breaks up into fibers, like a rope, before rupture.

The rupture of a material for which the tensile strength is less than the shearing strength takes place by separation in tension along a helicoidal surface, as shown in Fig. 10.10b. This type of break occurs when cast iron or plain concrete is subjected to torsion. The outline of the fracture makes a complete revolution of the bar, the ends of the helix being joined by an approximate straight line. The helicoidal type of fracture may be easily obtained by twisting a piece of chalk in torsion with the fingers.

Thin-walled tubular specimens of ductile material having a reduced section of length greater than the diameter fail by buckling, which is illustrated in Fig. 10.10c, but those having a short reduced section fail in torsion on a right section, as shown in Fig. 10.10d.

Modulus of rupture In a bar subjected to torsional loading above the proportional limit, if a straight-line variation of *strain* is assumed, the actual *stress* variation is something like that shown by the solid line in Fig. 10.9c. It is customary, however, for comparison purposes with similar materials, to compute a nominal extreme fiber stress at rupture by means of the torsion formula ($\tau = Tr/J$), which gives what is called the *modulus of rupture in torsion*. If proportionality between stress and strain were maintained up to the rupture point, the nominal straight-line stress distribution would be something like that shown by the dashed line in Fig. 10.9c. It can be seen that the modulus of rupture or maximum nominal stress is larger than the true maximum stress. For tubes, of course, the two are more nearly equal.

Upton shows a construction by means of which a corrected value of shearing

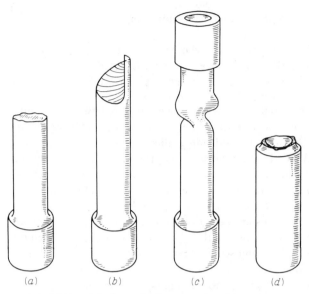

(a) (b) (c) (d)

(a) Solid bar of ductile material. Fracture on plane right section.
(b) Solid bar of brittle material. Helicoidal fracture.
(c) Tubular specimen of ductile material. Failure by buckling.
(d) Tubular specimen of ductile material; short reduced section.
    Failure on plane right section.

Figure 10.10 Types of failure in torsion. (*Report of ASTM Committee E-1, Ref. 76.*)

stress may be determined from a diagram of nominal stress vs. strain [97, 104]. His reasoning indicates that the ratio of the stress at rupture to the calculated modulus of rupture must lie between $\frac{3}{4}$ and 1 and approaches $\frac{3}{4}$ as ductility increases. This value of $\frac{3}{4}$ is roughly corroborated by direct shear tests on steels.

For materials that fail by tension under torsional loading, the torsion modulus of rupture roughly approximates the breaking strength in tension, although it is always somewhat higher than the true tensile strength, since above the proportional limit the torsion formula gives stresses that are too high. For cast iron, the torsion modulus of rupture as determined from solid rods appears to be about 1.2 times the tensile strength [97].

Ductility Ductility in a torsion test is determined by comparing the final fiber length L' (Fig. 10.9a) at rupture with the original fiber length or gage length L. The value of L' is computed knowing L and $r\theta$. The ductility is expressed as a percentage of elongation of the outer fiber and is equal to $[(L' - L)/L] \times 100$.

10.8 EFFECTS OF VARIABLES

In shear tests, the results obtained depend to no small measure on the type of test chosen. This is largely explained by the unavoidable fact that, in contrast to axial force tests, the stresses are not distributed uniformly over the cross section. Thus it is best to choose a test that fairly represents service conditions, such as torsion for a shaft or direct shear for a pin.

Direct shear test results further depend on whether the specimen is loaded in single or double shear. No doubt these differences can be attributed to the bending stresses inherently present. Yielding and failure of torsion specimens depend on the ductility of the material as well as on cross-sectional dimensions.

If the tensile strength of a material is less than its shearing strength, then failure under shear loading occurs by (tensile) separation along a plane at an angle of 45° to the plane of maximum shear. Under torsional loading, this results in a fracture with a helicoidal surface (see Fig. 10.10b). The ratio of shear strength to tensile strength appears to vary from perhaps 0.8 for ductile metals to values of about 1.1 or 1.3 for brittle ones such as cast iron. The shearing *elastic* strength of ductile and semiductile steels appears to be close to 0.6 of the tensile elastic strength [79].

Insofar as practical testing problems are concerned, the shearing stress-strain relations are of interest principally in connection with torsional loading. In the common theory of torsion, it is assumed that plane sections remain plane after twisting. The circular section is the only one that conforms to this condition, and hence the simple theory of torsion does not apply satisfactorily to sections other than those of circular form. However, in practical calculations for noncircular sections, the results of the simple torsion theory are often used in conjunction with suitable correction factors.

In a solid cylindrical bar in torsion, the interior fibers are less highly stressed than the surface fibers. Consequently, when the surface fibers reach the proportional limit or yield point, they are in a sense supported by the interior fibers. Thus the effect of yielding of the surface fibers during their early stage of plastic action is masked by the resistance of the remainder of the section. It is not until considerable yielding has taken place that any noticeable effect is apparent with instruments ordinarily used for measuring angle of twist. This difficulty is overcome by the use of properly designed tubular test specimens, which can be made to give more sensitive measures of the shearing elastic strength since all the fibers are at about the same stress. However, if a thin sheet is subjected to shear or a thin tube to torsion, before the shear strength of the material is reached failure may occur by buckling due to the *compressive* stresses that act at 45° to the planes of maximum shear (see Fig. 10.10c). Thus in tubular specimens for torsion tests the relative thickness of the wall must be greater than some critical value if shear failure is to be ensured.

PROBLEM 10 Torsion Test of Steel

Object: To determine the behavior of a ductile steel when subjected to torsion, and to obtain the following torsional properties:
1. Shearing proportional limit using the torsion formula
2. Probable true shearing proportional limit
3. Yield strength at an offset of 0.04 rad per meter of gage length (using the torsion formula), or as assigned
4. Shearing modulus of rupture
5. Probable true shearing strength
6. Modulus of rigidity

7. Average energy absorbed per unit volume at true proportional limit
8. Approximate percentage elongation in outer fiber at failure
9. Probable tensile proportional limit
10. Probable tensile strength

Specimen: Steel rod.

Special apparatus: Torsion testing machine and troptometer.

Procedure:

1. With a micrometer caliper determine the mean diameter of the specimen near its midlength. Assuming the shearing proportional limit to be 0.6 the tensile proportional limit and the shearing modulus of rupture to be equal to the tensile strength, compute loading increments that will give at least 10 observations below the proportional limit, several close together near the proportional limit, and at least 10 beyond the proportional limit. See Table 17.3 for property values.

2. Note the gage length and least reading of the troptometer. Securely clamp the instrument to the specimen, making certain that the axes of the troptometer and the test piece coincide and that the troptometer is in proper position for ease of reading.

3. Adjust the torsion machine to read zero and then insert the specimen into the two heads. See that each end is centered inside each head. Gradually bring the grips in the heads to a firm equal bearing, taking care not to displace the specimen. If in tightening the grips they produce some torque, operate the machine in forward or reverse so that it will be reduced to zero.

4. Remove the troptometer spacer bar and set the instrument to read zero.

5. Apply load at a slow speed. Take readings of torque and twist simultaneously without stopping the machine. After the specimen shows definite signs of yielding, apply the load at higher speed until failure occurs. Note the character of the fracture.

Report:

1. Plot two diagrams from the same origin, showing the relation between torque in kN · m/rad as ordinates against unit twist in rad/m as abscissas. One diagram will extend to the yield strength with a slope of not more than 60° with the strain axis. The second diagram will show the curve for the entire test.

2. Determine the torques at (a) the proportional limit and (b) the yield strength, and mark them on the first diagram.

3. Compute the quantities required. Tabulate them and compare with such values as are given in Chap. 17.

DISCUSSION

1. Discuss the feasibility of determining the shearing strength of brittle materials by the torsional method.

2. Are torsion specimens subjected to other than shearing stresses during the test? If so, what are these stresses, what causes them, and what is their probable effect on the results?

3. List the relative advantages and disadvantages of tubular and solid cylindrical torsion specimens for determinations of shearing strength.

4. Which of the properties determined in a torsion test is of most significance in the selection of steel for coil springs? Why?

5. Explain why ductile materials under torsional stress shear on a right section, whereas brittle materials fracture on a helicoidal section.

6. Why is the torsion formula not applicable to noncircular cross sections?

7. If a given round steel bar having proportional limits in shear and tension of 250 and 415 MPa respectively is to be used to absorb energy without undergoing plastic deformation, would the bar absorb more energy if it were used as a torsion member or as an axially loaded tension member? Explain.

8. Devise a method to test the shear strength of the glue in plywood (more about this in Chap. 20).

9. If a helical spring is loaded in compression, where and in what direction do the maximum shear stresses occur?

10. Find a small tube, such as a large drinking straw, and test it in torsion with your hands. Slice a longitudinal seam into it, and test it again. What differences do you notice? Can you explain them?

READING

Kollbrunner, C. F., and K. Basler: *Torsion*, Springer-Verlag, Berlin, 1966.
Shear and Torsion Testing, ASTM, Philadelphia, 1960.
Shear in Reinforced Concrete, American Concrete Institute, Detroit, 1974.

ELEVEN

STATIC BENDING

11.1 BENDING

If forces act on a piece of material in such a way that they tend to induce compressive stresses over one part of a cross section of the piece and tensile stresses over the remaining part, the piece is said to be in bending. The common illustration of bending action is a beam acted on by transverse loads; bending can also be caused by moments or couples such as may result, for example, from eccentric loads parallel to the longitudinal axis of a piece.

In structures and machines in service, bending may be accompanied by direct stress, transverse shear, or torsional shear. For convenience, however, bending stresses may be considered separately, and in tests to determine the behavior of materials in bending, attention is usually confined to beams. In the following discussion, it is assumed that the loads are applied so that they act in a plane of symmetry, so that no twisting occurs, and so that deflections are parallel to the plane of the loads. It is also assumed that no longitudinal forces are induced by the loads or by the supports. For more complicated cases of bending, texts on mechanics of materials should be consulted.

Figure 11.1 illustrates a beam subjected to transverse loading. The bending effect at any section is expressed as the bending moment M, which is the sum of the moments of all forces acting to the left (or to the right) of the section. The stresses induced by a bending moment may be termed *bending stresses*. For equilibrium to be achieved, the resultant of the tensile forces T must always equal the resultant of

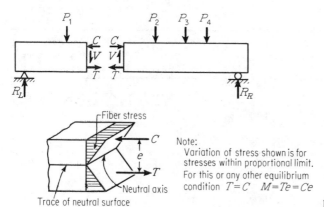

Figure 11.1 Bending of a beam.

the compressive forces C. The resultants of the bending stresses at any section form a couple that is equal in magnitude to the bending moment. When no stresses act other than the bending stresses, a condition of pure bending is said to exist. Pure bending is developed only under certain loading conditions; in the usual case, bending is accompanied by transverse shear. The resultant of the shearing stresses across a transverse section equals the total transverse shear V, which is computed as the algebraic sum of all transverse forces to the left (or to the right) of a section. Bending action in beams is often referred to as *flexure*.

The variations in total transverse shear and in bending moment along a beam are commonly represented by shear and moment diagrams, which are illustrated for several cases of concentrated loading in Fig. 11.2. It should be noted that symmetrical two-point loading given a condition of pure bending (constant moment) over the central portion of the span (see Fig. 11.2b).

In a cross section of a beam, the line along which the bending stresses are zero is called the *neutral axis*. The surface containing the neutral axes of consecutive sections

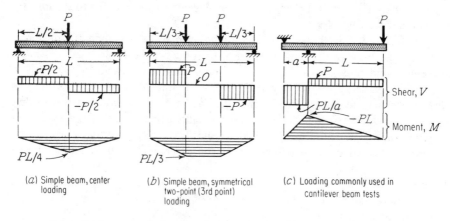

Note: Bending moment diagrams drawn on tension side.

Figure 11.2 Shear and moment diagrams.

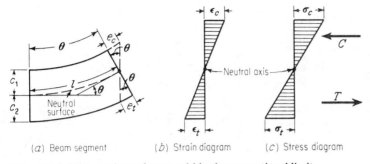

(a) Beam segment (b) Strain diagram (c) Stress diagram

Figure 11.3 Fiber strains and stress within the proportional limit.

is the *neutral surface*. On the compressive side of the beam the "fibers" of the beam shorten, and on the tensile side they stretch; thus the beam bends or deflects in a direction normal to the neutral surface, becoming concave on the compressive side (see Fig. 11.3).

It has been well established by theory and testing that in pure bending the *strains* are proportional to the distance from the neutral axis, and that appears to hold, at least to a good approximation, within the range of inelastic action as well as within the range of elastic action. This is referred to as a condition of "plane bending"; i.e., plane sections before bending remain plane sections after bending. The relative rotation of one cross section of an initially straight beam with respect to a reference cross section is illustrated in Fig. 11.3a. The elongation or shortening of the fibers in any given length of beam over which the moment is constant divided by that length gives unit fiber strain, as illustrated in Fig. 11.3b. If the stresses are proportional to the strains (say, within the proportional limit), the stress variation across a section is linear, as shown in Fig. 11.3a. The extreme fiber stresses σ then are

$$\sigma = \frac{Mc}{I}$$

where M is the bending moment, c is the distance from the neutral axis to an extreme fiber, and I is the centroidal moment of inertia.

Above the proportional limit, bending stresses do not vary linearly across a section because stress is not proportional to strain. Illustrations of common cases of this are shown in Fig. 11.4. In Fig. 11.4a, the solid line shows the stress variation in a homogeneous beam of symmetrical section for a material that has the same stress-strain characteristics in both tension and compression. The equivalent linear stress distribution that would yield the same moment is shown by the dotted line. The maximum value of the fictitious straight-line fiber stresses, the "modulus of rupture," is seen to be greater than the true maximum stresses.

If the material does not have the same stress-strain characteristics in tension as in compression, the neutral axis must shift toward the stiffer side of the beam in order to maintain equality of the resultants of the tensile and compressive forces, as shown in Fig. 11.4b. In this case the linear extreme fiber stress, computed with the flexure

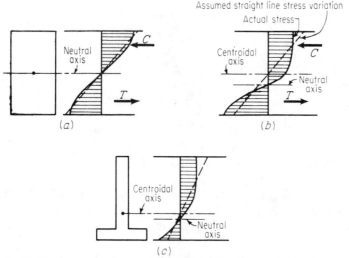

Figure 11.4 Fiber stress above the proportional limit.

formula, is less than the true fiber stress on the stiffer side of the beam and greater than the true fiber stress on the less stiff side.

If the beam has a cross section such as that shown in Fig. 11.4c, the stresses are lower on the side nearer the centroidal axis. This side of the beam is in effect stronger because of the concentration of material to resist stress. Yielding of the more highly stressed fibers on the opposite side then causes a shift of the neutral axis toward the stronger side, giving a stress distribution similar to that shown.

Increasingly large moments cause increasingly large strains until the beam either breaks (if the material is brittle) or forms a plastic hinge (if the material is ductile). The slopes and deflections can be obtained by considering Fig. 11.3a. The curvature is seen to equal e_t/c_2 and in turn is approximately equal to the first derivative of the slope and the second derivative of the deflection, with respect to the longitudinal axis of the beam. Solution of the differential equations yields the desired results, some of which are given in App. B. In the special case of uniform bending moment, the stresses, and therefore the strains and the curvature, are constant: the beam forms a circular arc.

In cases of nonuniform bending moment, shear exists. The shear stresses are generally highest near the neutral axis and zero at the extreme fibers. Only in very short, thick beams do shear stresses, shear strains, and shear deformations play an important role.

11.2 SCOPE AND APPLICABILITY

Most structures and machines have members whose primary function is to resist loads that cause bending. Examples are beams, hooks, plates, slabs, and columns under eccentric loads. The design of such structural members may be based on tensile, compressive, and shearing properties appropriately accounted for in various bending formulas. In many instances, however, bending formulas give results that only approximate the real conditions. While special analyses can often be made of stresses arising from unusual loading conditions and from local distortions and discontinuities, it is not always feasible to make such analyses, which may be very complicated. The bending test may serve then as a direct means of evaluating behavior under bending loads, particularly for determining the limits of structural stability of beams of various shapes and sizes.

Flexural tests on beams are usually made to determine strength and stiffness in bending; occasionally they are made to obtain a fairly complete picture of stress distribution in a flexural member. Beam tests also offer a means of determining the resilience and toughness of materials in bending.

Under the general designation of strength is included the proportional limit, yield strength, and modulus of rupture. These properties may be determined with a view to establishing appropriate load factors and allowable bending stresses for use in design. The modulus of rupture also may be used simply as a criterion of quality in control tests.

The stiffness of a material can be determined from a bending test in which the load and deflection are observed. The modulus of elasticity for the material in flexure is computed by using an appropriate deflection formula. The value of the modulus of elasticity may then be used to compute the elastic deflection of beams of the same material but under different loading conditions, although some error may be involved owing to (1) ignoring shearing deflections, which are of importance in short, deep beams; (2) deviations from the straight-line relation of stress and strain as expressed by Hooke's law; and (3) lack of uniformity of the material.

Because the loads required to cause failure may be relatively small and easily applied, bending tests can often be made with simple and inexpensive apparatus. Because the deflections in bending tests are many times the elongations in tension tests, a reasonable determination of stiffness or resilience can be made with less sensitive and less expensive instruments than are required in a tension test. Thus the bending test is often used as a control test for brittle materials, notably cast iron and concrete. It is less suited for determining the ultimate strength of ductile materials.

For wire and sheet metals, a simple bend test is sometimes used as an arbitrary measure of relative flexibility. For ductile materials in the form of rods, such as reinforcement bars for concrete, a cold-bend test is used to determine whether or not the rod can be bent sharply without cracking and serves as an acceptance test for this form of ductility.

11.3 SPECIMENS

If a beam specimen is to be tested in flexural failure, as is the case when the modulus of rupture of a material is to be determined, it must be so proportioned that it does not fail by lateral buckling or in shear before the ultimate flexural strength is reached. In order to avoid shear failure, the span must not be too short with respect to the depth. Values of $L = 6d$ to $L = 12d$ (the actual value depending on the material, the shape of beam, and the type of loading), in which L = length and d = depth, serve as an approximate dividing line between the short, deep beams that fail in shear and the long, shallow beams that fail in the outer fibers. A value of $L < 15b$, where b = width, usually safeguards against lateral buckling.

Although many forms of beams are used for special and research testing work, standardized specimens are used for routine and control testing of a number of common materials, such as cast iron, concrete, brick, and wood.

Test specimens of cast iron are cylindrical bars, cast separately but under the same sand-mold conditions and from the same ladle as the castings they represent (ASTM A 438). Standard cast iron specimens range in diameter from about 20 to 50 mm[†] and in length from 375 to 675 mm. They are tested as simple beams under center loading.

Concrete specimens are square or rectangular in cross section, a typical size being 150 by 150 mm. They may be tested under third-point loading (ASTM C 78) or under midspan loading (ASTM C 293). Building bricks are tested flatwise under center loading (ASTM C 67).

Small clear pieces of wood, 50 by 50 by 750 mm in size, are tested under center loading (ASTM D 143), but large timber beams having a length of about 5 m are often tested under third-point loading (ASTM D 198).

There are also standard specimens for various building materials such as gypsum board, building stone, slate, and the like.

11.4 APPARATUS

The principal requirements of the supporting and loading blocks for beam tests are as follows:

1. They should be of such shape that they permit the use of a definite and known length of span.
2. The areas of contact with the material under test should be such that unduly high stress concentrations (which may cause localized crushing around the bearing areas) do not occur.
3. There should be provision for longitudinal adjustment of the position of the supports so that longitudinal restraint will not be developed as loading rogresses.
4. There should be provision for some lateral rotational adjustment to accommodate beams having a slight twist from end to end so that torsional stresses will not be induced.
5. The arrangement of parts should be stable under load.

A number of specifications describe in detail the type of support to be used with particular materials. The principal features of representative supporting arrangements are shown in Fig. 11.5.

Many flexure tests are conducted in universal testing machines, with the supports

[†]1 mm = 0.039 in.

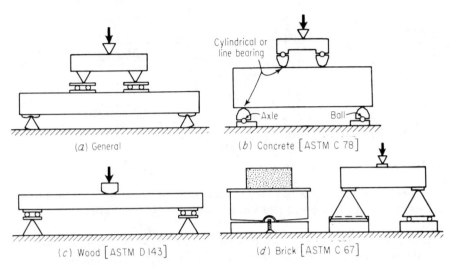

Figure 11.5 Supporting and loading devices for beam tests.

placed on the platen or on an extension of it and with the loading block fastened to or placed under the movable head. However, for control tests of some materials (e.g., foundry tests of cast iron and field tests of concrete) specially designed hand-operated machines are often employed.

Apparatus for measuring deflection should be so designed that crushing at the supports, settlement of the supports, and deformation of the supporting and loading blocks or of parts of the machine do not introduce serious errors into the results. One method of avoiding these sources of error is to measure deflections with reference to points on the neutral axis above the supports. Typical arrangements are shown in Figs. 11.6a and 11.6b. In general, deflections within the proportional limit should be read to at least 0.01 of the deflection at the proportional limit; for greater deflections they should be read to at least 0.01 of the deflection at rupture.

For determining fiber strains, a surface strainometer, resistance gages, or a portable strain gage may be used to measure strains along selected gage lines, or a special device may be used to indicate the relative rotation of plane cross sections at some given distance apart, as shown in Fig. 11.7. The dial indicators of this device give the movement of the two collars with respect to each other to the nearest 2.5 μm. By drawing to scale a simple sketch of the strainometer, including the location of the

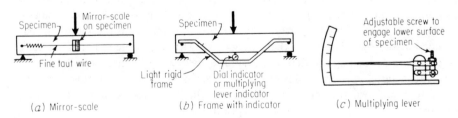

Figure 11.6 Deflection measuring devices (schematic).

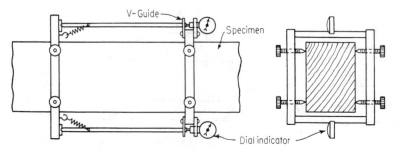

Figure 11.7 Strainometer for beam tests.

extreme fibers of the beam and the change in dial readings, the position of the neutral axis and the strains in the extreme fibers of the beam can be read from the drawing or computed.

Deflectometers should be designed in such a way that they will not be injured by failure of the test piece, or the precaution should be taken of removing them well in advance of final rupture.

11.5 PROCEDURE

Conducting routine flexure tests of the type used in quality-control work is quite simple. Often only the modulus of rupture is required, which can be obtained from the load at rupture, the location of reactions and loads, and the dimensions of the critical cross section. Even in this simple case, however, measurements must be made very precisely. This pertains particularly to the vertical dimensions of the beam, since the stresses, including the modulus of rupture, depend on the section modulus, I/c, which generally varies with the third power of the beam depth.

When deflections or strains are to be measured, as is the case when the modulus of elasticity is to be determined, advantageous locations for the instruments must be decided on. Suitable arrangements will depend on the purpose of the investigation, the instrumentation available, and the physical configuration of the specimen and apparatus. In general, locations along the centerline of the specimen and at its midspan are best, so as to reduce the effect of minor twisting and to maximize the precision of the results by measuring relatively large deflections.

For tests requiring fiber-strain measurements, the symmetrical two-point loading (Fig. 11.2b) is desirable in that the portion of the beam length between the loads is subject to constant moment and so to constant stress along any fiber. A reasonable gage length can thus be used for strain measurements. Further, in the section of constant moment the shear is zero, so that the observed strains are due to bending stresses only. This type of loading also has an advantage over center loading in tests of brittle materials to determine the modulus of rupture, because with center loading the failure is forced to occur near the center, and defects in the material would probably remain undiscovered unless they happen to occur close to the midspan.

In any case the assembly of specimen and supports must be properly aligned and centered in the machine. Deflectometers and strain gages, like the other components, should be checked for proper operation over the required range and must be located carefully. Dimensional sketches of the arrangement are essential.

In planning the test, some calculations are necessary in order to estimate the loads and displacements to be expected, as well as the rates of these quantities. Some useful formulas are included in App. B, but a review of the theory of beams in a text on mechanics of materials may be helpful.

When the cross section of a flexure specimen is measured in millimeters, computation of the moment of inertia usually leads to very large numbers, in units of mm^4. Rather than resorting to the use of cm^4, which would likely lead to computational difficulties later, it is best to express this quantity in units of 10^6 mm^4, which equals 10^{-6} m^4. Other cross-sectional properties may be treated in analogous ways.

Test speeds in general should be planned so that the readings can be taken accurately and precisely. Even in simple rupture tests, when no intermediate readings are necessary, the load rate should be slow enough so that dynamic effects are avoided; a minimum test duration under 30 s is seldom justified. Standard test specifications may be stated in terms of load rate, of crosshead speed, of extreme fiber stress or strain, or of total duration to yielding or fracture. Depending on the test conditions, especially the instruments available, these parameters may be monitored directly or indirectly. Normally the specified speeds are maximum values, but occasionally a minimum may also be stated in order to avoid creep effects. The actual test speeds should always be noted and recorded.

11.6 OBSERVATIONS

The general types of observation and record of tests in flexure are similar to those of tension, compression, and shear tests.

The conditions under which the modulus of rupture is determined (type of specimen, span length, type and rate of loading, and so on) should always be recorded, since these markedly affect the results.

In computing the modulus of elasticity from load-deflection data, the simplest procedure is to plot a load-deflection diagram and from the slope of the diagram compute the load-deflection ratio for substitution in the pertinent deflection formula.

Values of the proportional limit determined from beam tests are generally higher than values obtained from tension or compression tests because yielding of the extreme fibers is masked by the supporting effect of the less highly stressed fibers nearer the neutral axis.

In addition to measuring the primary test parameters, it is important to observe the type of failure. To this end it is helpful to know the behavior that can be expected.

Rupture The failure of beams of brittle material such as cast iron and plain concrete always occurs by sudden rupture. Although as failure is approached the neutral axis shifts toward the compression face and thus tends to strengthen the beam, failure finally occurs in the tensile fibers because the tensile strength of these materials is only a fraction of the compressive strength. The ratio of tensile to compressive strength is about 25 percent for cast iron and about 10 percent for concrete.

The failure of reinforced-concrete beams may be caused by (1) excessive elongation of the steel due to stresses above the yield point, resulting in vertical cracks on the tensile side of the beam; (2) failure of the concrete in compression at the outermost compressive fibers; and (3) failure of the concrete in diagonal tension, primarily due to excessive shearing stresses, resulting in the formation of cracks that slope downward toward the reactions, often becoming horizontal just above the main steel in simple span beams.

A wooden beam also may fail in a number of ways. (1) It may fail in direct compression at the concave compression surface. (2) It may break in tension on the convex tension surface. Since the tensile strength of wood parallel to the grain is usually greater than its compressive strength, the neutral axis shifts toward the tensile face so as to maintain equality of the tensile and compressive forces, as shown in Fig. 11.4b. Therefore, the first visible signs of failure may be in the tensile face even though the wood is stronger in tension than in compression. This type of failure occurs only in well-seasoned timbers, since green test pieces usually fail in compression before the tension fibers are ruptured. (3) It may fail by lateral deflection of the compression fibers acting as a column. (4) It may fail in horizontal shear along the grain near the neutral axis. This type of failure is sudden and more common in well-seasoned timbers of structural sizes than in green timbers or in small beams. (5) It may fail in compression perpendicular to the grain at points of concentrated load.

The fracture of materials such as cast iron and concrete is definite, usually occurring on an approximately plane surface at a section of maximum moment. The texture of the fracture may be of significance and should be noted. The designations of various modes of failure of wood are indicated in Fig. 11.8. In tests of beams that do not fail by sudden rupture, indications of impending failure, such as cracks, localized yielding, buckling, and so on, should be observed carefully. The causes of primary and secondary failure of structural beams should be noted.

For beams of brittle material, the nominal fiber stress at rupture as computed by the flexure formula (the modulus of rupture in bending) is usually appreciably greater than the true tensile strength of the material. The ratio of the modulus of rupture to the true tensile strength is about 1.8 for cast iron and about 1.5 to 2 for concrete. The ratio of the modulus of rupture to the compressive strength is about 0.5 for cast iron, about 0.15 to 0.20 for concrete, and about 2 for wood (considering the compressive strength parallel to the grain).

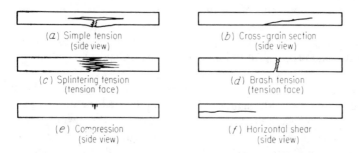

Figure 11.8 Various modes of failure of wood beams (ASTM D 143).

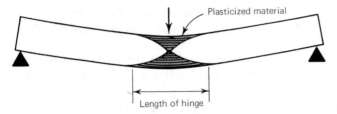

Figure 11.9 Plastic hinge in steel beam.

Yielding Beams of ductile steel do not rupture at all. When bending moment has reached a sufficient magnitude, called the "plastic moment," nearly the entire section is strained into the plastic range and hence stressed to the yield stress (in tension on one side of the neutral axis and in compression on the other). Under this constant moment the beam develops a "plastic hinge," as illustrated by Fig. 11.9, that is characterized by very large curvature. The length of the hinge and its other characteristics should be noted. If the yield stress is assumed to have equal magnitude in tension and compression, it may be computed by dividing the plastic moment by the plastic section modulus, a quantity equal to the sum of the moments of the areas above and below the neutral axis about the neutral axis, which bisects the cross section.

11.7 EFFECTS OF VARIABLES

In flexure tests of brittle materials, some of the more important factors that affect the results are the type and rate of loading, the length of span, and the cross-sectional dimensions of the beam.

Loading type The effect of the type of loading is illustrated by the results of numerous tests of concrete. These tests have indicated the relative magnitudes of the modulus of rupture for three common types of loading to be as follows [29, 44, 106] :

1. In a simple span, the largest value of the modulus of rupture is obtained from center loading. Values computed on the basis of the moment at the center of the span tend to be somewhat greater (about 7 percent) than values computed on the basis of the moment at the section of break.
2. Cantilever loading tends to give slightly higher results than center loading on a simple span, although on the average the difference is not great.
3. Third-point loading on a simple span invariably gives results somewhat less than center loading (roughly 10 to 25 percent). It seems reasonable to suppose that since the strength of the material varies somewhat throughout the length of the beam, in the third-point loading, the weakest section (of those subjected to constant moment) is sought out.

These relations probably hold, at least in principle, for other brittle materials. In general, the third-point-loading method appears to give the most concordant results.

Specimen dimensions Tests of both cast iron and concrete have shown that for beams of the same cross section, the shorter the span length, the greater the modulus of rupture [44, 57].

The shorter the span, the less is the computed value of the modulus of elasticity of cast iron, although the difference is not over about 10 percent for length-diameter ratios ranging from 10 to 30.

The shape of the cross section of a beam may appreciably affect the resistance of the beam. Tests of cast iron beams having a variety of shapes but of about the same cross-sectional area show that in general the modulus of rupture and the modulus of elasticity are lower for beams having a relatively large proportion of the cross-sectional area situated near the extreme fibers. This is the case with an I section, although the breaking *loads* are considerably greater for such sections [104].

Tests of both cast iron and concrete indicate lower strengths for beams of larger cross-sectional dimensions [44, 56, 106]. These results are in line with the results of tension and compression tests on cast iron and concrete. Test results for wood show that both the strength and the modulus of elasticity of large timbers are less than for small clear specimens on short spans; this is attributed to the presence of defects in members of structural size.

Stability The capacity of shaped specimens, such as I beams and channels usually made of steel, may be affected by instability of their elements. The following failure modes may result:

1. The beam may fail by yielding of the extreme fibers; if the beam twists or buckles, failure takes place before a plastic hinge forms.
2. The compression flange may act somewhat like a long column, and failure may occur by elastic instability. This lateral buckling may be a primary cause of beam failure, in which case the computed fiber stress, in general, does not reach the yield-point strength of the material before buckling occurs. Buckling often limits the strength of narrow, deep beams, especially I beams or channel-section beams with tension and compression flanges connected by a thin web. Whether it is a primary cause of failure or a final manner of failure, sidewise buckling results in a clearly marked and generally sudden failure. Hence it would appear desirable to maintain a length-width ratio of something less than 15 in tests of beams to determine properties of a material.
3. Failure in thin-webbed members, such as an I beam, may occur because of excessive shearing stresses in the web or by buckling of the web under the diagonal compressive stresses that always accompany shearing stress. If the shearing stress in the web reaches a value as great as the yield-point strength of the material in shear, beam failure may be expected, and the manner of failure will probably be by some secondary buckling or twisting action. The inclined compressive stress always accompanying shear may reach so high a value that the buckling of the web of the beam is a primary cause of failure. Danger of web failure as a primary cause of beam failure exists, in general, only for short beams with thin webs.
4. In the parts of beams adjacent to bearing blocks that transmit concentrated loads or reactions to beams, high compressive stresses may cause local failure of the web.

Test speed Speed of testing has the same general effect in the flexure test as in the tension and compression tests; that is, the greater the speed, the higher the indicated strength. Tests on timber beams in which the time from zero load to failure varied from 0.5 s to 5 h indicate that it requires a tenfold increase in rate of loading to produce a 10 percent increase in bending strength. Timber beams under continuous loading for years will fail under loads one-half to three-quarters as great as required in the usual static bending test, which ruptures the test specimen in a few minutes [105]. For concrete beams, increasing the rate of stressing from 140 to 7860 kPa/min resulted in an increase of about 15 percent in the modulus of rupture [106].

11.8 "BEND" TESTS

The "bend" tests (of which the "cold-bend" test is the most common) offer a simple, somewhat crude, but often satisfactory means of obtaining an index of ductility. Essentially the test consists in sharply bending a bar through a large angle and noting whether or not cracking occurs on the outer surface of the bent piece. Often the angle of bend at which cracking starts is determined. The severity of the test is generally varied by using different sizes of pins about which the bend is made.

Bend tests are sometimes made to check the ductility for particular types of service or to detect loss of ductility under certain types of treatment. Thus, cold-bend tests, which are made by bending a metal at ordinary temperatures, may serve to detect too high a carbon or phosphorus content or improper rolling conditions in steel. Cold-bend tests are required in the specifications for some steels, particularly those in the form of rod and plate, such as bars for concrete reinforcement (ASTM A 615M). In various steel specifications, for example, ASTM A 36 and A 285, bend tests are made optional by reference to ASTM A 6 or A 20. Figure 11.10a shows a diagram of a suitable device. In a guided weld test the specimen is placed on a jig similar in concept to that in Fig. 11.10b and bent into a U shape (ASTM E 190).

The specified angle of bend and size of pin around which the piece is to be bent without cracking depends on the grade of metal and the type of service for which it is to be used. In the case of concrete-reinforcement bars, which must be bent cold on the

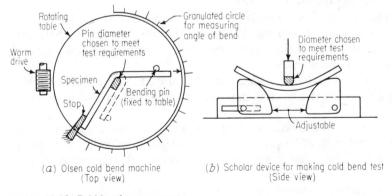

(a) Olsen cold bend machine (Top view) (b) Scholar device for making cold bend test (Side view)

Figure 11.10 Cold-bend test apparatus.

Table 11.1 Bend-test requirements for concrete-reinforcement bars[†]

Bar designation no.	Pin diameter for bend tests d = nominal diameter of specimen[‡]	
	Grade 300	Grade 400
10, 15	$4d$	$4d$
20	$5d$	$5d$
25	$5d$	$6d$
30, 35	$5d$	$8d$

[†]Based on ASTM A 615M.
[‡]Test bends 180°.

job, the requirements are shown in Table 11.1. The requirement for structural steel is that the specimen be bent through 180° in each case. It is required that a specimen of rivet rod stock be bent flat on itself.

A "hot-bend" test is sometimes made, for example, on wrought iron by heating it to a welding temperature (about 1000°C) and bending the heated piece on an anvil; the test serves to detect too high a sulfur content. A "quench-bend" test is occasionally used in connection with rivet steels for boilers and is made by heating, quenching, and then bending; the test in this case is used to detect too high a carbon content.

A "nick-bend" test is made when it is desired to make a rapid examination of a metal for coarse crystalline structure or for internal defects. Sometimes the specimen is nicked with a cold chisel, clamped in a vise, and bent with a hammer. In more carefully made tests, the nick or groove may be made by a hacksaw or in a milling machine, and after a slight bend is started with a hammer, it is completed by axial loading in a testing machine.

Similar to the nick-bend tests are tests in which a hole is made in the specimen by punching or drilling. The effect of such operations on ductility is then qualitatively determined by bending the metal at the restricted section.

Fiber-strain measurements are sometimes made in connection with cold-bend tests. What is to be the convex surface of the bend specimen is marked by scribe marks at intervals over a distance of perhaps 100 mm. The elongation of the outer fiber is then determined through the use of a flexible tape.

From a consideration of the bending action it may be seen that the elongation of the outer fiber varies directly with the thickness of the specimen and inversely with the radius of curvature. It is for this reason that metals of various thicknesses are bent around pins of different diameters.

In bend tests of some materials, such as wire (ASTM F 113) and plastics (ASTM D 747), ASTM specifies that the bending moment as well as the angle of bend be observed.

Figure 11.11 shows schematically an arrangement for obtaining simultaneous values of moment and angle of bend of the specimen. During the test the vise holding the specimen is rotated by a motor and the bending moment is measured by the pendulum system.

For particular uses a number of special bending tests have been developed, such as

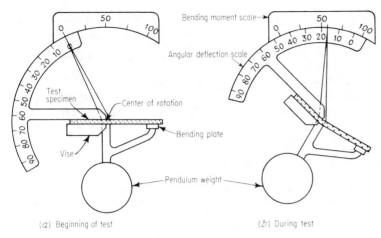

Figure 11.11 Mechanical system of stiffness tester (ASTM D 747). (*From Latour and Sutton, Ref. 116*)

those for investigating the ductility of thin sheet metals, flexing tests for ply separation and cracking of rubber products (ASTM D 430), and tests for measuring the "flexivity" of thermostat metals (ASTM B 106).

PROBLEM 11 Flexure Test of Wood

Object: To determine the mechanical properties of wood subjected to bending, to observe the behavior of the material under load, and to study the failure. The specific items to be determined are
1. Proportional limit stress in outer fiber
2. Modulus of rupture
3. Modulus of elasticity
4. Average work to proportional limit
5. Maximum shearing stress
6. Average total work to ultimate load
7. Type of failure

Preparatory reading: ASTM D 143.

Specimen: Clear wood, approximately 50 by 50 by 760 mm.

Special apparatus: Deflectometer, beam supports, and loading block.

Procedure:

1. Measure and weigh each specimen, count the number of annual growth rings per cm, and estimate the percentage of sapwood and of latewood. Make a sketch of each specimen in perspective, showing any defects and the direction of rings on end sections.

2. Mark the center and endpoints for a 700-mm span. Draw right-section lines through these points, using a square. Drive small brads in one side of the beam at mid-depth over each end support. Set the beam supports so that they can move away from the center of span as the lower fibers elongate due to their being in tension. Place the specimen in position so that the tangential surface nearest the pith will be face up. Place the deflectometer on the brads at the supports and attach it to the center of span. See that the deflectometer fits close to the beam, and bend up the two end brads slightly so that vibrations will not displace it.

3. Adjust the deflectometer and the testing machine to read zero.

4. Estimate the ultimate load and deflection by using property values listed in Table 20.3 or 20.4.

5. Apply the load continuously through a standardized wooden loading block at midspan at the rate of 2.5 mm/min or 1.75 N/min, taking simultaneous load and deflectometer readings for increments of load that will give at least 20 readings below the ultimate. Obtain a reading at the ultimate load if possible.

6. Sketch the appearance of the failure.

7. Cut a moisture sample about 25 mm in length from the specimen near the ruptured section. Remove all splinters and weigh to the nearest 0.1 g. Place in a drying oven controlled at 103 ± 2°C. After the moisture specimen has dried to constant mass, which may take two to four days, again weigh it.

Report:

1. Plot a diagram showing the relation between applied loads (ordinates) and center deflections (abscissas).

2. Determine the moisture content, approximate density both as received and when dry, and all other properties listed. Summarize the results in tabular form and compare them with similar values shown in Chap. 20.

DISCUSSION

1. Are the values obtained from tests on small clear wood specimens applicable to large beams?

2. If you had to walk across a plank of doubtful strength would you go rapidly or slowly? Why?

3. Concrete is strong in compression, weak in tension. Where would you expect the neutral axis to be when a rectangular plain concrete beam fails?

4. Explain the concept of the modulus of rupture. What is the effect of span length on the modulus of rupture of a cast iron flexure bar?

5. Discuss the relative importance of shear and direct stresses as span length increases.

6. Why is it preferable to measure the deflection of a beam specimen to the top rather than the bottom of the beam?

7. If each of the following factors is doubled in turn while all the others remain constant, show the effect on both the elastic strength and the elastic stiffness of a beam: (a) width, (b) depth, (c) span, (d) modulus of elasticity.

8. Name some manufactured products, other than those mentioned in the text, for which bend tests would be appropriate.

9. List the characteristics that are particularly important for each of the following items: (a) a diving board, (b) a chinning bar, (c) a baseball bat, (d) a fishing pole.

10. A ladder of minimum mass is to be designed, perhaps as part of a lunar vehicle. Why and how would flexure tests be helpful?

TWELVE

HARDNESS

12.1 HARDNESS

The general concept of hardness as a quality of matter having to do with solidity and firmness of outline is easily comprehended, but no single measure of hardness, universally applicable to all materials, has yet been devised. The fundamental "physics" of hardness is not yet clearly understood.

A number of different arbitrary definitions of hardness form the basis for the various hardness tests now in use. Some of these definitions are

1. Resistance to permanent identation under static or dynamic loads—indentation hardness
2. Energy absorption under impact loads—rebound hardness
3. Resistance to scratching—scratch hardness
4. Resistance to abrasion—wear hardness
5. Resistance to cutting or drilling—machinability

Such definitions generally develop with the necessity for some way of expressing quantitatively performance requirements under differing conditions of service. In spite of their apparent divergence in meaning, the method of test implied by each definition has a certain useful field of application.

An expert on the subject of hardness once stated [100]:

In the solid state the cohesive and adhesive forces are so strong that the atoms retain fixed positions relative to each other. This produces an aggregate that has and retains definite form. Resistance to change in form, i.e., to change of relative positions of the atoms composing a body, may be defined as the rigidity of that material. The stress necessary to produce permanent deformation of the structure of the solid is most intimately allied to the property of hardness, for, in the measurement of hardness by a penetration method, it will be necessary to vary permanently these fixed positions that the atoms bear with respect to each other.

All the hardness measures are, no doubt, functions of interatomic forces, but the various hardness tests do not bring these fundamental forces into play in the same way or to the same extent; thus no method of measuring hardness uniquely indicates any other single mechanical property. Although some hardness tests seem to be more closely associated than others with tensile strength, some appear to be more closely related to resilience, to ductility, and so forth. In view of this situation, it is obvious that a given type of test is of practical use only for comparing the relative hardnesses of similar materials on a stated basis. The results of ball-indentation tests on steel, for example, have no meaning when compared with results of such tests performed on rubber but serve nicely to evaluate the effectiveness of a series of heat-treatments on a given steel or even to classify steels of various compositions.

The fact that hardness is arbitrarily defined sets hardness tests apart from most others in an important respect: the standards must be scrupulously observed, measurements must be exact. This contrasts, for example, with a compression test on concrete: although major differences in specimen size have some minor effect, it makes no difference, for example, whether the specimen diameter is 6 in (152.4 mm) or 150 mm (provided, of course, that the correct value is used to compute the area), and ASTM allows either. If, however, a concrete test were designed to check what total force breaks a standard cylinder, all cylinders would have to have exactly the same diameter. That sort of thing is the case in hardness tests.

Most hardness tests predate the introduction of SI in 1960 by decades. In those days a common unit of force was the kilogram-force, that is, the weight of a mass of one kilogram. This force, which gives an acceleration of 9.806 650 m/s^2 to a 1-kg mass, was widely used in the standards rather than the newton, which gives the 1-kg mass an acceleration of 1.0 m/s^2. In some standards inches were used, but millimeters have generally predominated.

In spite of the transition to SI units it has been internationally agreed that the values of the various hardness scales shall not be changed. For calibration purposes, however, SI units must be used to measure the forces applied by the indenters. The ISO Technical Committee 17 (Steel) has agreed that the value for the gravitational acceleration should be rounded to 9.81 m/s^2. For the inverse, the factor 0.102 s^2/m may be used.

In this chapter, therefore, we will state indentation masses in kilograms with the understanding that in all cases the *weights* of the indicated masses are meant; newtons are obtained by multiplying the values by 9.81.[†] Indenter dimensions given in inches may be multiplied by 25.4 (exactly) to obtain millimeters.

12.2 SCOPE AND APPLICABILITY

Hardness tests have a wide field of use, although as commercial tests they are perhaps more commonly applied to metals than to any other class of material. The results of a hardness test may be utilized as follows:

[†]If a testing machine is calibrated in lbf, the appropriate kg value may be multiplied by 2.2 to obtain the corresponding dial reading. If the machine is calibrated in "kgf" or kp, no conversion is necessary.

1. Similar materials may be graded according to hardness, and a particular grade, as indicated by a hardness test, may be specified for some one type of service. The degree of hardness chosen depends, however, on previous experience with materials under the given service and not on any intrinsic significance of the hardness numbers. It should be observed that a hardness number cannot be utilized directly in design or analysis as can, for example, tensile strength.

2. The quality level of materials or products may be checked or controlled by hardness tests. They may be applied to determine the uniformity of samples of a metal or the uniformity of results of some treatment such as forming, alloying, heat-treatment, or case-hardening.

3. By establishing a correlation between hardness and some other desired property, for example, tensile strength, simple hardness tests may serve to control the uniformity of the tensile strength and to indicate rapidly whether more complete tests are warranted. One should be aware, however, that correlations apply only over a range of materials on which tests have previously been made; extrapolation from empirical relations should rarely be made and then only with great caution.

Most of the operations called hardness tests can be classified as shown in Table 12.1.

Table 12.1 Classification of hardness tests

Active element or tool	Line of action of load applicator	Fixed load; variable indentation or attrition		Fixed indentation or attrition; variable load	
		Static	Dynamic	Static	Dynamic
Two specimens, one pressed against the other	Normal to specimen	Réaumur (1722)			
Tool of material harder than specimen	Normal to surface of specimen	Brinell (1900) Rockwell (1920) Vickers (1925) Knoop (Tukon) (1939)	Shore scleroscope (1906) Ballentine Cloudburst Schmidt Various abrasion tests—e.g., sandblast type	Monotron Wood-hardness tool	
	Parallel to surface of specimen	Marten sclerometer (1889) Bierbaum sclerometer Herbert pendulum (1923) Machinability (cutting, drilling) tests Various wear or abrasion tests		Allcut and Turner sclerometer (1887) Various qualitative scratch-hardness tests—Mohs (1822)	

The fundamental idea that hardness is measured by resistance to indentation is the basis for a variety of instruments. The indenter, either a ball or a plain or truncated cone or pyramid, is usually made of hard steel or diamond and ordinarily is used under a static load. Either the load that produces a given depth of indentation or the indentation produced under a given load can be measured. The variable is, in either case, a function of the hardness. A similar choice exists in the case of dynamic or impact loads. With most of the dynamic machines, however, the force, whether developed by a drop or a spring load, is of fixed magnitude, as with the scleroscope, for which the height of rebound of the indenter is taken as a measure of the hardness.

Indentation and rebound-type tests, because of their simplicity, have become one of the important quality-control tests for metals. Because they are relatively inexpensive, do not require highly experienced personnel, and are nondestructive, some hardness tests may be employed for a "100 percent inspection" of finished parts.

Probably the most commonly used hardness tests for metals in the United States are the Brinell and Rockwell tests. However, the increasing use of very hard steels and hardened-steel surfaces has brought into use several other tests, such as those made with the Shore scleroscope and with the Vickers, Monotron, Rockwell superficial, and Herbert machines. Also, the need for determining the hardness of very thin specimens and very small parts and the hardness gradients over very small distances has led to the development of the so-called microhardness tests, such as those using the Knoop indenter.

A static ball-indentation test has been standardized for use with wood, although the test is not in common use. A dynamic indentation test using the Schmidt hammer is made to determine the hardness (and probable compressive strength) of concrete in place (see Art. 16.7). The Rockwell L-scale test is applied to hard rubber (ASTM D 530), and a special ball-indentation test is used for soft rubber (ASTM D 1415). The durometer is used for rubber and plastics.

Abrasion or wear tests have found their principal use in connection with paving materials, and a number of such tests have been standardized. Wear or abrasion tests have also been proposed and used experimentally for tests of metals and concrete-floor surfaces. In the case of mineral aggregates and brick, the sample (composed of a number of pieces) is tumbled in a drum—the "rattler" test (ASTM C 131, C7). Sandblast tests as well as rubbing or tumbling tests are in the class of abrasion tests.

For determining the machinability of metals, various special tests have been proposed. The hardness reported as the depth of hole made by a special drill in a given time while running at a constant speed and pressure is sometimes called the Bauer drill test, after the originator. It has been described as testing cutting hardness or machinability. There are other tests described as grooving and cutting tests, but all have a limited field of application.

For a qualitative classification of materials over a wide range, perhaps the most applicable type of test is the scratch test, in which an arbitrary scale is set up in terms of several common materials, each of which will just scratch the material of next lower hardness number. In familiar form, this is the mineralogist's or Mohs' scale. To take into account exotic materials of extreme hardness, a modification of this scale has been designed.

12.3 STANDARD TESTS

In this article a brief overview is given of the major test types classified in Table 12.1. More detailed descriptions of the widely used Brinell, Rockwell, and scleroscope tests will follow. Details of other tests may be obtained from cited ASTM standards and other references.

Static indentation hardness tests Several of the best-known hardness tests employ the principle of exerting a static load on an indenter, which in turn deforms the specimen. The *Brinell* test consists of pressing a hardened steel ball into a test specimen. In accordance with ASTM specifications (ASTM E 10), it is customary to use a 10-mm ball and a load exerted by a mass of 3000 kg for hard metals, 1500 kg for metals of intermediate hardness, and 500 kg (or even as low as 100 kg) for soft materials.

The *Rockwell* test is similar to the Brinell test in that the hardness number found is a function of the degree of indentation of the test piece by action of an indenter under a given static load [35]. Various loads and indenters are used, depending on the conditions of the test. The Rockwell test differs from the Brinell test in that the indenters and the loads are smaller (the masses being 60, 100, or 150 kg), and that the resulting indentation is smaller and shallower. It is applicable to the testing of materials having hardnesses beyond the scope of the Brinell test and is faster because it gives arbitrary direct readings. The Rockwell test is widely used in industrial work. The procedure has been standardized by ASTM (ASTM E 18).

The Rockwell superficial-hardness test uses a special-purpose machine intended exclusively for hardness tests where only very shallow indentation is possible and where it is desirable to know the hardness of the specimen close to the surface. This test was designed particularly for nitrided steel, razor blades, lightly carburized work, brass, bronze, and steel sheet. The masses producing the loads range up to 45 kg and so are even smaller than those used for the regular Rockwell tester.

The relative sizes of the indentations made by the Brinell, ordinary Rockwell, and Rockwell superficial-hardness testers for one material are illustrated in Fig. 12.1.

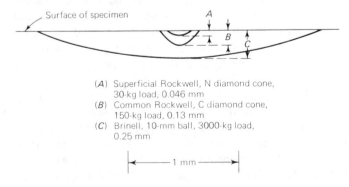

(A) Superficial Rockwell, N diamond cone,
30-kg load, 0.046 mm
(B) Common Rockwell, C diamond cone,
150-kg load, 0.13 mm
(C) Brinell, 10-mm ball, 3000-kg load,
0.25 mm

|← 1 mm →|

Figure 12.1 Comparative impressions in steel (Rockwell C 39) using Brinell and Rockwell testers. (*From Lysaght, Ref. 53.*)

The *Vickers* hardness tester is somewhat similar to the Brinell in that an indentation is made and the hardness number is determined from the ratio P/A of the load exerted by a mass P (in kilograms) to the contact surface area A of the indentation (in square millimeters). The indenter is a square-based diamond pyramid in which the angle between the opposite faces is $136°$ (ASTM E 92). The mass may be varied from 1 to 120 kg.

In conducting a test, place the specimen on the anvil and raise it by a screw until it is close to the point of the indenter. Tripping the starting lever unlocks a 20:1 ratio loading beam, and the load is slowly applied to the indenter and then released. Pressing a foot lever resets the machine. After the anvil is lowered, a microscope is swung over the specimen and the diagonal of the square indentation is measured to 0.001 mm. The machine is also designed to make tests with 1- and 2-mm ball indenters.

One advantage of the Vickers machine over the Brinell machine lies in the shape of the indentation: a much more accurate reading can be taken from the diagonal of a square than can be taken from the diameter of a circle, where the measurement must be made between two tangents. The Vickers machine provides a fairly rapid method of measuring hardness and can be used on metal as thin as 0.15 mm. It is said to be accurate for hardnesses as high as 1300 (about 850 Brinell) and to indicate the friability of nitrided-steel cases. The hardness so determined seems to be a good criterion of the wearing qualities of nitrided steel.

The *Monotron* hardness test is also based on the indentation principle. However, it is essentially a constant-depth indicator, since the hardness is arbitrarily taken as the pressure in kilograms per square millimeter necessary to give a fixed indentation depth of 45 μm. This corresponds to an indentation equal to 6 percent of the diameter of the 750-μm spherical-tipped diamond indenter used, and yields an indentation 360 μm in diameter. The load is applied by a hand lever and is difficult to control precisely. The load and depth are read from separate dials. The depth is measured under load and from the *original* surface of the specimen; it is the depth of the "unrecovered" indentation. The machine is well adapted to determining the hardness of thin materials or case-hardened surfaces. It is usable over the entire range of hardness of metals and operates quite rapidly.

Microhardness testers are used for very small specimens. Because of a real need for a device that would determine the hardness of a material over a very small area and that would produce a small indentation, the National Bureau of Standards (NBS) developed the *Knoop* indenter [46]. It is made of diamond and ground so that it produces a diamond-shaped indentation, the ratio of the long diagonal to the short being about 7:1.

The *Tukon* tester (see Fig. 12.2), with which the Knoop indenter is used, can apply masses of 0.25 to 3.6 kg to produce loads. It is fully automatic in making an indentation. The operator selects the position for test under high microscopic magnification, places the selected area under the indenter, and finally moves the specimen under the mciroscope for reading the length of the diagonal of the impression, from which the Knoop hardness number is calculated. This number is the ratio of the applied load (in kilograms) to the unrecovered projected area (in square millimeters).

The Tukon-Knoop device, or a somewhat similar *Wilson-Knoop* device, is useful

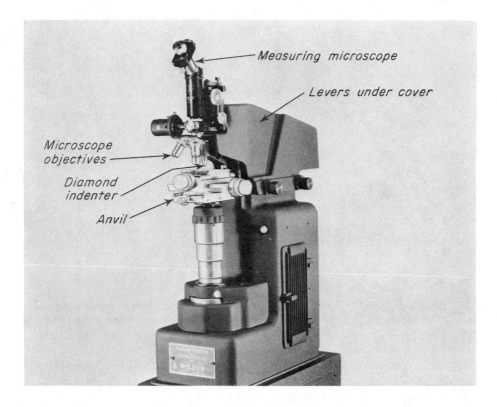

Figure 12.2 Tukon tester with Knoop indenter.

for hardness tests of small parts such as those in watches, thin materials, small wires, tips of cutting tools, single crystals or constituents of alloyed metals, and surface layers, as well as for exploring variations in hardness of small areas such as over the thickness of thin sheets or adjacent to a critical surface.

Although the Tukon tester is normally supplied with the Knoop indenter, it can easily be adapted to the 136° Vickers diamond-pyramid indenter.

Another type of microhardness tester is the *Eberbach*, which uses a spring-loaded Vickers indenter and an electronic device to indicate when the full indenting load is applied to the specimen.

Durometers come in various types, the selection depending on the material to be tested. The type A-2 durometer shown in Fig. 12.3 is used for soft rubber and non-rigid plastics, while the type D is used for harder rubber and plastics (ASTM D 2240). All durometers are quite similar, differing primarily in the sharpness of the point of the conical steel indenter and the magnitude of the load applied to the indenter by a calibrated spring, the type D having the sharpest and most heavily loaded indenter. The durometer measures hardness based on the depth of indentation; the hardness

Figure 12.3 Type A-2 Durometer for rubberlike materials. (*Courtesy of Shore Instrument and Manufacturing Co.*)

varies from 100 at zero indentation to 0 at an indentation of 0.100 in (2.54 mm) and is automatically indicated on a scale. The load acting on the indenter usually varies inversely with the depth of penetration, being a maximum at zero penetration and reducing to practically zero load at maximum penetration, although some durometers use a weight to apply a constant load. Results obtained with one type of durometer cannot be correlated with those obtained with another type.

The only standardized hardness test for *wood* is of the indentation type (ASTM D 143). The hardness is determined by measuring the load required to embed a 0.444-in (11.3 mm) steel ball to one-half its diameter into the wood. This test is of value for comparative purposes only. The approximate range in the hardness of air-dry wood is from about 2 kN[†] for poplar to 18 kN for persimmon. The hardness of Douglas fir is about 4 kN.

Dynamic-hardness tests Most dynamic-hardness tests are virtually indentation tests, irrespective of the manner of loading. As in most dynamic testing, since the methods of calculating the energy absorbed by the specimen are questionable, a test procedure is fixed and the results are therefore arbitrary. In order to secure comparable results, specified equipment and a specified procedure must be employed.

Perhaps the first dynamic-hardness tests were those by Rodman, who experimented with a pyramidal punch in 1861. Later investigations using a hammer with a spherical end verified Rodman's tests and showed that the work of the falling hammer is pro-

[†]1 kN = 225 lbf.

portional to the volume of the indentation. The hardness is expressed as the work required to produce a unit volume of indentation. This method is useful in determining the hardness of metals at high temperatures since the hammer does not stay in contact with the specimen long enough to be affected by the heat.

Among many modern dynamic-hardness testing devices, the *Shore scleroscope* may be the most widely used. The hardness measured by this instrument is often referred to as "rebound hardness." Scleroscope hardness is expressed by a number given by the height of rebound of a small pointed hammer after falling within a glass tube from a height of 10 in (254 mm) against the surface of the specimen. The standard hammer is about 6 mm in diameter and 19 mm long and weighs $\frac{1}{12}$ oz (2.36 g), with a diamond striking tip rounded to a 0.01-in (0.25-mm) radius (ASTM E 448).

The indications obtained by the use of this instrument depend on the resilience of the hammer as well as on that of the material tested, but the permanent deformation of the material is also an important factor. When the hammer falls onto a soft surface, it penetrates it to some extent before rebounding and produces a minute indentation. In so doing, part of the energy of fall is absorbed and the energy available for rebound is comparatively small. If the hammer is dropped on a hard surface, the size of the indentation is much smaller, so that less energy is absorbed in making it. The rebound of the hammer in this case is therefore much higher than before.

The *Herbert* hardness tester operates on a principle different from that of any other type of hardness testing device. The test may be considered to be in the class of dynamic tests, although it does not involve an impact load but rather the oscillations of an arch pendulum supported on a 1-mm ball. Procedures have been set up to give various measures of hardness [33]. The device appears to have many possibilities in the study of work-hardening problems that arise in connection with the machining of metals. It also affords an easy means of measuring hardness while a specimen is in a special furnace at a controlled, elevated temperature.

Scratch-hardness and wear-hardness tests A convenient and definite hardness scale similar to a temperature scale is lacking because materials are not available that have invariable hardness to serve as calibration points analogous to the various boiling and freezing points that are standards of comparison for thermometers and pyrometers.

An approach to this is the arbitrary mineralogical scale of hardness, in which a mineral will scratch other minerals that are lower on the scale (smaller hardness number) and will in turn be scratched by minerals higher on the scale. The well-known scale used by mineralogists is Mohs' scale, shown in Table 12.2. With the development of extremely hard abrasives has come the need for more adequately distinguishing between materials in the range of hardness from that of quartz to that of diamond. An extension of Mohs' scale devised for this purpose is also shown in Table 12.2.

A widely used test of the scratch type is the *file test*. In many modern industrial plants it is used as a qualitative or inspection test for hardened steel. In addition to the standard and more accurate methods of testing used on representative samples from each lot, each piece in the lot may be gone over with a file; by making one pass over the surface with a file of appropriate hardness the operator is able to cull unsatisfactory pieces.

Table 12.2 Scratch hardness—mineralogical basis

Hardness no.	Reference mineral	Hardness no.	Reference mineral	Metal equivalent
	Mohs' scale		Extension of Mohs' scale[†]	
1	Talc	1	Talc	
2	Gypsum	2	Gypsum	
3	Calcite	3	Calcite	
4	Fluorite	4	Fluorite	
5	Apatite	5	Apatite	
6	Feldspar (orthoclase)	6	Orthoclase	
		7	Vitreous pure silica	
7	Quartz	8	Quartz	Stellite
8	Topaz	9	Topaz	
		10	Garnet	
		11	Fused zirconia	Tantalum carbide
9	Sapphire or corundum	12	Fused alumina	Tungsten carbide
		13	Silicon carbide	
		14	Boron carbide	
10	Diamond	15	Diamond	

[†]R. R. Ridgeway, A. H. Ballard, and B. B. Bailey, "Hardness Value for Electrochemical Products," *Trans Electrochem. Soc.*, vol. 43.

One of the few *abrasion* tests, for which the Los Angeles abrasion machine is used, is briefly described in Chap. 19. It applies primarily to aggregates for asphalt mixes.

12.4 BRINELL TESTS

Brinell tests are static indentation tests using relatively large indenters. They are suitable for test pieces that are neither very small nor extremely hard. Highly specialized equipment is not required.

Specimens Although the Brinell test is a simple one to make, specimens must be chosen with care in order to obtain good results. It is not suitable for extremely hard materials, because the ball itself would deform too much, not is it satisfactory for thin pieces such as razor blades, because the usual indentation may be greater than the thickness of the piece. It is not adapted for use with case-hardened surfaces, because the depth of indentation may be greater than the thickness of the case and because the yielding of the soft core invalidates the results; also, for such surfaces, the indentation is almost invariably surrounded by a crack that may cause fatigue failure if the part is used in service. Obviously the Brinell test should not be used for a part if marring of the surface impairs its value.

The surface of the specimen should be flat and reasonably well polished; otherwise difficulty will be experienced in making an accurate determination of the diameter of the indentation. If the specimen is prepared from rough stock, the surface should be dressed with a file and then polished with fine emery cloth. For some materials, the edge of the indentation is very poorly defined, even when the surface finish is good. To increase the sharpness of definition of the edge of the indentation, ASTM suggests the use of a movable lamp for illumination, placing it so that the contrast of light and shade will bring first one edge of the indentation, then the other, into sharp definition. For some specimens, the indentation may be made more distinct by using balls lightly etched with nitric acid or by using some pigment, such as Prussian blue, on the ball.

Apparatus Various types of machines for making the Brinell test are available. They may differ as to (1) method of applying load: oil pressure, gear-driven screw, weights with lever; (2) method of operation: hand, motive power; (3) method of measuring load: piston with weights, bourdon gage, dynamometer, weights with lever; and (4) size: large (usual laboratory size), small (portable). The Brinell test may be made in a small universal testing machine by using a suitable adapter for holding the ball, as well as by using special machines designed for the purpose. For tests of thin sheet-metal products such as cartridges and cartridge cases, a small hand-plier device using a $\frac{3}{64}$-in (1.2-mm) ball and a 22-lb (98.1-N) spring pressure has been employed.

The principal features of a typical hydraulically operated Brinell testing machine are illustrated in Figs. 12.4 and 12.5. The specimen is placed on the anvil and raised

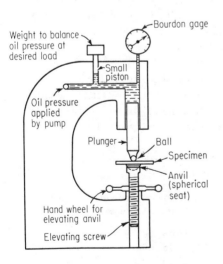

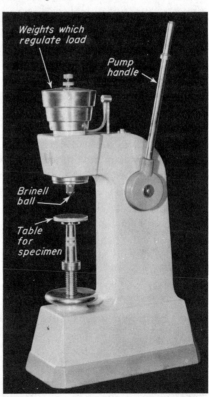

Figure 12.4 Features of hydraulic Brinell machine.

Figure 12.5 Hydraulic Brinell machines. (*Courtesy of Riehle Testing Machine Division.*)

to contact the ball. Load is applied by pumping oil into the main cylinder, which forces the main piston or plunger downward and presses the ball into the specimen. The plunger has a ground fit so that frictional effects are usually negligible. The bourdon gage is used only to give a rough indication of the load. When the desired load is applied, the balance weight on top of the machine is lifted by action of the small piston; this ensures that an overload is not applied to the ball.

It is required that the ball be within 0.01 mm of the nominal 10-mm diameter. This requirement is necessary in order to obtain concordant results with different machines. When used on very hard steels, it is required that the ball should not show a permanent change in diameter of more than 0.01 mm. For this reason, carboloy (tungsten carbide) balls are often used for testing the harder steels.

The load-measuring device may be calibrated (1) by the use of weights and proving levers, (2) by an elastic calibration device, or (3) by making a series of indentations on specimens of different degrees of hardness and comparing them with a second series of indentations made by the use of any standardized testing machine and standard test ball. According to ASTM standards, a Brinell machine is acceptable for use over a loading range within which its load-measuring device is correct to within 1 percent.

The size and uniformity of the ball are checked by measurements with a micrometer caliper of suitable accuracy.

The Brinell microscope is checked by comparing its readings with a standardized scale. The error of reading throughout the range should not exceed 0.02 mm.

Procedure In the Brinell test, the load must be chosen in accordance with the expected hardness of the material as shown in Table 12.3. In the standard test, the ball diameter is 10 mm.

For testing small or thin specimens, it is sometimes necessary to make Brinell hardness tests with a ball less than 10 mm in diameter. Such tests (which are not to be regarded as standard Brinell tests) will approximate the standard tests more closely if the relation between the applied load P (in kilograms) and the diameter of the ball D (in millimeters) is the same as in the standard tests.

Normally, the full load is applied for a minimum of 15 s for ferrous metals and 30 s for softer metals; then the load is released and the diameter of the indentation is measured to the nearest 0.02 mm with the microscope. Often, however, a 30-s interval is used for ferrous metals and a 60-s interval for other metals. In rapidly made control tests, sometimes the time interval is permitted to be less than the standard.

Table 12.3 Hardness range for standard Brinell loads[†]

Ball diameter, mm	Load, kg	Recommended range of Brinell hardness
10	3000	96 to 600
10	1500	48 to 300
10	500	16 to 100

[†]From ASTM E 10.

Observations For high precision, the diameter of the indentation is measured with a micrometer microscope that has a transparent engraved scale in the field of view. The scale has divisions corresponding to 0.1 mm, and the measurements are estimated to within at least 0.02 mm. The diameter is taken as the average of two readings measured at right angles to each other. Sometimes the depth of indentation is measured by means of a dial indicator fastened to the plunger and actuated by a gimbal ring held in contact with the surface of the specimen.

The Brinell hardness number is nominally the load mass per unit area, in kilograms per square millimeter, based on the area of the indentation that remains after the load is removed. It is obtained by dividing the applied load by the area of the surface of the indentation, which is assumed to be spherical. If P is the applied load (in kilograms), D is the diameter of the steel ball (in millimeters), and d is the diameter of the indentation (in millimeters), then

$$\text{Brinell hardness number (Bhn)} = \frac{\text{load mass}}{\text{indented area}} = \frac{P}{\frac{\pi D}{2}(D - \sqrt{D^2 - d^2})}$$

The hardness numbers obtained for ordinary steels (with 3000-kg load) range from about 100 to 500; the medium-carbon structural steels have hardness numbers on the order of 130 to 160. For very hard special steels the hardness numbers may be as high as 800 or 900, but the Brinell test itself is not recommended for materials having a Bhn over 630.

For a standardized test such as this, the projected circular area computed from the diameter of the indentation might just as well have been used. In fact, it seems as logical as using the assumed spherical area of the indentation, which was first used by Brinell. However, a change appears undesirable now because of the wide familiarity with and extensive use of the standard Brinell numbers.

The surface of the indentation is not truly spherical, because the ball undergoes some deformation under load and because there is some recovery of the test piece when the load is removed. Thus the indentations made by different-size balls and different loads are not geometrically similar. Because of this, it is essential that the ball size and the load be reported with the hardness number whenever the standard 3000-kg load and the 10-mm ball have not been used.

When the depth of indentation is to be measured, the observation is made just after the load is released. A hardness number is computed from the depth of the indentation by using the following equation:

$$\text{Brinell hardness number (Bhn)} = \frac{\text{load mass}}{\text{indented area}} = \frac{P}{\pi D t}$$

where t = depth of indentation, mm
$\quad\quad D$ = diameter of ball, mm

However, the observed depth of indentation t_1 (usually determined from the relative motion of the ball plunger and the specimen) and the actual depth t corresponding to the diameter of indentation d do not always agree, owing to the possible formation of

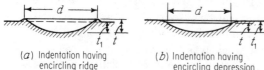

(a) Indentation having   (b) Indentation having
    encircling ridge   encircling depression   **Figure 12.6** Indentations in Brinell test.

an encircling ridge (Fig. 12.6a) or depression (Fig. 12.6b). Soft materials such as copper and mild steel show the former effect, and hard materials such as manganese steel and some bronzes develop a depression. Although the observed depth of indentation t_1 appears to be a logical basis of comparison, it is not the basis of the Brinell hardness number according to definition. However, it is possible for routine control work to establish by test the relation between depth and diameter of indentation for each kind of material in different stages of hardness. These relations may be plotted on a graph or recorded in tabular form so that rapid routine tests in mass production may be made to yield standard Brinell values.

Effects of variables The material of the specimen is permanently deformed for an appreciable distance below the surface of the indentation. If this deformation extends to the lower or opposite surface, the size of the indentation may be greater for some materials and less for others than for a thicker specimen of the test material. Tests made at NBS on a large variety of materials showed that under each indentation where the thickness was less than the "critical" value, a spot of altered surface was visible on the underside of the specimen [75]. ASTM specifies that no marking shall appear on the side of the piece opposite the indentation and also requires the thickness of the specimen to be at least 10 times the depth of the indentation. To satisfy this requirement the minimum hardness for a given thickness of a specimen should be as shown in Table 12.4.

If an indentation is made too near the edge of the specimen, it may be both too large and unsymmetrical. If the indentation is made too close to a previous one, it may be too large owing to lack of sufficient supporting material or too small owing to work-hardening of the material by the first indentation. However, tests have shown that the errors may be neglected if the distance of the center of the indentation from the edge of the specimen or from the center of adjacent indentations is equal to or greater than $2\frac{1}{2}$ times the diameter of the indentation.

Table 12.4 Thickness of Brinell specimen†

Thickness of specimen, mm	Minimum Brinell hardness for which a Brinell test may safely be made		
	500-kg load	1500-kg load	3000-kg load
2	79	238	476
4	40	119	238
6	26	79	159
8	20	60	119
10	16	48	95

†From ASTM E 10.

If the compressive properties of a flat specimen are not uniform, perhaps because of the direction of rolling or the cooling stresses, a noncircular indentation will result. In this case, the average Brinell hardness of the material may be obtained if the diameter is taken as the average of measurements in four directions, roughly 45° apart.

It is frequently necessary in practice to measure the Brinell number on a curved surface rather than on a plane one, but an indentation on a curved surface of a specimen having uniform properties will not have a circular boundary unless the specimen is spherical. Provided the radius of the specimen is not less than 25 mm, the diameter of the impression may be taken to be the average of the maximum and minimum diameters. For smaller radii a flat spot may be prepared on the surface of the specimen.

A rapid rate of applying the load affects the diameter of the Brinell indentation in two ways: (1) the effect of inertia of the piston and weights and the friction of the plunger cause a momentary rise of the load above 3000 kg and consequently enlarge the indentation, and (2) a rapid rate of penetration allows less time for the plastic flow of the material, resulting in a decreased size of the indentation.

Tests have shown that the error due to inertia of the plunger can be, and often is, very much greater than that due to an insufficient period of sustaining the load to permit plastic flow to occur. The inertial effect is of importance not only when the load is first applied but also when, during the maintenance of the required load, the downward drift of the weights is reversed into an upward motion by a stroke of the pump in order to maintain the weights in floating equilibrium. By exercising care in operating the pump, the inertial effect may be reduced to a negligible factor. If a UTM is used and the load cannot be held quite steady during the load interval, it is better to err on the low side occasionally than to apply excessive loads.

From tests at NBS on 29 steels and nonferrous metals, it was found that the flow for most materials is quite rapid during the first 30 s under required load; it is much less rapid in the interval from 30 to 120 s [75]. For most metals the Brinell number varies less than 1 percent for loading intervals between 30 and 120 s.

The error in the Brinell number is less than 1 percent as long as the error in diameter does not exceed 0.01 mm. Errors in reading the diameter of the indentation may be ascribed to two causes: (1) to an error in reading the microscope and (2) to indefiniteness of the boundary of the indentation. The error in reading a modern Brinell microscope should not exceed 0.02 mm, provided it is in proper adjustment as determined by comparison with the calibrated scale furnished with the microscope. The indefiniteness of the boundary of the indentation may cause considerable uncertainty in the magnitude of the diameter; some of the precautions that may be taken to obtain a clear indentation have already been outlined.

The variations from standard size and shape of modern well-made balls are usually too small to introduce appreciable errors. However, the flattening of the ball, particularly when the hardness of the specimen approaches that of the ball, may lead to serious errors. For testing materials with Brinell hardness numbers greater than about 400, the ball should be frequently checked for distortion, and for testing materials having hardness numbers over 450, balls of harder material than steel should be used. Carbide balls may be used for values to 630, the maximum hardness for which the Brinell test should be used. In quoting Brinell numbers greater than 200, the material of the ball should be stated, since the results depend to some extent on what kind of material it is.

12.5 ROCKWELL TESTS

Rockwell tests, like Brinell tests, are of the indentation type, but they can be used for smaller specimens and harder materials. A special tester is required, however.

Specimens Because of the smallness of the indentation and because of the way it is measured, there are some differences in selecting and preparing test pieces for the Rockwell test as compared with the Brinell test. Certain precautions are necessary, of which the following are the most important.

The test surface should be flat and free from scale, oxide films, pits, and foreign material that may affect the results. A pitted surface may give erratic readings, owing to some indentations being near the edge of a depression. This permits a free flow of metal around the indenting tool and results in a low reading. Oiled surfaces generally give slightly lower readings than dry ones because the friction is reduced under the indenter. The bottom surface should also be free from scale, dirt, or other foreign materials that might crush or flow under the test pressure and so affect the results.

The thickness of the piece tested should be such that no bulge or other marking appears on the surface opposite the indentation, since then the depth of the indentation is noticeably affected by the supporting anvil. For very hard materials, the thickness may be as little as about 0.25 mm [36]. Selector charts for the proper scale to be used with thin sheets of various ranges of hardness are shown in ASTM E 18.

All hardness tests should be made on a single thickness of the material, regardless of the thickness of the piece. The use of more than one piece of thin material to give adequate thickness does not yield the same result as for a solid piece of the same thickness. Relative movement takes place on the surfaces between the various pieces, and lack of flatness of the separate pieces will lead to compression of the pile under the load.

The hardness number determined by indenting a curved surface is in error because of the shape of the surface. In industrial applications, this problem is often encountered, commonly with shafts. If feasible, a small flat spot may be filed on the rod before making the indentation. However, corrections to be added to observed Rockwell values on cylindrical specimens having diameters from 5 to 40 mm are given in ASTM E 18.

Apparatus The test is conducted in a specially designed machine that applies load through a system of weights and levers. A photograph of one model is shown in Fig. 12.7. The indenter or "penetrator" may be either a steel ball or a diamond cone with a somewhat rounded point. The hardness value, as read from a specially graduated dial indicator, is an arbitrary number that is related to the depth of indentation.

The accuracy of Rockwell hardness testers is checked with special test blocks that are available for all ranges of hardness. If the error of the tester is more than ±2 hardness numbers, it should be reconditioned and brought into proper adjustment.

Levers under cover

Depth and hardness indicator

Dash pot

Diamond indenter

Anvil

Adjustable weights

ROCKWELL
HARDNESS TESTER
WILSON

Figure 12.7 Rockwell tester. (*Courtesy of Wilson Mechanical Instrument Co.*)

Procedure In the operation of the machine, which is explained diagrammatically in Fig. 12.8, a minor load of 10 kg is first applied, which causes an initial indentation that sets the indenter on the material and holds it in position. The dial is set at the "set" mark on the scale, and the major load is applied. The major load is customarily 60 or 100 kg when a steel ball is used as the indenter, although other loads may be used when found necessary, and the load is usually 150 kg when the diamond cone is employed. The ball indenter is normally $\frac{1}{16}$ in (1.6 mm) in diameter, but others of larger diameter may be employed for soft materials. After the major load is applied and removed, the hardness reading is taken from the dial while the minor load is still in position.

The "superficial" tester operates on the same principle as the regular Rockwell tester but employs lighter minor and major loads and has a more sensitive depth-measuring system. Instead of the 10-kg minor load and the 60-, 100-, or 150-kg major loads of the regular Rockwell, the superficial tester applies a minor load of 3 kg and major loads of 15, 30, or 45 kg. One point of hardness on the superficial machine corresponds to a difference in depth of indentation of 0.001 mm.

Observations The dial of the machine has two sets of figures, one red and the other black, which differ by 30 hardness numbers. The dial was designed in this way to

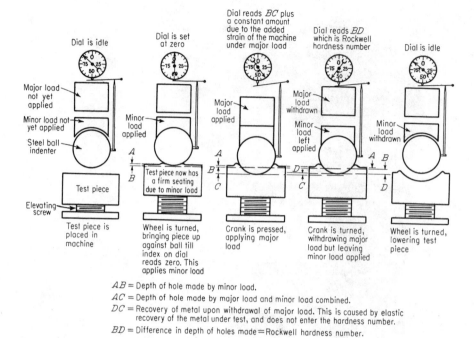

AB = Depth of hole made by minor load.

AC = Depth of hole made by major load and minor load combined.

DC = Recovery of metal upon withdrawal of major load. This is caused by elastic recovery of the metal under test, and does not enter the hardness number.

BD = Difference in depth of holes made = Rockwell hardness number.

Figure 12.8 Procedure in using Rockwell tester.

accommodate the B and C scales, which were the first ones standardized and are the most widely used. Two advantages were gained by the shifting of the zero points: (1) negative numbers are avoided on soft material such as brass, when tested on the B scale; and (2) this established B 100 as the upper practical limit of hardness that might be tested with the 100-kg major load and the $\frac{1}{16}$-in (1.6-mm) ball without deforming the ball.

There is no Rockwell hardness value designated by a number alone because it is necessary to indicate which indenter and load have been employed in making the test. Therefore, a prefix letter, as shown in the first column of Table 12.5, is used to designate the test conditions. The letter designations for the various Rockwell scales are assigned as follows:

	Major load mass, kg		
Indenter	60	100	150
Cone	A	D	C
$\frac{1}{16}$"-ball	F	B	G
$\frac{1}{8}$"-ball	H	E	K
$\frac{1}{4}$"-ball	L	M	P
$\frac{1}{2}$"-ball	R	S	V

Table 12.5 Rockwell hardness scales and prefix letters[†]

Group	Scale symbol and prefix letter	Indenter	Major load, kg	Dial numerals	Typical applications of scales
1 Common scales	B	$\frac{1}{16}$-in ball (1.6-mm)	100	Red	Copper alloys, soft steels, aluminum alloys, malleable iron
	C	Diamond cone	150	Black	Steel, hard cast iron, pearlitic malleable iron, deep case-hardened steel
2	A	Diamond cone	60	Black	Cemented carbides, thin steel, shallow case-hardened steel
	D	Diamond cone	100	Black	Thin steel, medium case-hardened steel
	E	$\frac{1}{8}$-in ball (3.2-mm)	100	Red	Cast iron, aluminum and magnesium alloys, bearing metals
	F	$\frac{1}{16}$-in ball (1.6-mm)	60	Red	Annealed copper alloys, thin soft sheet metals
	G		150	Red	Phosphor bronze, beryllium copper, malleable iron
	H	$\frac{1}{8}$-in ball (3.2-mm)	60	Red	Aluminum, lead, zinc
	K		150	Red	
3	L	$\frac{1}{4}$-in ball (6.4-mm)	60	Red	
	M		100	Red	Bearing metals and other very soft or thin materials. Use smallest ball and heaviest load that does not give anvil effort
	P		150	Red	
	R	$\frac{1}{2}$-in ball (12.7-mm)	60	Red	
	S		100	Red	
	V		150	Red	

[†]Based on ASTM E 18.

In the designation of scales (Table 12.5), it will be noted that red figures are used for readings obtained with ball indenters regardless of the size of the ball or the magnitude of the major load, and that black figures are used only for the diamond cone.

The B scale is for testing materials of medium hardness such as low- and medium-carbon annealed steels. The working range of this scale is from 0 to 100. If the ball indenter is used to test material harder than about B 100, there is danger that is will be flattened. Furthermore, because of its shape, the ball is not as sensitive as the rounded conical indenter to differences in hardness of hard specimens. If the $\frac{1}{16}$-in (1.6-mm) ball is used on material softer than B 0, there is danger that the cap of the indenter that holds the ball in place will make contact with the specimen or that the weight arm will descend too far and rest on its stop pin. Below B 0, the ball, owing to its shape, becomes supersensitive and the readings are erratic.

The C scale is the one most commonly used for materials harder than B 100. The hardest steels run about C 70. The useful range of this scale is from C 20 upward. Any inaccuracies that occur in grinding the diamond cone to its proper shape have a proportionately greater effect on small indentations, and it should therefore not be used below this lower value. Also, on soft materials, the spherical apex (0.2-mm radius) of the cone is driven into the material a considerable distance by the minor load, and unless the speed with which the indenter makes contact against the work and the time interval between applying minor and major loads are standardized, there will be considerable variation in the readings. These effects are negligible when using the cone on harder material.

In general, a scale should be selected to employ the smallest ball that can properly be used, because of the loss of sensitivity as the size of indenter increases. An exception to this is when soft nonhomogeneous material is to be tested, in which case it may be preferable to use a larger ball that makes an indentation of greater area, thus obtaining more of an average hardness.

The Rockwell scales are divided into 100 divisions, and each division or point of hardness is the equivalent of 2 μm in indentation; therefore the difference in indentation between dial readings of B 53 and B 56 is 3 $\times$ 2, or 6 μm. Since the scales are reversed, the higher the number, the harder the material, as shown by the following expressions, which define the Rockwell B and C numbers:

$$\text{Rockwell B number} = 130 - \frac{\text{depth of penetration } (\mu m)}{2 \ \mu m}$$

$$\text{Rockwell C number} = 100 - \frac{\text{depth of penetration } (\mu m)}{2 \ \mu m}$$

Since the diamond cone in the superficial machines is intended especially for use on "nitrided" work and the $\frac{1}{16}$-in (1.6-mm) steel ball for testing "thin" sheet, the letters N and T have been selected for these two scale designations. The W, X, and Y scales are used for very soft materials. Although these machines have but one set of dial graduations, scale symbols must be assigned as given in Table 12.6 to indicate the indenter and major load used.

Table 12.6 Rockwell superficial-hardness scales[†]

	Scale symbols				
Major load, kg	N scale, diamond cone	T scale, $\frac{1}{16}$-in ball (1.6-mm)	W scale, $\frac{1}{8}$-in ball (3.2-mm)	X scale, $\frac{1}{4}$-in ball (6.4-mm)	Y scale, $\frac{1}{2}$-in ball (12.7-mm)
15	15 N	15 T	15 W	15 X	15 Y
30	30 N	30 T	30 W	30 X	30 Y
45	45 N	45 T	45 W	45 X	45 Y

[†]Based on ASTM E 18.

Effects of variables If the table on which the Rockwell hardness tester is mounted is subject to vibration, the hardness numbers will be too low, since the indenter will sink farther into the material than when such vibrations are absent. If, when the operating handle is being returned to its normal position, the latch operates with such a snap as noticeably to change the position of the dial pointer, felt or rubber washers should be placed under the trip mechanism in order to cushion this blow. If the snap is severe, a difference in reading of several hardness numbers may result.

Specimens should be prepared with care. If curved plates are tested, the concave side should face the indenter. If such specimens are reversed, an error will be introduced owing to the flattening of the piece on the anvil. Specimens that have sufficient overhang so that they do not balance themselves on the anvil should be properly supported. To prevent injury to the anvil and indenter, the two should not be brought into contact without a test specimen between them.

The speed and time of application of the major load should be established, accurately adhered to, and reported when comparing results. The dashpot should be adjusted so that the operating handle completes its travel in 4 to 5 s both with no specimen in the machine and with the machine set up to apply a major load of 100 kg. An interval of full application of the major load of not more than 2 s is specified. For soft metals, plastic flow may cause variations as high as 10 hardness numbers. With such materials, the operating lever should be brought back the moment it is seen that the major load is fully applied.

It is advisable to check the ball indenters regularly to see that they have not become flattened and the diamond cone to see that it has not become blunted or chipped.

12.6 SCLEROSCOPE TESTS

Scleroscope tests are dynamic in nature, using rebound energy as a measure of hardness. They are applicable even to very thin specimens made of very hard materials.

Specimens For scleroscope tests the minimum thickness of the specimen that can be tested depends on its hardness. For hard steel, as in safety-razor blades, the thickness should be at least 0.15 mm; for cold-rolled unannealed brass and steel it should be 0.25 mm; and for annealed sheets, 0.38 mm. The scleroscope is very useful for testing the hardness of case-hardened surfaces provided they are at least 0.04 mm thick. The surface must be flat, smooth, and clean.

Apparatus There are two types of scleroscopes, one a direct-reading or visual type, as shown in Fig. 12.9a, in which the height of rebound must be caught by eye, and the other an improved dial-recording instrument, as shown in Fig. 12.9b, in which the dial hand remains at the height of rebound until reset. Both instruments are portable, and tests can be made quite rapidly with them. In the direct-reading type, the instrument is provided with an ingenious automatic head by means of which the hammer is lifted and released by air pressure from a rubber bulb.

A magnifier hammer is available for use on soft materials. This has a larger point

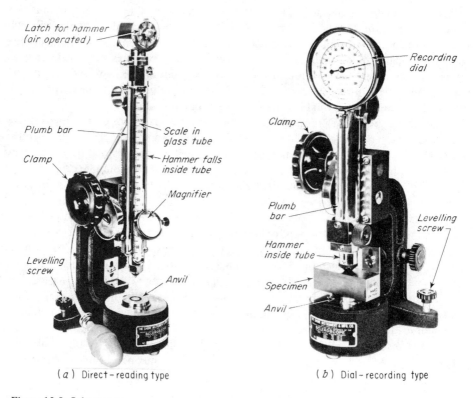

(a) Direct-reading type   (b) Dial-recording type

Figure 12.9 Scleroscopes.

area than the standard hammer and gives higher readings, thus magnifying small but significant variations of hardness.

The condition of the diamond point is very important and should be checked frequently by using hardened-steel reference blocks. The indicated hardness should not vary more than ±5 percent from that stamped on the blocks; larger variations suggest possible cracking or chipping of the diamond.

Procedure Various precautions must be observed if reliable results are to be obtained in a Shore scleroscope test.

The glass tube through which the hammer falls must be carefully plumbed to avoid rubbing of the hammer on its inner surface. All specimens should be securely clamped to the anvil to avoid inertial effects.

Several readings should be taken to get a fair average, but the hammer should not be dropped more than once on any one spot because of the possible effect of strain-hardening.

Observations The scale is graduated into 140 divisions, a rebound of 100 being equivalent to the hardness of martensitic high-carbon steel. For this material, the area of contact between the hammer and specimen is only about 0.25 mm^2 and the stress

Table 12.7 Approximate hardness relations for steel†

Diameter, mm	Brinell, 3000 kg — Standard 10-mm ball	Brinell, 3000 kg — Tungsten carbide ball	Rockwell‡ A Cone 60 kg	Rockwell‡ B 1/16" ball 100 kg	Rockwell‡ C Cone 150 kg	Rockwell‡ D Cone 100 kg	Rockwell‡ E 1/8" ball 100 kg	Rockwell‡ 30 N Cone 30 kg	Rockwell‡ 30 T 1/16" ball 30 kg	Sclero-scope	Vickers diamond pyramid	Mohs	Tensile strength MPa	Tensile strength ksi
2.50	...	601	79.8	...	57.3	68.7	...	75.1	...	77	640	7.8	2130	309
2.75	...	495	76.7	...	51.6	64.3	...	69.9	...	68	539	7.3	1765	256
3.00	415	415	72.8	...	44.5	58.8	...	63.5	...	59	440	6.8	1450	210
3.25	352	352	69.3	...	37.9	53.8	...	57.6	...	51	372	6.4	1215	176
3.50	302	302	66.3	...	32.1	49.3	...	52.2	...	45	319	6.0	1035	150
3.75	262	262	63.6	...	26.6	45.0	...	47.3	...	39	276	5.6	890	129
4.00	229	229	60.8	...	20.5	40.5	...	41.9	...	34	241	5.0	765	111
4.25	202	...	58.0	93.0	...	37.0	111	37.0	78.0	30	...	5.1	675	98
4.50	179	...	56.0	88.0	...	33.0	108	32.0	78.0	27	...	4.8	605	88
4.75	159	...	53.0	83.0	...	31.0	106	28.0	72.0	24	...	4.6	545	79
5.00	143	...	...	78.0	...	...	103	...	69.0	22	...	4.5	495	72
5.25	128	...	...	71.0	...	...	100	...	64.0	...	...	...	450	65
5.50	116	...	...	65.0	...	...	96	...	60	...	...	...	415	60
5.75	105	...	...	58.0	...	...	92	...	55	...	...	...	370	54

†Based on *Metals Handbook–1961*, American Society for Metals, Cleveland, Ohio, 1961. See ASTM E 140 for additional relations.

‡ 1/8 in = 3.175 mm; 1/16 in = 1.5875 mm (exactly).

developed is over 2750 MPa. Notwithstanding the lightness of the hammer, the forces exerted in all cases are sufficient to overcome the surface resistance of the hardest materials used in engineering practice; however, the indentations produced are so minute that they do not seriously impair a finished surface.

It should be noted that the scleroscope hardness numbers are arbitrary and that they are comparable only when determined on similar material. Obviously there would be little relation between the scleroscope hardness numbers of two such dissimilar materials as rubber and steel.

12.9 CORRELATIONS

Because of the rather arbitrary nature of the several different methods of measuring hardness, precise correlations cannot be expected either among the results of the various hardness tests or with other properties of materials.

Some approximate relations have been determined by tests of the same material using the various devices. Since these relations vary with the different materials and with the mechanical and heat-treatment given them, too much reliance on them must be avoided. Table 12.7 presents comparative values for steel as determined by the more common types of equipment. Additional relations covering alloys are shown in ASTM E 140.

No correlation exists between any indentation hardness and the yield strength determined in a tension test, since the amount of inelastic strain involved in the hardness test is much greater than in the test for yield strength. However, because of the greater similarity in inelastic strain involved in the test for ultimate tensile strength and indentation hardness, empirical relations have been developed between these two properties, at least in the case of certain steels. For example, in the commonly used carbon and alloy steels, the tensile strength in megapascals is approximately 3.5 times the Brinell hardness number. The approximate tensile strength of steel for the several types and degrees of hardness is indicated in the last column of Table 12.7.

PROBLEM 12 Hardness Tests of Steel

Object: To study the Brinell, Rockwell, and Shore-scleroscope hardness testers and to determine the hardness numbers of assigned specimens of steel.
Preparatory reading: ASTM E 10, E 18, and E 448.
Specimens: Steel specimens as assigned. Both top and bottom surfaces of specimens should be clean and smooth.
Special apparatus: Brinell tester with measuring microscope and possibly with depth gage, Rockwell machine, and Shore scleroscope.
Procedure for Brinell test:
1. Examine the depth gage on the Brinell machine. Note the value of the smallest dial division in terms of depth of ball impression.
2. Place the specimen on the anvil so that its surface will be normal to the direction of the applied load. Raise the anvil with the handwheel until the specimen just makes contact with the ball. See that the ball is at least 5 mm from the edge of the specimen.
3. Read the depth gage or set the dial to read zero.

4. Apply load by means of the hand pump until the yoke and attached weights rise and float. Be sure that the yoke does not rise more than 10 mm. Use a slow, steady stroke in lifting the yoke. Maintain the full load for 20 s.

5. Release the load but make certain that the ball still contacts the specimen. Record the depth-gage reading. Remove the specimen from the machine.

6. Check the scale in the field of view of the measuring microscope with the standard scale provided for that purpose. Measure the diameter of the impression left by the ball to the nearest 0.01 mm.

7. Make five independent hardness determinations on each specimen.

8. After the test has been completed, place a dummy specimen on the anvil to protect it.

Procedure for Rockwell test:

1. Examine the Rockwell hardness tester and get it ready for the test. Use a $\frac{1}{16}$-in-diameter ball indenter. Move the operating handle as far forward as it will go and see that the weight marked "100 kg" is in position.

2. Place the specimen on the anvil of the machine. Raise the anvil and the test specimen by means of the elevating screw until the specimen comes in contact with the ball. Continue to raise the work slowly until the initial load is applied and the pointer is within ± 5 scale divisions of its upper vertical position. Turn the bezel of the gage until the mark B 30, which is also designated by a red arrow and by the word "set," is directly behind the pointer.

3. Release the operating handle so as to apply the major load. Allow the operating handle to move without interference until the major load is fully applied. This is accomplished when the weight arm is completely free from the control of the dashpot. Immediately after the major load has been fully applied, gently bring back the operating handle to its latched position.

4. Read the position of the pointer on the red or B scale of the dial, which gives the Rockwell hardness number.

5. Make five tests on each specimen.

Procedure for scleroscope test:

1. Examine the scleroscope, and to facilitate the reading of the rebound of the hammer, place it on a bench so that light will be reflected from the scale. Adjust the leveling screws so that the plumb bar will hang freely in the center of the ring at its lower end.

2. Using the central portion of the specimen so that it will be practically balanced on the anvil, clamp it in position and maintain a slight pressure on the clamp handle during the test.

3. Release the hammer and note the scale reading at the top of the hammer on the first rebound. Lift the hammer back into position. Make five determinations on each specimen. Do not let the hammer fall twice on the same spot.

Report:

1. Compute the average of the five values of diameter of impression of each specimen using the Brinell method and compute the hardness number. Also compute the hardness from the depth of the Brinell impression.

2. Determine the probable tensile strengths of the materials from which the specimens were made.

3. Determine the average Rockwell hardness number of each specimen and the probable corresponding Brinell number.

4. Determine the average scleroscope hardness number of each specimen and the probable corresponding Brinell number.

5. In a single table present the results of tests by all three methods.

DISCUSSION

1. Is the depth of impression or the diameter of impression measured in the ASTM standard Brinell hardness test?

2. Why do the Brinell hardness numbers obtained from the depth of indentation usually differ from those computed from the diameter of indentation?

3. Do the hardness numbers of unlike materials give a satisfactory basis for comparing their hardness?

4. Would you expect the Brinell hardness numbers for loads of 500, 1500, and 3000 kg to vary or to remain constant for any given specimen? Explain.

5. How long should the load be maintained in the Brinell test for hardness of steel according to ASTM specifications? Why is a definite loading period essential?

6. Why is a minor load applied before setting the Rockwell depth-measuring dial?

7. How would the lack of verticality of the glass tube affect the scleroscope hardness numbers?

8. Discuss the probable effect of (a) surface roughness and (b) an oily surface of the specimen on the hardness numbers by each method.

9. How would compressible material between the specimen and the anvil affect the Rockwell and Brinell hardness numbers? Explain.

10. If a Brinell or Rockwell impression were made close to the edge, causing a side bulge, would the hardness number be greater or less than its true value?

READING

Lysaght, V. E., and A. DeBellis: *Hardness Testing Handbook*, American Chain and Cable Co., 1969.

Westbrook, J. H., and H. Conrad (eds.): *The Science of Hardness Testing and Its Research Applications*, American Society for Metals, Metals Park, Ohio, 1971.

THIRTEEN

IMPACT

13.1 IMPACT

An important type of dynamic loading is that in which the load is applied suddenly, as from the impact of a moving mass. This chapter is concerned with some of the aspects of the behavior of materials under such impact loads.

As the velocity and thus the kinetic energy of a striking body are changed, a transfer of energy must occur; work is done on the parts receiving the blow. The mechanics of impact involve not only the question of stresses induced but also a consideration of energy transfer and of energy absorption and dissipation.

The energy of a blow may be absorbed in a number of ways: through elastic deformation of the members or parts of a system, through plastic deformations in the parts, through hysteresis effects in the parts, through frictional action between parts, and through effects of inertia of moving parts. The effect of an impact load in producing stress depends on the extent to which the energy is expended in causing deformation. In dealing with problems involving impact loading, the predominant way in which the load is to be resisted obviously determines the type of information that is needed.

In the design of many types of structures and machines that must take impact loading, the aim is to provide for the absorption of as much energy as possible through *elastic* action and then to rely on some kind of damping to dissipate it. In such structures the resilience (i.e., the elastic energy capacity) of the material is a significant property, and resilience data derived from static loading may be adequate.

Satisfactory performance of certain types of machine parts, such as parts of percussion drilling equipment, parts of automotive engines and transmissions, parts of railroad equipment, track and buffer devices, highway guardrails, and so on, depends on the toughness of the parts under shock loadings. Although a direct approach to this problem would seem to be the use of tests that involve impact loads, the solution to the problem is not simple. Without doubt the results of impact tests have contributed indirectly to the improved design of certain types of parts, but in general such tests are of only limited signifance in producing basic design data.

In most tests to determine the energy-absorption characteristics of materials under impact loads, the object is to utilize the energy of the blow to cause *rupture* of the test piece. There is thus a distinction to be made between problems that largely involve elastic energy absorption and problems to which data on energy capacity at rupture are pertinent. This difference contributes to a basic limitation to the general applicability of the results of the ordinary impact test.

In some experimental studies of the properties of materials under impact loading, detailed determinations have been made of stress-strain-time relations. Generally, such studies of dynamic properties require very special and complicated experimental procedures that are not adapted to normal testing, although notable contributions have been made to the understanding not only of the impact problem but of the mechanics of plastic strain and fracture. The development of electric strain gages, electronic measuring equipment, and high-speed oscillographs has greatly extended the scope of research into the behavior of materials under very rapidly applied loads.

In some tests limited measurements of strain or deflection under impact loading are made, but in the most commonly used impact tests, the so-called "notched-bar" tests, only the energy required to produce rupture is determined. In these tests one objective is to obtain a relative measure of the tendency to exhibit brittleness with a decrease in temperature, especially as affected by the presence of minor constituents or small variations in composition or structure of a particular metal or other material.

For many uses, knowledge of the overall or macroscopic behavior of a material provides adequate information. However, the failure of materials under some conditions can be adequately explained only in terms of the behavior of their microstructure. An understanding of the behavior of metals in the form of notched bars under impact loading requires a knowledge of the mechanism of fracture as influenced by a polycrystalline structure [19].

The property of a material relating to the work required to cause rupture is termed "toughness." Toughness depends fundamentally on strength and ductility and appears to be independent of the type of loading. It is a fact, however, that the rate at which the energy is absorbed may markedly affect the behavior of a material, and thus different measures of toughness may be obtained from impact loadings than from static loadings.

All materials do not respond in the same way to variations in speed of load application; some materials display what is termed "velocity sensitivity" to a much more marked extent than others. Among striking examples of materials that display radically different behavior under slow- and high-speed loadings are ordinary glass, which is punctured with a fairly clean hole by a high-speed bullet but shatters under slowly applied point loading, and sealing wax, a stick of which breaks as if it were brittle under a sharp blow but slowly sags plastically under its own weight if supported as a beam.

Over the range of low- and medium-carbon steels, the *relative* toughness of a series of steels determined from impact and static tests of *plain* (*unnotched*) *tension* specimens appears to be more or less the same, although the *actual* work required to cause rupture under impact loadings (where the striking velocities are less than some critical value, as is the case with the usual impact tests) probably runs 25 percent greater than the work as obtained from the usual static stress-strain diagram. But the toughness as determined by impact loading is not necessarily greater than that determined by static loading; in the case of chrome-nickel steel, for example, the impact toughness is *less* than the static toughness [89]. However, it has been pointed out by Mann [58] that, for velocities obtained with the ordinary impact machines, good correlation between (unnotched) tension test results under impact and static loading is obtained if the area under the true stress—conventional strain diagram is used to calculate the energy to rupture.

With a given material the toughness does not vary greatly over a considerable range in striking velocity, but above some critical speed (different for various materials) the energy required to rupture a material appears to decrease rapidly with increase in speed [14, 59]. This critical velocity is found to be associated with the rate of propagation of plastic strain and is affected by the length of the piece subjected to impact loading [14, 21].

In addition to the velocity effect, the form of a piece may have a marked effect on its capacity to resist impact loads. At ordinary temperatures a plain bar of ductile metal will not fracture under an impact load in flexure. In order to induce fracture to take place under a single blow, test specimens of a ductile material are notched. The use of a notch causes high localized stress concentrations, restricts the drawing-out action (i.e., artificially tends to reduce ductility), causes most of the energy of rupture to be absorbed in a localized region of the piece, and tends to induce a brittle type of fracture. The tendency of a ductile material to act like a brittle material when broken in the form of a notched specimen is sometimes referred to as "notch sensitivity." Materials that have practically identical properties in static tension tests, or even in impact tension tests when unnotched, sometimes show marked differences in notch sensitivity. It has been appropriately suggested that impact testing and notched-bar testing really belong in different categories [89].

From various illustrations cited above, it appears that all materials do not respond in the same way to impact loadings. An analogous situation was pointed out in connection with the results of hardness tests.

13.2 SCOPE AND APPLICABILITY

In research on the behavior of materials under dynamic loading, many devices and techniques have been used, and many more will probably be developed in attempts to learn the detailed mechanism of deformation and fracture as affected by the many variables of composition, temperature, velocity of loading, and geometry of specimen. We will take a brief look here at the results of this kind of research, in order to explain some of the behavior of materials under test. However, the remainder of this chapter will be devoted primarily to the conduct and use of the more common and more or less standardized impact tests.

The ideal impact test would be one in which all the energy of a blow is transmitted to the test specimen. Actually this ideal is never realized; some energy is always lost through friction, through deformation of the supports and of the striking mass, and through vibration of various parts of the testing machine. In some tests, it is impossible to obtain a truly accurate measure of the energy absorbed by a specimen. Further, the particular values obtained from an impact test depend very much on the form of specimen used. These facts necessitate close attention to standardization of details in any given type of test if concordant results are to be obtained, and usually preclude direct comparisons of results from different types of impact tests. Each type of impact test has its own specialized field of use, and its applicability depends largely on satisfactory correlation with performance under service conditions. In this connection, it may also be observed that the applicability of a test may not necessarily be confined to materials for use in parts that are to be subject to impact.

ASTM's Committee A-3 on steel described the impact tests of steel for production purposes as follows (ASTM A 370-61T):

> An impact test is a dynamic test in which a selected specimen, machined or surface-ground and usually notched, is struck and broken by a single blow in a specially designed testing machine and the energy absorbed in breaking the specimen is measured. The energy values determined are qualitative comparisons on a selected specimen and cannot be converted into energy figures that would serve for engineering design calculations. The notch behavior indicated in an individual test applies only to the specimen size, notch geometry, and testing conditions involved, and cannot be applied to other sizes of specimens and conditions. Minimum impact requirements are generally specified only for quenched and tempered, normalized and tempered or normalized materials, as provided in the appropriate product specifications.

In making an impact test, the load may be applied in flexure, tension, compression, or torsion. Flexural loading is the most common; tensile loading is less common; compressive and torsional loadings are used only in special instances. The impact blow may be delivered through the use of a dropping weight, a swinging pendulum, or a rotating flywheel. Some tests are made so as to rupture the test piece by a single blow; others employ repeated blows. In some tests of the latter type, the repeated blow is of constant magnitude; in others, the "increment-drop" tests, the height of drop of the weight is increased gradually until rupture is induced.

Perhaps the most commonly used impact tests for steels in the United States are the Charpy and the Izod tests, both of which employ the pendulum principle. These tests are made on small notched specimens broken in *flexure*. In the Charpy test, the specimen is supported as a simple beam, and in the Izod test it is supported as a cantilever. In such tests a large part of the energy absorbed is taken up in a region immediately adjacent to the notch, and a brittle type of fracture is often induced. It should be observed that these tests do not, and are not intended to, simulate shock loading in service; they simply give the relative resistance of a particular notched metal specimen to fracture under a particular type of blow. It has been found that the results indicate differences in condition of a metal that are not indicated by other tests. The results appear to be particularly sensitive to variations in the structure of steel as effected by heat-treatment; by certain minor changes in composition that tend to cause "embrittlement," such as variations in the sulfur or phosphorus content; and by various alloying elements. Also, these tests, when made on specimens at low temperatures, have proved useful in indicating whether or not adequate toughness is maintained at those temperatures. While Charpy or Izod tests may not directly predict the ductile or brittle behavior of steel as used in large structural units, they find use as acceptance tests or tests of identity for different lots of the same steel or in choosing between different steels when correlation with reliable service behavior has been established.

Procedures for the Charpy and Izod tests as applied to metals have been standardized (ASTM E 23), and formal specification of impact-strength limits has been made in the case of materials for a number of products such as airplane-engine parts, transmission gears, parts for tractor belts, turbine blading, many types of forgings, and pipe and steel plate for low-temperature service. The Charpy and Izod tests are also used to determine the impact resistance of plastics and electrical insulating materials (ASTM D 256).

Cast iron is not often used in parts that must have high shock resistance, and the results of the static-flexure test may give most of the information needed for estimating the relative energy capacity of cast irons [4, 63]. However, a number of investigations on the impact resistance of cast iron have been made, and a variety of procedures have been employed, including single-blow pendulum tests, increment-drop tests, and repeated-blow tests. A repeated-blow method and both Charpy and Izod test procedures for cast iron have been standardized in ASTM A 327. All these test methods employ flexural loading; impact-tension tests of cast iron do not appear to give reliable results. Unnotched specimens are standard, since they seem to be more satisfactory than notched specimens. Cast iron is not as "notch sensitive" as steel; this may be due to the notchlike effect of the graphite flakes in cast iron, which is not greatly increased by the additional effect of an external notch.

Impact loading is used in tests for the acceptance of a number of metal products such as rails and axles. The American Railway Engineering Association specification for steel rails calls for an increment-drop test (ASTM A 1) by use of a special machine, the details of which are also specified. Likewise, for axles a repeated-drop test is specified in ASTM A 383. Tests of this kind have significance in relation to performance in service.

For wood, the Hatt-Turner test–a flexural impact test of the increment-drop type—is used, and the procedure is standardized (ASTM D 143). In this test the height of drop at which failure occurs is taken as a measure of toughness, but from the data of the test the modulus of elasticity, the proportional limit, and the average elastic resilience may also be found.

Impact tests of metals in *tension* have been made largely for experimental purposes. They have been made as single-blow tests in a drop-weight machine, in a pendulum machine when suitably modified to accommodate a tension specimen, and in a flywheel machine. The impact-tension test affords the opportunity to study the behavior of ductile materials under impact loading without the complications introduced by the use of a groove or notch, although notched tensile specimens have also been used in some tests. The flywheel type of machine, in particular, is capable of providing very high impact velocities; studies concerned with the "transition" or "critical" velocity have been made with this type of machine [14, 21, 59].

A falling weight test of the *compression-impact* type is used for some plastics (ASTM D 3029). At a drop height of 0.66 m the mass of the tup is adjusted until visible damage occurs in 50 percent of the sample. A similar test is used for thermoplastic pipe (ASTM D 2444).

For testing tool steels, a *torsion-impact* test, the Carpenter test, has enjoyed some prominence [50, 51, 52]. The test appears to offer a method for investigating and controlling optimum heat-treatment conditions of products such as drills, taps, and rock-drill parts.

13.3 TEST PRINCIPLES

The effect of a blow depends in general on the mass of the parts receiving the blow as well as on the energy and mass of the striking body. Items that require standardization are the foundation, anvil, specimen supports, specimen, and the striking mass and its velocity.

The principal features of a *single-blow pendulum* impact machine are (1) a moving mass whose kinetic energy is great enough to cause rupture of the test specimen placed in its path, (2) an anvil and a support on which the specimen is placed to receive the blow, and (3) a means for measuring the residual energy of the moving mass after the specimen has been broken.

The pendulum should be supported so that it falls in a vertical plane without possibility of lateral play or lateral restraint, and the bearings should be such that friction is small. The pendulum should be sturdy enough so that excessive vibrations do not cause serious variations in the results. The release mechanism should not influence the free-fall movement of the pendulum by causing any binding, accelerating, or vibrating effects.

The anvil should be heavy enough in relation to the energy of the blow so that an undue amount of energy is not lost by deformation or vibration. The device for supporting the specimen should be such that the specimen is held accurately in position prior to the instant of impact.

The striking edge of the pendulum should coincide with a vertical line through the center of the rotation when the pendulum is hanging unrestrained. The line of action of the reactive force between the specimen and the pendulum should pass through the center of percussion at the instant of impact. It is considered desirable that the center of percussion be as close as possible to the striking edge.

The machine should be so constructed that the space between the anvils does not decrease in width in the direction of motion of the pendulum. In a standard machine the width should increase, in fact, to prevent drag between the specimen and the anvils. This is illustrated in Fig. 13.8*a* (see Art. 13.6).

To indicate the swing of the pendulum of the Charpy- and Izod-type machines after the specimen has been broken, an arm attached to the pendulum moves a "friction pointer" over an arc graduated in angular or energy units. The friction pointer, the axis of rotation of which coincides with that of the pendulum, is simply an arm that can rotate on a pin bearing of such tightness that the pointer is prevented from changing position under its own weight. The bearing pressure should be adjusted to a minimum that will prevent overcarrying or dropping down of the pointer. At the beginning of each test, this pointer is positioned to make contact with the pendulum and to indicate the proper reading when the latter is hanging vertically.

In the Oxford machine the anvil is mounted on a pendulum that can swing independently of the striking pendulum and thus measure the energy transmitted to the anvil [89]. The energy relations are obtained from the motions of both pendulums.

In *drop-weight* machines the principal features are the moving mass of known kinetic energy, but not necessarily of such magnitude that rupture is caused by a single blow, and an anvil. Many drop-weight machines do not have a device for measuring the residual kinetic energy of the weight, hammer, or tup after it has ruptured the specimen (one exception is the Fremont machine). In an increment-drop test, in which the height of drop is increased gradually, an approximate measure of the least energy load required to cause rupture is obtained. In some machines the variation in velocity of the tup before and after impact may be found from electronic time-displacement data, from which the energy relations may be calculated. In repeated-blow machines there must be some provision for keeping flexure specimens from being displaced without restraining or fixing the ends.

In drop-weight machines it is necessary that the axes of the tup and the guides be vertical and in alignment and that the supports and anvil be so placed that the blow can be delivered squarely to the specimen. Friction in the guides should be minimized by keeping them free from grease or rust; they may be lubricated with powdered graphite.

The general requirements for *flywheel* machines are similar to those for pendulum machines, although the mechanical details are, of course, different. The Guillery machine has a fixed anvil, and the energy of the blow is determined from the change in velocity of rotation of the flywheel, before and after impact [7]. In the Mann-Haskell machine, the anvil is carried on a pendulum, the displacement of which is determined in order to obtain the energy of rupture [59]. In both these machines the blow is delivered through the use of a retractible striker arm, which is carried flush with the flywheel until the wheel is brought up to speed. By means of a tripping device the striker can be released, and it is then forced to its projecting position by centrifugal action.

For the purpose at hand, detailed discussion of impact-test procedures is confined to the Charpy and Izod tests of metals and the Hatt-Turner test of wood. These may be considered fairly representative of impact-test work.

13.4 APPARATUS

The following discussion concerns two basic types of machines, namely the pendulum and drop-weight devices. Of these, the former type is definitely more common than the latter. The principles of both were described in the preceding article. In each case a mass is released from rest some distance above the impact point and strikes the specimen.

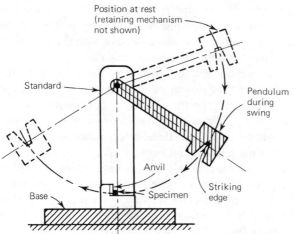

Position at rest
(retaining mechanism
not shown)

Standard

Pendulum
during
swing

Anvil

Base

Specimen

Striking
edge

Figure 13.1 Diagram of Charpy impact machine.

Figure 13.2 Universal pendulum inpact tester.
(*Courtesy of Tinius Olsen Testing Machine Co.*)

The kinetic energy of the tup or head at the time of impact is $mv^2/2$, equal to the relative potential energy of the tup before its release (mgh), where m is the mass of the tup and v ($=\sqrt{2gh}$) is its tangential velocity at impact (g is the gravitational acceleration of 9.806 m/s² and h is the height of the drop). Energy losses due to friction generally amount to less than 1 percent in quality machines. Note that the drop height determines the velocity and the drop height and mass jointly determine the energy.

Pendulum machines can be used to perform Charpy, Izod, or tension-impact tests. Some machines are made for only one or two of these tests, such as the Charpy impact machine shown diagramatically in Fig. 13.1. More commonly, modern machines, like the one shown in Fig. 13.2, have heads and specimen supports that allow them to be used for any of these tests and for others as well.

For different tests and materials, different levels of kinetic energy and tangential velocity at the time of impact are required. These can be changed by adjusting one or more of three variables: the radius from the center of mass to the center of rotation of the pendulum, the mass of the head, and the position of the pendulum arm before release.

If a particular machine is to be made adjustable, it would be impractical to change the length of the arm. Changing the mass of the head, however, or the initial position of the arm is feasible and some machines feature this flexibility.

A typical machine, like the one in Fig. 13.2, might have a head (with arm) weighing 27.25 kg and a drop of 1.34 m; the potential energy is then (27.25) (9.806) (1.34) = 358 J.† The impact velocity would be $\sqrt{2(9.806)(1.34)}$ = 5.126 m/s or slightly less because of frictional losses, which generally are well below 1 percent.

Special features allow the energy to be adjusted down to about 0.2 J and the velocity to 0.1 m/s.

The simplest type of instrumentation consists of a pointer that indicates on a circular scale the position of the pendulum. The scale is usually calibrated to indicate potential energy, rather than angular rise. The pointer has just enough friction to remain in position after the test without being affected by either gravity or its own momentum. This friction, of course, causes a slight energy loss in the system, but usually it amounts to only a fraction of 1 percent.

In addition, a tester may be equipped with electronic instrumentation. Electric strain gages mounted in the head sense the load during impact, which generally lasts from 0.1 to 4.0 ms. The velocity of the tup during this interval is measured by a light beam and photosensor combination, which also triggers the data-acquisition system. A microprocessor computes travel distance, energy, and other output desired. The resulting functions may be displayed on a storage oscilloscope and photographed or be printed on a plotter. Commonly load and energy are plotted vs. time, as illustrated in Fig. 13.3. The processor system may also be capable of storing and recalling the results for further analysis and printout.

---

†1 J = 0.738 ft-lbf.

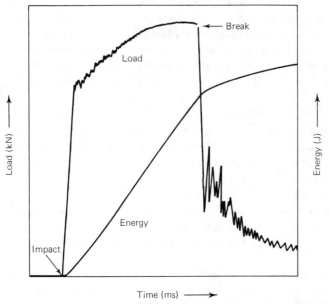

Figure 13.3 Plot of impact functions. (*Courtesy of Effects Technology, Inc.*)

Drop-weight testers like the one shown in Fig. 13.4 are principally used to conduct flexure tests such as the Charpy test and the Hatt-Turner test for wood, but they may accommodate other specimens as well.

The tup is raised by an electric motor to any desired, and possibly predetermined, release position. Machines with drop heights up to about 2.5 m are available, but 0.6 m or so is more usual. Thus velocities may be as high as 7 m/s but are typically around 3 m/s.

The tup masses usually range from about 10 to 200 kg, but optional equipment can increase the mass up to perhaps 750 kg. The energy achieved may be as high as 18 kJ, although more typical values are from 400 to 500 J.

Hence it can be seen that drop heights and velocities of drop-weight testers tend to be similar to or less than those of pendulum machines, but tup masses and energy levels are greater, sometimes much greater.

In contrast to pendulum-type tests, the drop-weight tests do not necessarily break a specimen. There may be an automatic rebound brake to limit the impact to a single blow. Testers may include other special features, including alarm systems that alert the operator to unsafe conditions.

Minimum instrumentation consists of a hollow drum free to rotate about its vertical axis and mounted on the base near the guide columns. A pencil carried by the tup is pressed by a spring against a piece of paper secured to the surface of the drum, giving a graphic record from which can be computed the height of drop of the tup and the corresponding deflection of the specimen.

Modern testers, however, feature instrumented tups and associated analysis and recording systems like those described for the pendulum machines.

Figure 13.4 Drop-weight impact system. (*Courtesy of Tinius Olsen Testing Machine Co.*)

13.5 SPECIMENS

Since the tests are comparative, providing relative rather than absolute results, specimens must be carefully specified. Even if standard specimens are not available or used, the dimensions for a test series or program must be identical.

The standard metal specimen for the Charpy test is a square prism 10 by 10 by 55 mm, notched as shown in Fig. 13.5*a* (ASTM E 23). Other sizes are used in special cases. In many specifications a keyhole notch or a U notch is required, as shown in Fig. 13.5*b* and *c*. For the test, the specimen is arranged as a simple beam with a span of 40 mm, the notch being on the tension side.

For the Izod test, prismatic specimens 10 by 10 by 75 mm, notched as illustrated in Fig. 13.6, are clamped to act as vertical cantilevers, with the notch on the tension side.

In the Hatt-Turner test for wood, clear specimens in the shape of a prism 50 by 50 by 760 mm[†] are used. They are arranged as simple beams over a 700-mm span, arranged so that the tangential surface nearest the pith will be face up. Figure 13.7 shows a specimen.

[†]Or 2 by 2 by 30 in (28-in span).

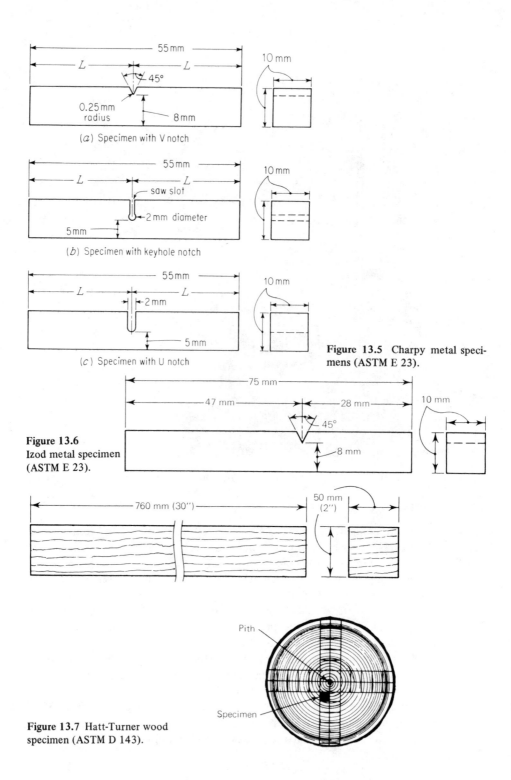

(a) Specimen with V notch

(b) Specimen with keyhole notch

(c) Specimen with U notch

Figure 13.5 Charpy metal specimens (ASTM E 23).

Figure 13.6 Izod metal specimen (ASTM E 23).

Figure 13.7 Hatt-Turner wood specimen (ASTM D 143).

13.6 PROCEDURE

This discussion of procedure must be prefaced with a warning. Most impact testers can easily break a finger-sized steel specimen in a millisecond. It is easy to imagine what harm they can do to an actual finger or, for that matter, to an arm or a head! Extreme care is needed when working with any test apparatus, but particularly with an impact machine.

To prepare for a test, the dimensions of the specimen should be measured to an accuracy of 0.1 percent. This applies especially to the dimensions at the notch, if any.

If a multipurpose machine is used, the anvil and the striker are installed corresponding to the test planned: Charpy, Izod, and impact-tension. The bolts must be tight and the machine properly leveled and generally in good working order.

If the drop height is adjustable, which is the case for some pendulum machines and all drop-weight machines, the level that will give the tup an impact velocity appropriate to the specimen must be estimated. To obtain the desired impact energy, the mass of the tup is adjusted if possible. In the absence of standards or instructions, this may require some experimentation.

The pointer is properly set. If electronic instrumentation is to be used, it should be checked, connected, and turned on.

Next, with the tup still in the lowered position, the specimen is inserted. For the Charpy test, the specimen is arranged as shown in Fig. 13.8a, with the notch on the side away from the striking edge. An Izod specimen is arranged as shown in Fig. 13.8b, with the notch toward the striking edge.

An impact-tension specimen is attached to the rear of the tup, and the hammer block is attached to the specimen.

For the Hatt-Turner test, a low-capacity drop-weight machine is used. The first

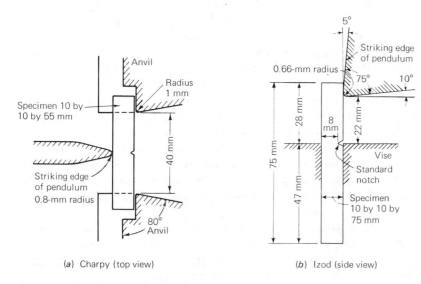

(a) Charpy (top view)   (b) Izod (side view)

Figure 13.8 Specimen arrangements.

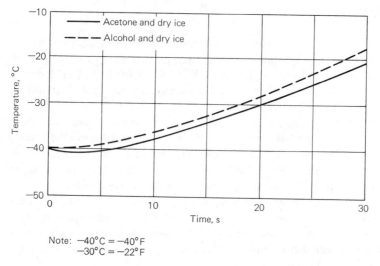

Figure 13.9 Temperature change in steel specimens when removed from bath at −40°C. (*From Driscoll, Ref. 20.*)

drop is 1 in (25 mm) and succeeding drops are increased by 1-in (25-mm) increments until a height of 10 in (250 mm) is reached, after which the drops are increased by 2-in (50-mm) increments until either complete rupture or a 6-in (150-mm) deflection is reached.

Tests to determine the impact resistance of metals at low temperatures are commonly conducted by immersing the specimens in some cool liquid in a widemouthed vacuum jar, with at least 2 to 3 cm of liquid above and below the specimens. For temperatures from ambient to −78°C, the liquid is usually alcohol or acetone, cooled to the desired temperature by the addition of small lumps of dry ice. For lower temperatures the cooling agent is usually liquid nitrogen (−195°C), and the liquid for immersion is usually alcohol to −123°C, isopentane to −157°C, and liquid nitrogen itself to −195°C.

Thermometers suitable for determining the temperature of the coolant are the mercurial type to −40°C, alcohol or bimetallic types to −100°C, and copper-constantan thermocouples or pentane-type thermometers for lower temperatures.

Specimens should be held at temperature for 15 min and the bath temperature should be held constant within +0°, −2°C during the last 5 min before testing. The test should be completed within 5 s after removing the specimen from the coolant. As shown in Fig. 13.9, at a test temperature of −40°C no appreciable temperature changes occur within 5 s when alcohol is used.

If available, a thermocouple sensor with remote readout can be attached to the specimen before it is immersed in the bath. The cooled specimen can then be placed into the tester at a temperature somewhat lower than the desired one and the test performed as soon as the specimen warms up to the test temperature.

13.7 OBSERVATIONS

For impact tests, the pertinent facts to be noted and recorded include the type, size, and temperature of the specimen; the type of device used; the mass of the tup; and the drop height. The impact velocity is computed from the height.

In the case of low-temperature tests, the bath temperature may be determined, as well as the duration of immersion and the elapsed time from removal of the specimen to impact. Then the actual test temperature can be estimated by means of diagrams similar to Fig. 13.9. If the thermocouple is used, the specimen temperature can be determined directly.

In any case, impact-test results of many steels are very sensitive to temperature changes within the normal atmospheric range. For this reason the actual temperature of test may be quite important and should be reported in conjunction with all test results; preferably the transition temperature range for the given steel should be determined.

After the test, the specimen should be examined. The shape, texture, and inclination of the fracture surface, as well as its relation to the notch, are important. A sketch or photograph will document the fracture.

In *pendulum tests*, calculation of the energy required for fracture is of primary importance. From Fig. 13.10 it can be seen that the pendulum's potential energy before release (point A) is mga. After release, the potential energy decreases and the kinetic energy increases, until just before impact (point B) the former is zero and the latter maximum. At B the amount of energy necessary to fracture the specimen is dissipated. As the pendulum continues its swing, the remaining kinetic energy is again converted to potential energy, the process being complete when the pendulum reaches its farthest excursion at C, where the potential energy is mgb. The difference between the potential energies at A and C is the fracture energy, which is equal to $mg(a - b)$. In most testers, this is the value indicated by the pointer.

If the scale is calibrated in energy units, it should be read to the nearest joule in a

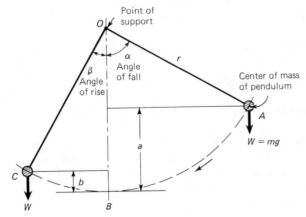

Figure 13.10 Space relations for pendulum machine.

normal machine. If the scale is in degrees, readings should be estimated to $0.2°$. The heights a and b equal $r(1 - \cos \alpha)$ and $r(1 - \cos \beta)$ respectively. If m is given in kg, g as 9.806 m/s^2, and a and b in m, the energies will be expressed in joules (J), without reference to the volume of material involved.

Energy losses can mainly be attributed to four causes:

1. Friction within the pivot bearing, which is greatest when the pendulum is at the nadir, the total force then being greatest

2. Air resistance (windage) of the pendulum, which is a function of velocity and therefore also greatest at the nadir

3. The frictional resistance of the pointer, which is approximately constant

4. The kinetic energy imparted to the specimen, or the part of the specimen broken off, which depends on its mass and velocity

Normally these losses total less than 1 percent and in some tests can be ignored.

For precision work, even the small errors present can be substantially reduced by means of a bit of experimentation and estimation. The magnitude of bearing friction and windage can be estimated by letting the pendulum swing freely for several cycles and observing the energy losses; assuming a constant energy loss rate per unit angle will yield a fairly accurate correction. The effect of pointer friction can be substantially reduced by presetting the pointer to a position only slightly short of the expected test value. The kinetic energy of the specimen can be closely estimated by determining its mass and assuming its velocity to equal that of the pendulum after rupture of the specimen, the result being another correction to the apparent fracture energy.

In *drop-weight tests*, such as the Hatt-Turner, the impact energy is of primary importance, especially the amount that breaks the specimen. As before, the kinetic energy of the tup equals the potential energy before release, less minor losses due to friction and air drag.

In cases that do not result in failure of the specimen, the deflections are recorded and used to compute the proportional limit, modulus of elasticity, and resilience, as explained in ASTM D 143.

If electronic instrumentation is available, much more detailed test records and analytical results are possible. Normally, graphical load-time curves and energy-time curves are obtained, which allow chronological study of impact behavior during the elastic and inelastic phases. In general, all quantities derived from force, time, and distance may be measured or calculated. Further, statistical analysis routines, such as curve fitting, may be possible.

The data of major interest include the impact velocity and energy, the slope of the load-deflection curve in the elastic range, elapsed times, the maximum and fracture loads, and the deflections and energies during the various phases of loading. Some modern data-processing systems allow all these results to be printed out, as well as summary tabulations for entire test series.

13.8 EFFECTS OF VARIABLES

The results of impact tests of metals made with various machines differ, owing to (1) variation in amounts of energy transformed at impact into vibrations of parts of the machines, (2) variations in striking velocity of the hammers, and (3) size and form of the specimen.

Provided the same form of notch is used, the results from ordinary designs of Charpy and Izod machines are fairly comparable, although the Charpy results tend to be somewhat higher than the Izod results. The tougher the material, the greater the spread appears to be [7].

Velocity Over the range in velocity developed in most ordinary machines (about 3 to 5 m/s),[†] the velocity at impact does not appear to appreciably affect the results. However, experiments conducted with a machine capable of developing velocities up to 300 m/s indicate that, above some critical velocity, impact resistance appears to decrease markedly. The magnitude of the critical velocity and the rate of decline in impact resistance with increase in velocity differ for different metals. In general, the critical velocity is much less for annealed steels than for the same steels in a hardened condition [14, 21, 59]. The range of velocities developed in the usual Charpy and Izod tests appears to be well below the critical velocity for ordinary carbon steels; some alloy steels appear to have critical velocities near, if not within, the range of velocity of the ordinary Charpy and Izod tests [89].

Specimens In some cases it is not possible to obtain a specimen of standard width from the stock available. Decreasing either the width or the depth of the specimen decreases the volume of metal subject to distortion and thereby tends to decrease the energy absorption when breaking the specimen. However, any decrease in size also tends to decrease the degree of restraint and by reducing the tendency to cause brittle fracture may increase the amount of energy absorbed. This is particularly true where a standard specimen shows a brittle fracture, and a narrower specimen may require more energy for rupture than one of standard width. Actual tests have shown that the Charpy values for specimens at room temperature, having widths of one-fourth and two-thirds that of the standard specimen, are roughly proportional to the width of the subsize specimen, but at low temperatures the narrow specimens of some steels may show up to three times the total energy resistance of a standard specimen [32].

Test results for mild steel as presented in Table 13.1 for a notch depth of 5 mm and a root radius of 0.67 mm show that the angle of notch does not appreciably affect results until it has exceeded 60°.

The sharpness of the root of the notch may have an appreciable influence on the energy of rupture of the test piece [7, 43]. As shown in Table 13.2, the energy of rupture decreases as the sharpness of the notch increases owing to the increase in stress concentration. It has been shown that the sharper the notch, the greater the difference

[†]1 m/s = 3.28 ft/s.

Table 13.1 Effect of angle of notch on energy of rupture of mild steel[†]

Angle of notch, °	Sketch of specimen	Charpy impact value	
		J	ft-lb
0		30.0	22.1
30		33.1	24.4
60		31.3	23.1
90		35.1	25.9
120		56.7	41.8
150		89.8	66.2
180		85.6	63.1

[†]J. J. Thomas, "The Charpy Impact Test on Heat-treated Steels," *Proc. ASTM*, vol. 15, pt. II, 1915.

between test results for brittle and tough materials, and that the tougher the material, the less the effect of the root radius of the notch [7]. Since a dead-sharp notch is difficult to produce, a 0.25-mm radius has been adopted as standard for the V notch.

The use of a shallow notch in place of a deep notch gives a greater spread of impact values for tough and brittle metals; in addition, the shallow notch appears to be more sensitive to differences in either composition or temperature (see Fig. 13.13). For this reason, in Charpy impact testing some prefer the 2-mm V notch instead of the keyhole notch having a 5-mm depth, although the latter is commonly used for tests at low temperatures.

Temperature In contrast to the relatively small effect of temperature on the static strength and ductility of metals, at least within the atmospheric range, temperature has a very marked effect on the impact resistance of notched bars. Figure 13.11 illus-

Table 13.2 Effect of root radius of 45° V notch on energy of rupture of 0.65 percent carbon steel[†]

Root radius of notch, 2 mm deep, mm	Charpy impact value	
	J	ft-lb
Sharp	5.4	4.0
0.17	9.5	6.9
0.34	11.3	8.3
0.68	18.6	13.7

[†]R. G. Batson, and J. H. Hyde, *Mechanical Testing*, Vol. I: *Testing of Materials of Construction*, Dutton, New York (Chapman & Hall, London), 1922.

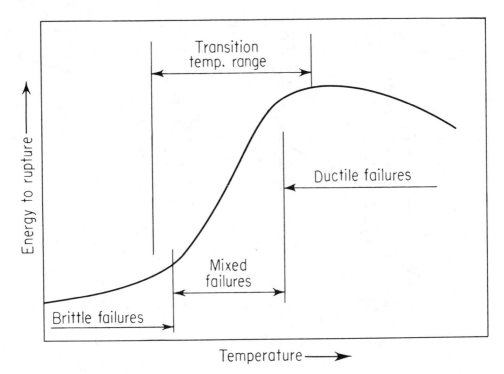

Figure 13.11 Variation with temperature of energy to rupture in impact tests of metals.

trates in generalized form the nature of the variation of the energy to produce rupture in the impact test over a considerable range in temperature. For a particular metal and type of test, below some critical temperature the failures are brittle, with low energy absorption. Above some critical temperature the failures are ductile, with energy absorption that may be many times that in the brittle-fracture range. Between these temperatures is what has been termed the "transition-temperature range," where the character of the fracture may be mixed. The transition-temperature range may be very short or abrupt for some steels (see Fig. 13.12) or may extend over some appreciable range for others (see Fig. 13.13) [16, 41, 43]. A significant fact to be noted is that in or near this critical-temperature range, a variation in testing temperature of only a few degrees may cause very appreciable differences in impact resistance. Above the transition-temperature range, the impact resistance decreases more or less slowly with increase in temperature until a temperature of perhaps 650°C is reached.

With the standard V notch the critical range for many steels appears to occur between the freezing point and room temperature; in some metals it may extend to temperatures well below the freezing point. The critical range appears to be higher with more highly "notch sensitive" steels and with test pieces having sharper or deeper notches.

Coarse grain size, strain hardening, and certain embrittling elements tend to raise the transition range of temperature, whereas fine grain size, ductilizing and refining

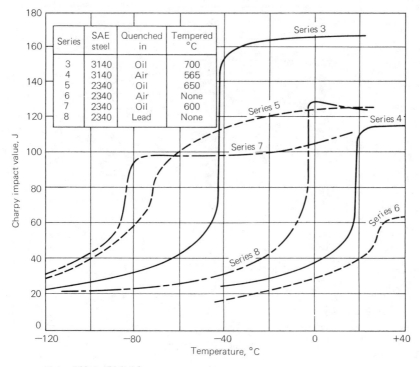

Series	SAE steel	Quenched in	Tempered °C
3	3140	Oil	700
4	3140	Air	565
5	2340	Oil	650
6	2340	Air	None
7	2340	Oil	600
8	2340	Lead	None

Note: 100 J = 74 ft-lbf.

Figure 13.12 Variation in transition-temperature range for steel in the Charpy test. (*From Dolan and Yen, Ref. 19.*)

heat treatments, and the addition of certain alloying elements tend to enhance notch toughness, even at fairly low temperatures [89]. Grain size resulting from the work done in forming the steel and from the heat-treatment operations is an important factor in determining impact resistance. Steels having a fine-grained structure may be expected to show superior impact values, especially at low temperatures. In Fig. 13.13 is shown the effect of temperatures as low as –120°C and of the type of notch on the Charpy impact resistance of a carbon steel and a nickel steel of about the same carbon content (0.18 percent), both steels being normalized and drawn. The general beneficial effect of nickel on impact resistance at low temperatures is apparent.

Failure Two modes of fracture may take place in a metallic or crystalline material—a separation fracture and a shear or sliding fracture (see Art. 2.5). In the separation fracture a cleavage may occur within the crystals along critical planes of the crystal lattices, or it may occur by separation along crystal boundaries or through intercrystalline material. The shear type of fracture results from extreme detrusion; it may take place across crystals or in the intercrystalline material.

The final fracture may occur after little or much plastic deformation. When little or practically no plastic deformation occurs before rupture, the failure is said to be "brittle." It is generally considered that a brittle failure is accompanied by a separation fracture (although this is not necessarily the case). In a brittle failure with a separation

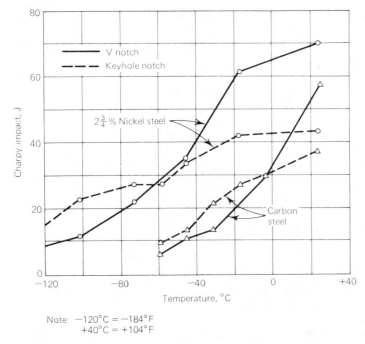

Note: $-120°C = -184°F$
$+40°C = +104°F$

Figure 13.13 Influence of low temperature and type of notch on impact strength of two steels. (*From Armstrong and Gagnebin, Ref. 3.*)

fracture, the fractured surface usually has a granular or crystalline appearance; the energy absorbed is relatively small because little work has been done in producing plastic strain.

When considerable plastic strain takes place before rupture occurs, the failure is said to be "ductile." While the fracture that terminates in a ductile failure is usually a shear fracture, this is not necessarily the case. The fractured surface following ductile action and shear fracture often has a fine, silky appearance; the energy absorbed in a ductile failure is relatively high, sometimes very high, because of plastic strain that takes place throughout the specimen.

Factors that inhibit plastic flow and hence tend to cause a brittle type of fracture in otherwise ductile metals are (1) a state of stress that holds the shearing stresses to a small magnitude relative to the tensile stresses, (2) a localization of the deformation by the presence of discontinuities or notches, (3) a very rapid application of stress (or high rate of strain), (4) lowered temperatures, and (5) certain types of structure or composition.

In the notched-bar impact test, the transition from ductile to brittle failure is usually brought into the atmospheric temperature range because of the effect of the notch and of the rapid strain rate in reducing plastic deformation. As the condition of the material at the base of the notch becomes a critical factor in determining the type of failure, the transition temperature as well as the relative amount of energy absorbed are indexes of the effect of composition and structure on toughness or brittleness.

PROBLEM 13 Impact-Flexure (Charpy) Test of Steel

Object: To study an impact-testing machine of the pendulum type and to determine the relative impact resistance of a steel, in the form of notched-bar Charpy specimens.
Preparatory reading: ASTM E 23.
Specimens: Standard notched flexure specimens (Charpy).
Special apparatus: Pendulum impact machine, low-temperature bath.
Procedure:
1. Measure the lateral dimensions of the specimens at a full section and at the notch. Weigh the specimens and place some of them into the bath.
2. If necessary, adjust the anvils of the machine by means of the template so that they are centered with respect to the pendulum and the clear distance between them is 40 mm.
3. Note the mass of the pendulum and the radius of its center of mass (these numbers may be stamped on the pendulum, although for some machines energy values are shown directly on the graduated scale). With the pendulum freely suspended, adjust the lever actuating the friction pointer (if necessary).
4. If instrumentation is available, prepare and connect it.
5. Place a specimen accurately in position on the anvils. (The notch should be on the side of the specimen farthest from the striking edge of the pendulum and directly in line with it.) Raise the pendulum to its upper position and let it fall to rupture the specimen.
6. Record the angle of rise of the pendulum or observe the energy to rupture.
7. Note the shape of the fractured surface, its inclination with respect to the axis of the piece, its texture, and its relation with respect to the notch.
8. Repeat with other specimens.
9. Without a specimen in the machine obtain data for determining friction losses, and estimate the kinetic energy of the broken specimens.
Report:
1. Calculate the effective angle of fall of the pendulum and compute the corrected angle of rise (if the scale does not give energy values directly). Make corrections for friction and windage, as well as for the energy used in imparting a velocity to the specimen.
2. Compute the energy of rupture of each specimen.
3. Prepare a tabulation of results, which should minimally consist of
    a. Type, model, and capacity of machine used
    b. Type and size of specimen used
    c. Maximum linear velocity of the hammer, m/s
    d. Energy losses, J
    e. Energy of blow with which specimen was struck, J
    f. Energy absorbed by specimen in breaking, J
    g. Temperature of specimen, °C
    h. Appearance of fractured surface
    i. Number of specimens tested and number failing to break
Any additional data provided by instrumentation should, of course, also be reported.

DISCUSSION

1. Discuss the significance and advantages of impact tests compared with static tests.
2. Why are impact-flexure specimens notched?
3. Discuss the effect of the following factors on the results of impact tests:
    a. Characteristics of the notch
    b. Velocity of the hammer
    c. Type of testing machine
    d. Temperature of specimen
4. Why is it necessary that impact tests be standardized?

5. Discuss the relative impact tensile resistance of plain and notched specimens of the same grade of steel that have the same breaking area. Is the relation affected by the characteristics of the steel?
6. Can the absolute impact resistance of a specimen be determined by the test procedure used? Explain.
7. Discuss the relative advantages of long vs. short bolts of a given size subject to impact loading.
8. What physical property of a material is determined by means of an impact test?
9. Devise a machine that can impart torsional impact to a shaft.
10. What are the relative advantages of drop-weight and pendulum testers?

READING

Impact Testing of Metals, ASTM Philadelphia, 1970.
Instrumented Impact Testing, ASTM, Philadelphia, 1973.
Johnson, W.: *Impact Strength of Materials*, E. Arnold, London, 1972.

FOURTEEN

FATIGUE

14.1 FATIGUE

Most structural assemblages are subject to variation in applied loads, causing fluctuations in stresses in the parts. If the fluctuating stresses are large enough, even though the maximum applied stress may be considerably less than the static strength of the material, failure may occur when the stress is repeated often enough. A failure induced in this manner is called a *fatigue* failure.

Metals are composed of aggregations of small crystals with random orientations. The crystals themselves are usually nonisotropic. Experiments indicate that some crystals in a stressed piece of metal reach their limit of elastic action sooner than others, owing, no doubt, to their unfavorable orientation, which permits slip to occur. Also, the distribution of stress from crystal to crystal within a piece of stressed metal is probably nonuniform, and when a piece is subjected to cyclic stress variation, the constituent particles tend to move slightly with respect to one another. This movement finally weakens some minute element to such an extent that it ruptures. In the zone of failure a stress concentration develops, and with successive repetitions of stress the fracture spreads from this nucleus across the entire section. For this reason fatigue failures are often referred to as "progressive fractures."

High localized stress is also developed at abrupt changes in cross section, at the base of surface scratches, at the root of a screw thread, at the edge of small inclusions of foreign substances, and at a minute blowhole or similar internal defect. These are typical conditions that accentuate the susceptibility to failure by fatigue. It should be noted, too, that failure of machine parts is often the result of a combination of fatigue and damage by occasional overstrain.

The relative movement of the elements of minute steel crystals, when under stress, was first observed by Ewing and Rosenhain in 1899. The movement became evident as parallel lines, called "slip lines," across the face of individual crystal grains microscope when illuminated by oblique lighting. Four years later, Ewing and Humphrey observed that the slip lines developed in steel by subjecting it to repeated cycles of stress would develop into microscopic cracks that in turn spread and cause failure of the piece [31].

Fatigue failures occur suddenly without any appreciable deformation, and the fracture is coarsely crystalline as in the case of a static failure of cast iron or brittle steel. The appearance of the fracture and the characteristic suddenness with which failure occurs have led to a theory of cold crystallization of metal—there are still those who refer to metal that has failed by fatigue as having become "crystallized." Many experiments have shown that this idea is incorrect, although it is fostered by the fact that sudden failures display the crystalline structure of a metal, which, in the case of a ductile failure, as in a static test, is disguised by the distortion of the piece owing to its ductility.

Metals crystallize when they solidify from the liquid state, and there is no change in their internal structure due to the action of repeated stresses. The large facets in a fatigue fracture are formed while the deterioration is in progress by a continuous cleavage plane extending through two or more adjacent crystals, which tends to exaggerate the apparent size of the crystals on the surface of the break. However, a microscopic examination of the metal behind a fatigue fracture shows no change in its structure or increase in size of the individual crystals. The idea of "crystallization" undoubtedly arose from the fact that many parts, ruptured under repeated loading, showed a coarsely crystalline fracture that may have been due to overheating, defective chemical composition, or some maltreatment in fabrication. The parts broke in many cases because these defects made them particularly weak in resisting repeated stresses.

Once a crack has formed in a part, even static loads producing average tensile stresses well below the material's nominal strength may produce fracture, especially in relatively brittle metals. The reason lies in the formation of high stress concentrations at the leading edge of the crack. "Fracture mechanics" investigations have shown that the fracture toughness of a material at a given temperature is proportional to a stress level and to the square root of a crack dimension. The fracture toughness can thus be expressed by a single parameter, the critical stress intensity factor K, which has the units $MPa \cdot m^{1/2}$. It can be determined by growing a crack in a notched specimen subjected to cyclic loading, measuring the crack, and then fracturing the specimen in a static tension test. A subscript denoting the loading and stress conditions is attached to K. Terms relating to fracture testing are defined by ASTM E 616.

Two methods of designating the nature of stress variations are (1) a statement of the numerically maximum stress, together with the ratio of the minimum to the maximum stress, called the *range ratio;* and (2) a statement of the mean value of the fluctuating stress, together with the alternating stress that must be superimposed on the mean stress to produce the given variation in stress conditions. A classification of types of "repeated" stresses is given in Table 14.1. In addition to designating the degree of stress *variation*, the *kind* of stress (tension, compression, or shear) must also be stated for complete definition of a stress condition. The stresses may be caused by axial, shearing, torsional, or flexural loadings or by combinations of them.

Table 14.1 Classification of types of repeated stresses[†]

In stating numerical values of stresses, the kind of stress should always be designated as tension, compression, or shear. The kind of loading should also be designated as axial, torsional, direct shear, or bending

Type of stress variation		Range-ratio nomenclature		Mean-stress nomenclature	
Description	Diagram	Maximum stress	Range ratio	Mean stress	Alternating stress
Steady stress, σ_1		σ_1	$\dfrac{\sigma_1}{\sigma_1} = 1.0$	σ_m	0
Pulsating stress, between σ_1 and σ_2		σ_1	$0 < \dfrac{\sigma_2}{\sigma_1} < 1$	σ_m	$\pm \sigma_a$
Pulsating stress, between σ_1 and 0		σ_1	$\dfrac{0}{\sigma_1} = 0$	σ_m	$\pm \sigma_a$
Partly reversed stress, between σ_1 and σ_2, where $-\sigma_1 < \sigma_2 < 0$		σ_1	$-1 < \dfrac{\sigma_2}{\sigma_1} < 0$	σ_m	$\pm \sigma_a$
Completely reversed stress, between σ_1 and σ_2, where $\sigma_2 = -\sigma_1$	One cycle	σ_1	$\dfrac{\sigma_2}{\sigma_1} = -1.0$	0	$\pm \sigma_a = \sigma_1$

Note: $\sigma_m = (\sigma_1 + \sigma_2)/2$ and $\sigma_a = (\sigma_1 - \sigma_2)/2$, with due regard to signs.

[†]"Report of the ASTM Research Committee on Fatigue of Metals," *Proc. ASTM*, vol. 37, pt. I, 1937.

For determinations of the fatigue characteristics of metals, one type of repeated loading is completely reversed bending.

The stress at which a metal fails by fatigue after a certain number of cycles is termed the *fatigue strength*. It has been found that for most materials there is a limiting stress below which a load may be repeatedly applied an indefinitely large number of times without causing failure. This limiting stress is called the endurance limit or *fatigue limit*. Its magnitude depends on the kind of stress variation to which the material is subjected. Unless qualified, the fatigue limit is usually understood to be that for completely reversed stress fatigue. For most constructional materials, the fatigue limit in completely reversed bending varies between about 0.2 and 0.6 of the static strength, although for a given kind of material the ratio of fatigue limit to static strength, called the endurance ratio or fatigue ratio, will range between narrower limits.

From a test series, a record may be produced that relates the number of cycles N to which the specimen had been subjected before failure occurred to the fatigue

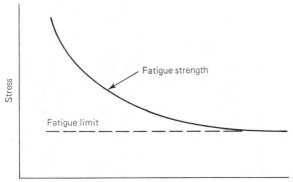

Figure 14.1 Schematic S-N curve.

strength S. These data are usually plotted in the form of an S-N curve, as illustrated by Fig. 14.1. The number of cycles is commonly plotted to a logarithmic scale, and sometimes the fatigue strength as well.

Simultaneous corrosion and repeated stress is called *corrosion fatigue*. Cycles of repeated stress and simultaneous corrosion by so mild an agent as fresh water may reduce the fatigue strength of a metal to a value below one-half its value for fatigue tests in air. Corrosion-resistant metals are somewhat superior in their resistance to corrosion fatigue. The actual surface corrosion of a specimen that fails by corrosion fatigue seems to consist in the mechanical breaking down by repeated stress of the more or less protective film that is formed as a result of reaction between the metal and the corroding agent, and consequent progressive spread of corrosion pits followed by the formation of a spreading fatigue crack. The time of exposure to the corroding agent plays an important part in corrosion fatigue because deeper pitting and consequent higher stress concentrations result from longer periods of exposure of the metal to corrosive agents while subject to repeated stress. Certain inhibitors, notably sodium chromate, have been effective in reducing corrosion fatigue.

Most elements of a structure or machine that are subjected to repeated stresses are made of metal—principally steel—so that this discussion is concerned primarily with the fatigue testing of metallic materials.

14.2 SCOPE AND APPLICABILITY

Only in certain types of structures does the question of fatigue require consideration. In general, the fluctuations in the stresses in bridges and buildings (except for elements that may be subject to vibration) are not large enough nor do they occur often enough to produce failure. It has been estimated that the stresses in the members of an ordinary railway bridge are repeated fewer than 2 million times in a period of 50 years. Yet some materials, such as high-strength steels, are quite vulnerable to fatigue failure. In 1979 a large arena roof in Kansas City collapsed because a few high-strength bolts apparently failed in fatigue a few years after construction.

In some instances, notably in rapidly moving machines and in parts subject to severe vibrations, appreciable stress fluctuations may occur, running to billions of repetitions during the useful life of the machine or structure. The crankshaft of a piston-type airplane motor is subjected to about 20 million reversals of stress in less than 200 hours of flying. Furthermore, the stresses are relatively high, since the motor is constantly operated at nearly maximum power and the size of the shaft is kept as small as possible to reduce its weight. The stresses in the shaft of a stream turbine, if operated continuously for 10 years, would be reversed about 16×10^9 times, and the stresses in the blades would be reversed about 250×10^9 times. Fatigue must be considered in the design of many parts subjected to cycles of stress such as motor shafts, bolts, springs, gear teeth, turbine blades, airplane and automobile parts, steam- and gas-engine parts, railway rails, wire rope, car axles, and many machine parts subjected to cyclic loading.

One of the simplest, still widely used type of test for determining the fatigue limit of a *material* employs completely reversed flexural loading on rotating-beam specimens, the maximum stress being computed with the simple flexure formula. When carefully prepared smoothly finished specimens without abrupt changes in cross section are used, fairly concordant results are obtained. In a very different type of flexural fatigue test, "triangular" specimens are loaded as cantilever beams (ASTM B 593 and D 671).

In some tests the specimen is loaded axially, often with complete reversal of stress. A constant amplitude test for metals is described by ASTM E 466. There are also compressive fatigue tests for materials such as rubber (ASTM D 623).

When grooved or notched specimens are used, the fatigue limits have been found to be functions of the true maximum concentrated stresses developed. In fact, repeated-load tests have been used for evaluating stress-concentration factors.

Although the effect of range in stress on the fatigue limit has not been fully explored, sufficient information is available to make what seems to be a fair (though rough) estimate of the fatigue limit of metals under types of loading other than those giving complete reversal of stress. Data are also available that make possible an estimate of fatigue strengths under axial and under torsional loadings. For each of a number of given types of materials these show some correlation with static ultimate strengths.

Because of uncertainties in the stress analysis it is difficult to apply the results of fatigue tests on small specimens to the design of complicated parts and built-up units. To obtain direct information on the behavior of such parts or units under repeated loadings, fatigue tests have been made on a number of kinds of full-sized specimens, notably axles and riveted, bolted, and welded joints [6, 37, 93, 102, 103]. A fatigue test is generally unsuitable for an inspection test, or quality-control test, owing to the time and effort required to collect the data.

As mentioned earlier, during their normal service life many structures do not experience even close to the number of stress cycles that would reduce the strength of the material to the fatigue limit. To design such structures on the basis of the fatigue limit would hardly be cost-effective. In order to account for the effects of stress reversals, low-cycle fatigue tests may be performed. The usual objective is to study the stress-strain behavior after repeated reverse plastic deformations (ASTM E 606).

Terms relating to fatigue testing and the statistical analysis of results are given in ASTM E 206, and those for low-cycle testing by ASTM E 513. All fatigue tests may be conducted at low or high as well as at normal temperatures.

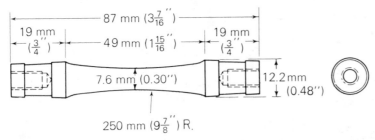

Figure 14.2 Specimen for rotating-beam test of metals. (*From* Metals Handbook, *American Society for Metals, Cleveland, Ohio, 1948.*)

14.3 SPECIMENS

In some classical flexure tests, the beam specimen is loaded by gravitational force and the specimen rotated; in others a stationary flexure specimen is acted on by loads rotating around it. If the specimen is circular in cross section, the flexural direct stresses will vary sinusoidally. Such a specimen is shown in Fig. 14.2.

In some tests triangular specimens of flat stock are arranged as cantilevers. The configuration specified for plastics is indicated by Fig. 14.3 (ASTM D 671). Specimens of the same general type, although slightly smaller, are used in flexural-fatigue tests of some copper alloy specimens (ASTM B 593). Specimens of this type are clamped at the larger end and loaded at the smaller one in such a way that the effective line of action of the load passes through the apex of the triangle. Because of the linearity of the bending moment diagram and of the section modulus, the fiber stresses will be uniform throughout the specimen portion bounded by the triangular sides.

Specimens for axial fatigue tests of metals (ASTM E 466) are illustrated by Fig. 14.4. The specimens may be circular or rectangular in cross section; in the latter case the width of the test section should be from two to six times the thickness. The grip area (cross section) should be at least half again as large as the test area, which may

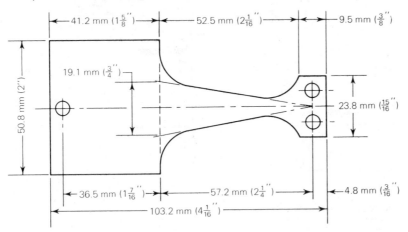

Figure 14.3 Specimen for cantilever-beam test of plastics (ASTM D 671).

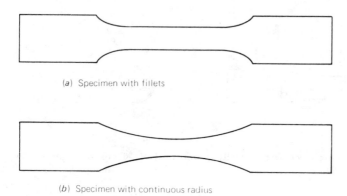

(a) Specimen with fillets

(b) Specimen with continuous radius

Figure 14.4 Round or flat specimens for axial test of metals (ASTM E 466).

range from 20 to 650 mm^2. Either specimen may have fillets, as shown in Fig. 14.4*a*, or a continuous radius, as shown in Fig. 14.4*b*. It is essential that all specimens be prepared with great care, and for any test series in a uniform manner.

For compression-fatigue tests of rubber (ASTM D 623) specimens in the shape of either a cylinder or a frustrum of a rectangular pyramid are specified.

A *compact tension specimen* (*CTS*) is often used for fracture-mechanics investigations (ASTM E 399). Figure 14.5 shows one type. It should be noted that the purpose of fatigue testing of these specimens is the propagation of a crack from the notch, as it would be impossible to achieve such a thin crack by machining. Thus the fatigue test is really an extension of the preparation of a specimen for a static-tension test.

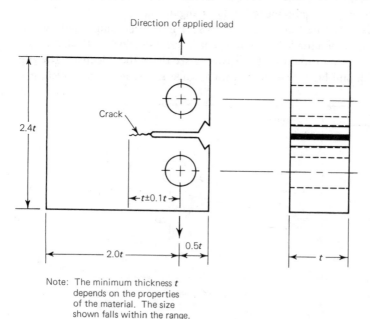

Note: The minimum thickness *t* depends on the properties of the material. The size shown falls within the range.

Figure 14.5 Compact tension specimen (ASTM E 399).

14.4 APPARATUS

Machines for making fatigue tests under cycles of repeated or reversed stress may be classified according to the type of stress produced:

1. Machines for cycles of axial stress (tension, compression)
2. Machines for cycles of flexural stress
3. Machines for cycles of torsional shearing stress
4. Machines for axial, flexural, or torsional shearing stresses or combinations of them

All repeated-stress testing machines must be provided with a means for applying load to a specimen and with a means for measuring the load. Also, there must be a counter for recording the number of cycles applied and some device that automatically disengages the counter when the specimen breaks. Frequently the disengaging device is also designed to stop the testing machine itself.

A major machine used in modern fatigue investigations is the dynamic hydraulic universal testing machine described in Art. 6.4. With the aid of a closed-loop control system an entire fatigue test can be programmed by choosing appropriate command inputs. Available frequencies range up to a kilohertz or so. The results can be displayed in almost any manner desired. Some machines have torsional as well as axial capabilities of 90° or more.

Bending fixtures permit flexural-fatigue testing. Self-aligning grips allow threaded and button-end specimens to be loaded in compression, which is important especially in low-cycle tests. For high-temperature tests the grips should be water-cooled. In fracture-mechanics investigations it is convenient to have the same machine perform precycling and the pull test.

There are several ingenious mechanical fatigue-testing devices, simple yet accurate. Mechanical fatigue-testing machines generally predate the hydraulic machines. The earliest flexural-testing machines were of the rotating-beam type, one example of which is shown in Fig. 14.6. For this machine a specimen is held at its ends in special holders and loaded through two bearings equidistant from the center of the span. Equal loads on these bearings are applied by means of weights that produce a uniform bending moment in the specimen between the loaded bearings. To apply cycles of stress the specimen is rotated by a motor; since the upper fibers of the rotated beam

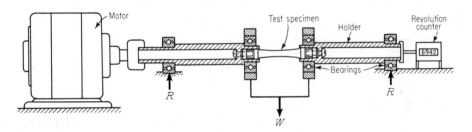

Figure 14.6 R. R. Moore rotating-beam machine.

are always in compression while the lower fibers are in tension, it is apparent that a complete cycle of reversed stress in all fibers of the beam is produced during each revolution.

Two types of testing machines are in use in which a cantilever specimen carries a load at its free end. In one type, the specimen is rotated while a gravity load is applied to the free end. In another, a bearing at the free end carries the load from a compression spring acting in a plane normal to the longitudinal axis of the specimen; thus the specimen is deflected, and as the spring rotates around it, cycles of reversed stress are produced.

Several types of fatigue-testing machines are broader in application. They have one stationary head or fixed platen and one vibratory platen, with provision for connecting various fixtures with attached specimens between them. The vibratory platen exerts a controlled motion or force on the specimen: if exerted axially, tensile and compressive stresses will be developed. By use of various fixtures, torsional or flexural stresses can be developed. Several basic methods are used to generate and control the force.

In the Haigh machine, which is of the ac magnet type, the specimen is subjected to cycles of axial stress by being attached to an armature that moves rapidly back and forth between two electromagnets energized by two-phase alternating current, one phase being connected to each magnet. Hence, when axially loaded, the specimen is alternately stretched and compressed by the action of the magnets [67].

Another type of machine uses a variable-throw crank, as shown in Fig. 14.7, to generate and control the force acting on the specimen. For any one test the throw is constant and the load on the specimen is measured by the deflection of the transmission beam. Static preload is applied by moving the fixed platen with either a hydraulic or screw mechanism. Some machines have special controls that sense the load and automatically maintain constant preload and vibratory forces to compensate for creep and slippage. The flexplates eliminate all transverse motion of the vibratory platen and ensure a sinusoidal motion in one plane only.

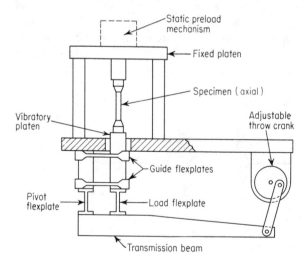

Figure 14.7 Constant-amplitude machine. (*From Breunich, Ref. 11.*)

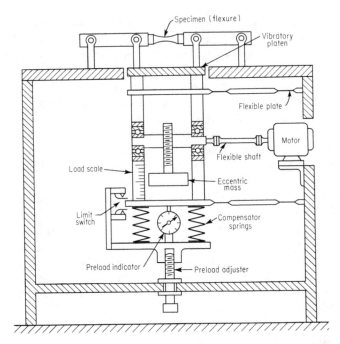

Figure 14.8 Sonntag inertia-type machine. (*From Breunich, Ref. 11.*)

A third type of universal machine, which is shown schematically in Fig. 14.8, uses a constant force instead of a motion of constant amplitude. The load on the specimen is produced by a mechanical oscillator that is, essentially, an eccentrically mounted mass rotated by a synchronous motor. The unbalance of this mass can be changed from zero to a maximum, and its position is calibrated to indicate the dynamic load. By adjusting the preload springs, any desired ratio of maximum to minimum stress can be produced.

The horizontal component of the centrifugal force is absorbed by restraining members, leaving a sinusoidal force in a vertical direction that is transmitted to the test piece.

14.5 PROCEDURE AND OBSERVATIONS

To determine the fatigue limit of a metal, it is necessary to prepare a supply of identical specimens that are representative of the material. The first specimen is tested at a relatively high stress so that failure will occur after a small number of applications of stress. Succeeding specimens are then tested, each one at a lower stress. The number of repetitions required to produce failure increases as the stress decreases. Specimens stressed below the fatigue limit will not rupture. In order to minimize the effect of random errors, usually many specimens are tested, often several at the same stress level. A grouped frequency analysis can then be performed on the results.

An *S-N* diagram, as described earlier, is next prepared. Examples using semilogarithmic plotting are shown in Fig. 14.9, which presents the results for various typical materials. For all ferrous metals tested, and for most nonferrous metals, the *S-N* diagrams become horizontal, as nearly as can be determined, for values of *N* ranging from 1 000 000 to 50 000 000 cycles, thus indicating a well-defined fatigue limit. The *S-N* diagrams for duralumin and monel metal do not indicate well-defined fatigue limits.

Certain test parameters must be selected before the test is started:

1. *Type of load variation.* A sine function is normally used, but many function generators also make haversine, square, sawtooth, ramp, dual ramp, triangular, and trapezoidal functions possible.
2. *Mean stress.* For endurance tests the mean stress is normally zero, but another selection may be desired (see Table 14.1).
3. *Alternating stress.* When the alternating stress is added to the (nonnegative) mean stress, the maximum stress results. In a test series, the latter should be close to the maximum static stress for the first specimen, and then be reduced by about 10 percent for the second one. Thereafter the previous data points will provide guidance.
4. *Frequency.* Usually a frequency of several hundred hertz is appropriate. ASTM E 466 specifies a maximum of 170 Hz.

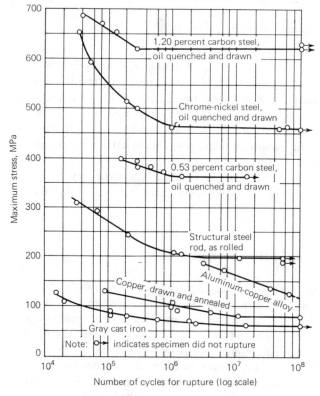

Note: 100 MPa = 14.5 ksi

Figure 14.9 Typical *S-N* diagrams for metals under reversed flexural stress. (*From Moore, Ref. 68.*)

In low-cycle fatigue testing self-aligning grips should be used to avoid instability. The test is started by pulling the specimen into the inelastic range and then cycling the load under strain control. Usually a double-ramp command pattern is chosen. The output is a stress-strain diagram showing the hysteresis loops. After a predetermined number of cycles, which simulates the expected number of lifetime stress reversals of the material, the specimen is subjected to a normal tension test. The result of a test series might be, for example, breaking strength as a function of the number of cycles.

14.6 EFFECTS OF SPECIMEN VARIABLES

The fatigue limit of metals varies with composition, grain structure, heat-treatment, and mechanical working. The fatigue limits as well as the static strengths of a few representative metals are shown in Table 14.2.

Various alloying elements increase both the static strength and the fatigue limit of steel. No alloying element is known that distinctly improves the fatigue limit without also increasing the static strength. Alloying elements (other than carbon) do not affect the fatigue limit as much as the microstructure of the steel does [25].

Proper heat-treatment improves both the static strength and the fatigue limit of steel, especially in the case of high-carbon steel and alloy steels, for which the percentage of increase may be as high as 200. However, as in the case of composition of the steel, no heat-treatment is distinctly beneficial solely to the fatigue limit. In general, heat-treatment and chemical composition are closely related, inasmuch as the use of

Table 14.2 Fatigue limit and fatigue ratio of various metals[†]

Metal	Static tensile strength, MPa[‡]	Fatigue limit in reversed flexure, MPa[‡]	Fatigue ratio
Steel, 0.18% carbon, hot-rolled	432	213	0.49
Steel, 0.24% carbon, quenched and drawn	465	203	0.44
Steel, 0.32% carbon, hot-rolled	453	216	0.48
Steel, 0.38% carbon, quenched and drawn	631	231	0.37
Steel, 0.93% carbon, annealed	580	210	0.36
Steel, 1.02% carbon, quenched	1382	724	0.51
Nickel steel, SAE 2341, quenched	1944	772	0.40
Cast steel, 0.25% carbon, as cast	463	186	0.40
Copper, annealed	223	69	0.31
Copper, cold-rolled	359	110	0.31
70-30 brass, cold-rolled	505	121	0.24
Aluminum alloy 2024, T36	496	124	0.25
Magnesium alloy AZ63A	276	76	0.27

[†]Adapted from H. F. Moore and J. B. Kommers, *Fatigue of Metals*, McGraw-Hill, New York, 1927.

[‡]1 MPa = 145 psi.

certain alloying elements is largely responsible for the depth of penetration and degree of heat-treatment possible. If a machine part is loaded only in flexure or torsion, then surface-hardening treatments may improve fatigue resistance, but if a machine part is to be used in axial loading, then the uniformity of properties throughout the whole cross section is important, and a heat-treated alloy steel that has good depth-hardening properties is superior to one that does not. Specimens refrigerated before tempering have a higher fatigue limit than unrefrigerated specimens, since refrigeration tends to reduce the amount of retained austenite [25]. Some heat-treatments may increase the static yield strength but decrease the fatigue limit because of minute quenching cracks or decarburization of the surface.

Cold-working of ductile steel increases its fatigue limit to about the same degree as it increases the static strength. For nonferrous metals the percentage of increase of fatigue limit due to cold-working is sometimes less than the percentage of increase in tensile strength.

The surface finish of a test specimen has a definite effect on the fatigue limit; the rougher the finish, the lower the fatigue limit. In fact, a very rough surface finish, particularly if the scratches or machine marks are normal to the applied force, may lower the fatigue limit of a metal by as much as 15 to 20 percent in comparison with the fatigue limit for a polished surface. However, a part that is machine-polished may fail more rapidly than one with a good tool finish, because of the tension set up in the skin by the heat of the polishing.

Shot-blasting the surface of unpolished machined specimens of some kinds of steel has a beneficial effect on the fatigue limit, in comparison with the result for polished but not shot-blasted specimens, although the tensile strength is not correspondingly improved [48]. For wire specimens 4.11 mm in diameter that were not machined or polished, the fatigue limits were as follows:

Untreated . . . . 510 MPa
Shot-blasted . . . . 662 MPa
Shot-blasted and tempered at 200°C . . . . 696 MPa

There does not appear to be any advantage, as regards fatigue limits, in shot-blasting surfaces where high residual stresses due to quenching and insufficient drawing are present.

Abrupt changes in cross section definitely lower the nominal fatigue limit owing to the high concentration of stress at such transitions. A V notch may reduce the nominal fatigue limit of a test specimen in reversed bending by about 65 percent, even though the net cross-sectional area remains constant. In actual machine parts subject to cycles of reversed stress any abrupt change in section due to holes, grooves, notches, screw threads, and shoulders must be given special consideration. However, the effect of such stress raisers is not so serious as it appears from the results of computations made on the basis of the theory of elasticity or photoelastic analysis.

Some of the alloy steels, such as chrome nickel, seem to be more sensitive to stress-concentration effects, a property that has been called the *tenderness* of a material. Cast iron is considerably less sensitive to imposed stress concentrations in fatigue tests than are steels, the reason apparently being that the inherent irregularities in the microstructure of cast iron produce considerable stress concentration, so that the effect of an imposed discontinuity is somewhat masked.

Chrome plating decreases the fatigue limit, but nickel plating will either increase or decrease the limit depending on the bath used.

14.7 EFFECTS OF TEST VARIABLES

It is difficult to carry out repeated tests of specimens under cycles of alternating direct tension and compression owing to the possibility that any slight eccentricity of load may cause serious flexural stresses and that high localized stresses are likely to occur at the shoulders of axially loaded specimens. These stress concentrations in items subjected to repeated stresses are of considerable importance even for ductile materials, although they have very little effect on the static tensile strength. In general, carefully performed tests have shown that the fatigue limit for cycles of alternating direct tension and compression is practically the same as the fatigue limit for cycles of reversed flexural stress.

The fatigue limit for shearing stresses is usually determined from tests in repeated or reversed torsion. Most of these determinations have been made on carbon and alloy steels. For tests of carbon steels the ratio of the fatigue limit in reversed torsion to the fatigue limit in reversed flexure has been found to vary from about 0.48 to 0.64, with an average of 0.55. For alloy steels the ratio varies from 0.44 to 0.71, with an average of 0.58. The average ratio for a few nonferrous metals is about 0.5 [67].

For specimens of metal subjected to repeated stresses involving a range of stress less than complete reversal, the smaller the range of stress, the higher the fatigue limit. The limiting value is, of course, the static strength. The general nature of the variation of strength with range in stress is shown in Fig. 14.10. Three methods of representing fatigue data involving the variable of range in stress are shown. Figure 14.10a shows the Goodman-Johnson type of diagram in which the minimum stress is plotted to give a straight line (the horizontal scale being without significance), and the fatigue limit corresponding to any minimum stress is plotted vertically above, giving the upper curved line. The range of stress is represented by the vertical ordinate between the lower and upper solid lines, whereas the mean stress is represented by the curved dashed line. Thus, for any minimum stress AC, the fatigue limit is BC, the mean stress is DC, and the range of stress is AB.

Figure 14.10b, the Schenck-Peterson diagram, is drawn in much the same manner except that the curved line representing mean stresses is drawn as a straight line at an angle of 45° with the horizontal axis. This makes the minimum stress line a curve and permits the horizontal axis to represent the mean stresses to the same scale as on the vertical axis.

Figure 14.10c, the Haigh-Soderberg diagram, represents the upper half of Fig. 14.10b, but with the 45° line representing mean stresses turned to the horizontal position. In this diagram ordinates to the curve represent the maximum value of the alternating stress BD, which can be applied simultaneously with a mean stress OD without causing failure by fatigue. For the purpose of estimating values of fatigue

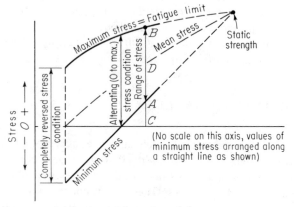

(a) Goodman-Johnson type of diagram

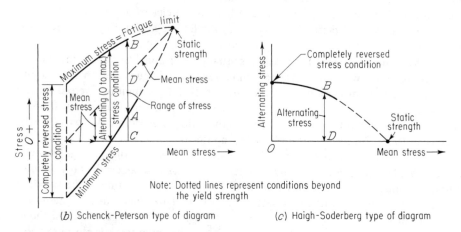

Note: Dotted lines represent conditions beyond
the yield strength

(b) Schenck-Peterson type of diagram   (c) Haigh-Soderberg type of diagram

Figure 14.10 General variation of strength with range of stress.

limit for use in design, a number of formulas, based on idealizations of the data, have been derived [67, 84].

The repeated loading of metals to stresses below the fatigue limit appears to produce a mild cold-working of the material resulting in a slight increase in its original or normal fatigue limit, but the strengthening varies widely for different metals; for progressively increasing cyclic understress, the increase in the limit may be considerable. On the other hand, many preliminary cycles of stress above the normal fatigue limit appear to start the formation of microscopic cracks that cause a reduction of the fatigue limit; the higher the preliminary stress and the greater the number of preliminary cycles of stress, the greater the damage done and the lower the resulting modified fatigue limit. Also, the static load capacity will be decreased due to the high stress developed at the root of the microscopic fatigue crack. Some materials appear to be damaged more than others by preliminary overstressing. Since many machines or structures are likely to be subjected to temporary overloads, the possible effect of such overloads on the fatigue limit should be considered [47].

In machine design, although there may be no intention of using allowable stresses above the fatigue limit, inability to analyze the stress condition accurately often results in higher stresses than those calculated. Many repetitions of this overstress may cause the part to fail. However, if during the life of a structure an occasional overload occurs owing to carelessness in operation or another unforeseen circumstance, such overstressing, unless it is excessive, will probably not materially affect the ability of the structure to carry its normal working loads.

The influence of the frequency of stress reversal on the fatigue limit has been under discussion for a long time. It seems clear that in the range of ordinary testing speeds of, say, 30 to 250 Hz there is little if any variation in the fatigue limits obtained, although there appear to be inconsistencies in results when speeds as high as 500 Hz are used.

14.8 CORRELATION

Correlation between the fatigue limit and one of the more readily determinable mechanical properties is desirable because of the expense involved and the time required to determine the fatigue limit by a long-time test. However, there is no direct relation between the fatigue limit and any of the other physical properties that will apply to all metals. Elastic strength, ultimate strength, and ductility seem to influence the fatigue strength. Fatigue failure is a progressive fracture of a metal. This may account for a closer correlation between fatigue limit and tensile strength than there is between fatigue limit and any other single physical property.

The results of many tests indicate that the fatigue limit under reversed flexure for rolled or forged steel and iron is usually between 45 and 55 percent of the tensile strength, averaging about 50 percent, although in some cases it may be as low as 35 or as high as 60 percent. For steel and iron castings this fatigue ratio averages about 40 percent. For nonferrous metals the fatigue ratio varies over a wide range, from less than 25 percent for cold-drawn copper and certain aluminum-copper alloys to 50 percent for annealed bronze. For representative values from various individual tests, see Table 14.2.

The Brinell hardness test seems to furnish a fair index of the fatigue limit for rolled and forged steel. For such materials the fatigue limit under reversed flexure (in megapascals) is about 1.7 times the Brinell hardness number.[†] This relation fails to hold when the Brinell number is greater than 400.

PROBLEM 14 Axial-Fatigue Test of Steel

Object: To conduct a constant-amplitude axial-fatigue test of structural steel and obtain one point on the *S-N* curve.
Preparatory reading: ASTM E 466.
Specimen: Flat or round fatigue specimen of high-strength steel.
Special apparatus: Axial-fatigue tester.

[†]Or, in psi, 250 times Bhn.

Procedure:
1. Carefully measure the dimensions of the specimen, obtain the cross-sectional area, and calculate the likely yield load.
2. Securely fasten the specimen in the chucks of the testing machine, making certain that the grips are properly aligned.
3. Set the load so that the maximum tension and compression equal each other and about 80 percent of the yield load (unless a different stress level is assigned).
4. Adjust the frequency to about 150 Hz, zero the cycle counter, and start the machine.
5. Continue the test until the specimen fractures.

Report: Report all circumstances surrounding the test (including temperature, specimen data, machine type, etc.) and the actual test data (including frequency, load amplitude, elapsed time, and number of cycles to failure). With the aid of a sketch describe the appearance of the fractured specimen.

DISCUSSION

1. What is an inherent advantage of fracture mechanics over the fatigue limit approach?
2. Why were early fatigue tests conducted in flexure?
3. Did you ever purposely cause a fatigue failure?
4. For axial fatigue specimens loaded into the compression range, the length-to-thickness ratio of the test section is limited. Why?
5. Would it be good engineering practice to use the fatigue limit as a design criterion for all structures?
6. What unusual test result is obtained in compression-fatigue tests of rubber? Why is it of interest?
7. What precautions must be observed in using an *S-N* curve for design purposes?
8. Contrast low-cycle and normal fatigue tests. What is the purpose of each?
9. Estimate the number of stress cycles in the suspension of an automobile during the automobile's life.
10. What differences in their reactions to fatigue loading do you expect in the case of two metals with identical strengths, one brittle and the other ductile?

READING

Brown, W. F., and J. G. Kaufman (eds.): *Developments in Fracture Mechanics Test Method Standardization*, ASTM, Philadelphia, 1976.

Cyclic Stress-Strain Behavior—Analysis, Experimentation and Failure Prediction, ASTM, Philadelphia, 1971.

Experimental Techniques in Fracture Mechanics, Society for Experimental Stress Analysis, Westport, Conn. 1973.

Mindlin, H., and R. W. Landgraf (eds.): *Use of Computers in the Fatigue Laboratory*, ASTM, Philadelphia, 1975.

Rolfe, S. T., and J. M. Barsom: *Fracture and Fatigue Control in Structures*, Prentice-Hall, Englewood Cliffs, N.J., 1977.

Sandor, B. I.: *Cyclic Stress and Strain*, The University of Wisconsin Press, Madison, Wis., 1972.

Weibull, W.: *Fatigue Testing and Analysis of Results*, Pergamon, New York, 1961.

FIFTEEN

CREEP

15.1 CREEP

At high temperatures many materials act somewhat like a very viscous liquid, so that a stress often much less than that required to cause failure in a few minutes may cause failure if sufficient time is allowed. The physical process that brings about failure is a comparatively slow but progressively increasing strain called *creep*. As in other contexts, here failure may mean either rupture or excessive distortion.

While most materials are subject to creep, structural differences among metals, plastics, rubber, concrete, and other materials cause considerable dissimilarities in the creep mechanisms. It is not surprising that the effects of temperature, as well as stress, vary widely. In steel, temperatures of several hundred degrees Celsius may make creep a problem, whereas some plastics, concretes, and lead may undergo creep at normal atmospheric temperatures. The discussion in this chapter will primarily concern metals, particularly steels, for which creep effects tend to start at about 40 percent of the melting temperatures (in K).

Most structural metals and alloys possess elastic properties at room temperature. As illustrated in Fig. 15.1, if a stress σ below the proportional limit is applied, an elastic strain OA occurs immediately on application of the load, and regardless of the duration of application the strain remains constant. The strain characteristics under this condition are represented by the curve OAB.

At elevated temperatures a strain OC occurs on application of the same stress σ, OC being greater than OA owing partly to the lower value of the modulus of elasticity at the higher temperature. The strain OC may be entirely elastic, or elastic and plastic, depending on the material, the temperature, and the stress. Under suitable stress-temperature conditions, however, the strain increases as the time of load applications is extended, the strain following the curve $CDEF$. This continued increase in strain with time while under a constant temperature and stress is represented by a *creep curve*.

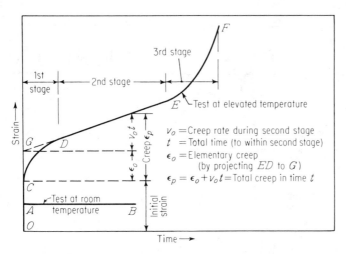

Figure 15.1 Strain-time curves at room and at elevated temperatures.

As indicated in Fig. 15.1, the creep sequence consists of three stages. In the first stage, creep continues at a decreasing rate that becomes approximately constant during the second stage; the beginning of the third stage is marked by a rapid increase in creep rate that continues until fracture occurs. In the majority of tests the stress and temperature conditions are selected so as to preclude the development of the third stage. Although broad generalizations are not permissible, it is generally found that the duration of each stage tends to vary inversely with the stress.

The characteristic failure of metals at low temperatures is by fracture through the crystals themselves, whereas failure at high temperatures occurs at crystal boundaries. The temperature at which the manner of fracture changes from intra- to intercrystalline is called the *equicohesive* temperature. This temperature is not accurately known; for ordinary mild steel it is about 450°C, but for certain alloy steels it is somewhat higher.

The mechanism of strain also appears to change as the testing temperature is increased. At temperatures below the equicohesive temperature, strain apparently occurs largely as an elastic intracrystalline movement, so that the metal exhibits elastic properties. At temperatures above the equicohesive temperature, strain may occur through quasi-viscous intercrystalline movement. However, the characteristics that render a material resistant to extension under high-temperature loading are not completely known.

The magnitude of the creep resulting from the application of a given stress over a given time period appears to depend on the opposing effects of the yielding of the material and the strain hardening caused by such yielding. At or below the equicohesive temperatures, strain hardening tends to predominate, and continuous measurable creep will not occur unless the stresses are large enough to overcome the resistance caused by strain hardening. If strain hardening predominates, the diagram representing the second stage of Fig. 15.1 becomes a horizontal line. For example, the yielding of steel above the proportional limit at room temperature is a creep phenomenon, but

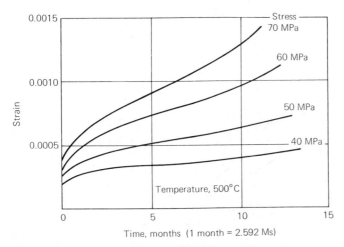

Figure 15.2 Creep curves for a typical carbon steel.

measurable creep soon stops as a result of strain hardening. At temperatures above the equicohesive temperature, however, the yielding rate exceeds the strain-hardening rate, and creep will proceed even under low stresses. Creep can be detected at these higher temperatures even under very low stress if sufficiently sensitive apparatus is used. Figure 15.2 shows a family of creep curves for various stress levels.

If a specimen undergoing creep is unloaded, some of the strain is recovered, as illustrated in Fig. 15.3, but an appreciable plastic strain has become permanent, its amount depending on the material and the test conditions (time of loading, temperature, and stress).

A measurement closely related to creep is that of *stress relaxation*. The specimen is first subjected to a fixed strain at an initial stress σ_o, and the load required to maintain the strain is observed progressively with time. A typical relaxation curve for a carbon steel is shown in Fig. 15.4. The importance of this stress is very evident. Various studies have been made to permit the prediction of relaxation from creep data [78].

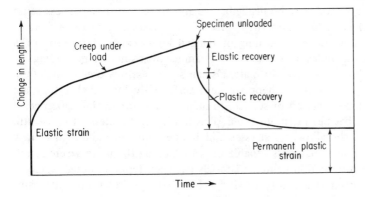

Figure 15.3 Recovery of strain after unloading.

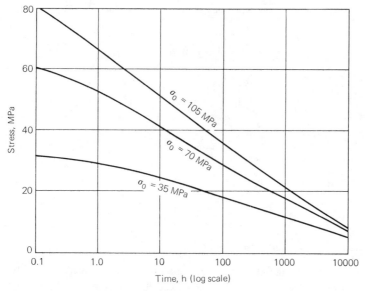

Note: 20 MPa = 2.9 ksi

Figure 15.4 Relaxation of 0.35% carbon steel at 455°C. (*From Roberts, Ref. 78.*)

15.2 SCOPE AND APPLICABILITY

Many types of engineering machines and structures are subjected to stress while at relatively high temperatures, a few common examples being internal-combustion and jet engines, high-pressure boilers and steam turbines, nuclear reactors, and cracking stills such as those used in the chemical and petroleum industries. Some types of equipment are operated for considerable periods of time at temperatures as high as 700°C and sometimes at higher temperatures.

At any temperature, creep may be rapid or slow; its rate decreases rapidly as the stress is lowered. Although at some relatively low stress the possibility of rupture due to creep within a long period may be negligible, such a stress may nevertheless cause undesirable structural *distortion* over the course of time. Examples of the effects of such excessive distortion are the loosening of flanged joints caused by creep in the connecting bolts, and the undesirable changes in clearance of steam turbine blading.

Experience has shown that certain materials can be successfully used at high temperatures provided the stresses in them are kept sufficiently low or that they are replaced before too long a period of service. In many actual applications absence of appreciable distortion has been demonstrated. This has led to the desire to raise still further either or both the allowable stresses and temperatures in the design of new units and to search for new materials capable of withstanding the more severe conditions. Progress in the production of new materials has been rapid; although some of these materials have given satisfactory service, others have proved inadequate, thus necessitating frequent replacements. The need for adequate high-temperature tests is obvious. Some of these tests are conducted over long periods of time to determine the resulting strain for a given temperature and stress. In other tests the time to rupture is determined at a given temperature and stress. Both types of test for metals are covered in ASTM E 139. Tensile- and compressive-creep tests for plastics are specified by ASTM D 2990 and compressive-creep tests for concrete by ASTM C 512.

Creep tests at high temperatures appear to be the only satisfactory guide to the performance of metals for high-temperature service, although creep testing as a scientific technique should be considered to be still in the developmental stage. While the general phenomena associated with creep are fairly well known, there is much detailed information yet to be obtained in order to achieve a well-rounded picture. Creep tests are inherently long-time tests, but the test periods may nevertheless be short in comparison with periods of high-temperature service in actual structures, so that extrapolation of creep-test data must be made with judgment. Creep tests require too much time to be used as acceptance tests, but they are the basis of data to be used in design.

Extrapolation of creep data from tests that have not been carried into the second stage (uniform creep rate, see Fig. 15.1) tends to give excessive values of predicted creep, although this procedure admittedly results in the selection of more conservative allowable stresses than those based on tests conducted over longer periods. On the other hand, extrapolation from data of tests that have been carried well into the second stage does not in itself give assurance that under very long-time service the metal will not enter the third or failure stage, but the use of a small permissible creep (commonly 1 percent) ordinarily gives adequate protection against such an occurrence.

In view of the importance of having some common basis for making creep tests, the Joint ASTM-ASME Research Committee on Effect of Temperature on Properties of Metals prepared a standard test procedure (ASTM E 139). In general, the method involves maintaining a standard test specimen at a constant temperature under a fixed tensile load and observing from time to time the strains that occur. Such a test may cover several months or even years, although some tests cover only two to three months. Suitable test periods depend on the reasonable life expected from the material in service. Test periods of less than 1 percent of the expected life are not considered to give significant results. Tests extending to 10 percent of the expected life are preferred, and even longer durations may be necessary, especially for new materials, although few tests extend for much more than a year. ASTM also recommends a procedure for tensile creep tests of metals under conditions of rapid heating and short duration (ASTM E 150).

Four variables are always involved in a creep test: stress, strain, time, and temperature. The test may be conducted on individual specimens at each of several loads and several temperatures. The results for a given material are often summarized in the form of composite diagrams, from which can be read the limiting stress for a given percentage of creep under stated temperatures and periods of time.

For use under moderately high temperature service but below the equicohesive temperature, or for materials highly resistant to creep, it may be that strength is of more significance than creep alone. In such cases, static elevated-temperature tests, involving the determination of yield or tensile strengths, are pertinent. A standardized procedure is available (ASTM E 21). It should be obvious that if the yield strength of a steel as determined by a static test is less than the creep limit for any given temperature, the yield strength is used in establishing the allowable stress.

15.3 APPARATUS

The determination of the creep characteristics of metals at high temperatures requires the use of three pieces of major equipment: (1) an electric furnace with suitable temperature-control, (2) an extensometer, and (3) a loading device.

Various types of devices are used for creep tests, the one shown in Fig. 15.5 being typical of many installations for one specimen; some furnaces are arranged to accommodate as many as 12 or more specimens. As shown in greater detail by Fig. 15.6, a winding of nickel-chromium wire is spaced around a tube that is usually of fused silica or Alundum. The winding can be held in position by a heat-resisting cement or may be laid in grooves in the tube. A large-diameter tube affords greater uniformity of temperature from the center to the outside of the test piece, and a long tube helps in reducing the temperature gradient between the center of the test length and the pull rods. A fair thickness of heat-insulating material is advisable for economy of heat.

It is common practice to space the turns along the tube in such a way that more heat is supplied to the pull rods attached to the specimen than to the specimen itself. This is done to reduce temperature gradients along the test piece. Recognizing the undesirability of nonuniform temperatures, ASTM specifies that the maximum variation of temperature over the gage length should not exceed ±2°C from the average test temperature for temperatures up to about 1000°C, ±3°C from the average test

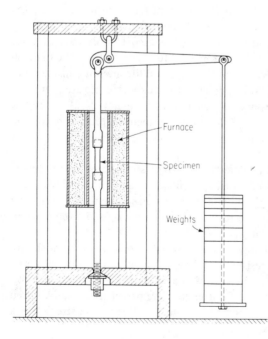

Figure 15.5 Loading apparatus. (*From Moore and Moore, Ref. 69.*)

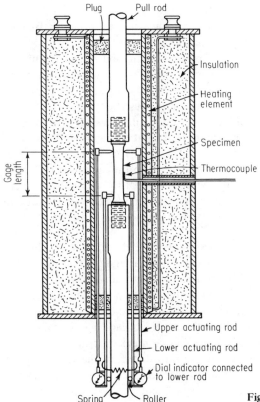

Figure 15.6 Electric furnace and extensometer.

temperature for higher temperatures. Also, the temperature variation of the specimen from the indicated nominal test temperature should not exceed ±2°C. The extent of the variation in temperature must be stated in reporting each test.

The ends of the furnace should be closed to prevent free circulation of air that would produce temperature variations and might cause oxidation of the surface of the specimen. In some cases an inert gas such as nitrogen is used inside the furnace.

Temperature is measured with a thermocouple whose junction is in direct contact with the specimen. Ordinarily, asbestos is used to protect the junction from direct radiation from the walls of the furnace tube.

An automatic controller, connected to the thermocouple, is used to ensure a constant temperature in the furnace. Relatively small fluctuations of temperature will not only introduce thermal strains that cause errors in the observed creep but also, near some critical high temperature, cause large changes in the creep characteristics of the specimen. Therefore, the extent of the fluctuations of furnace temperature should be stated in reporting the tests. Since fluctuations of the room temperature affect the extensometer readings as well as the furnace temperature, thermostatic control of the room temperature within close limits is desirable.

The load is applied to a specimen by means of direct weights or by a simple multiplying lever and weights as shown in Fig. 15.5. The applied load should be measured to within 1 percent. The outer ends of the pull rods are connected to the frame and loading lever through spherically seated nuts or other devices to permit freedom of rotation in all directions, thus ensuring axial loading of the test piece.

High-temperature furnaces are available for some static and dynamic universal testing machines, but the use of such machines for ordinary creep tests, convenient though it may be, has the obvious disadvantage of tying up a very expensive piece of equipment for perhaps months in order to conduct a single test.

Various types of extensometer are in common use to measure the strains in the specimen. One type makes use of two telescopes that sight through windows in the furnace on gage marks provided on the test piece. The strains are measured by means of a filar micrometer attached to one of the telescopes. The time required for setting a telescope and reading a micrometer prevents rapid readings of the elastic strain. Since the early stages of creep occur many times faster than the later stages, delay in obtaining measurements of the elastic strain tends to give too high a value of the *initial* strain and too low a creep value.

Two other types of extensometer, the mirror and the dial indicator, employ two pairs of actuating rods that are clamped to the specimen at the ends of the gage length, as shown in Fig. 15.6. These rods should have good strength and show little scaling at high temperatures; a chrome-nickel alloy is generally used. In the simple type shown in Fig. 15.6, the actuating rods are connected to dial indicators that give direct readings of the strains. The ease and rapidity with which readings may be taken with the dial-indicator type give it a distinct advantage in that quick determinations of the initial elastic strain can be made before errors are introduced by the very early but appreciable plastic strains. Thus elastic and plastic strains are more clearly defined. Readings should be taken on opposite sides of the specimen.

In place of the dial indicators other types of deformeters may be used. Differential transformers and strain-gage transducers have the advantage of allowing automatic readout.

Finally, it is possible to attach electric resistance strain gages to the specimen, at least for tests of moderate temperature and duration. Even special high-temperature gages and adhesives, however, cannot be used when temperatures exceed about 500°C or durations are greater than a few hundred hours.

15.4 PROCEDURE AND OBSERVATIONS

In a standard creep test the tensile or compressive load and the temperature are held constant; the resulting strain-time data form a creep curve. For round metal specimens, diameters ranging from about 2.5 to 15 mm are recommended, with a gage length of at least four diameters; the surface should be smooth and unscratched.

In making tests at a given temperature, the unloaded specimen is first heated to the required temperature. When the temperature of the specimen is steady, the gage length is observed and the predetermined load applied quickly without shock. The resulting instantaneous extension is largely an elastic strain. Measurements of the subsequent creep are observed at sufficiently frequent intervals to define the strain-time curve; daily or weekly observations usually suffice. At the conclusion of the test there should be at least 50 observations of specimen temperature, the average of which should be reported as the actual test temperature.

Care must be taken to avoid eccentricity of loading. The use of two (or three) extensometers allows eccentricity to be detected, but not to be corrected. As an inspection of Fig. 15.2 shows, at any particular temperature and time the stresses are by no means proportional to the creep strains. Therefore the average strain recorded for an eccentrically loaded specimen will not at all equal the uniform strain obtained for a concentrically loaded specimen under otherwise identical conditions. This caveat

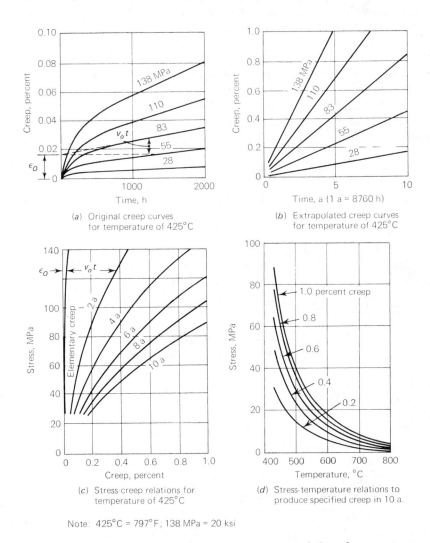

(a) Original creep curves for temperature of 425°C

(b) Extrapolated creep curves for temperature of 425°C

(c) Stress-creep relations for temperature of 425°C

(d) Stress-temperature relations to produce specified creep in 10 a.

Note: 425°C = 797°F; 138 MPa = 20 ksi

Figure 15.7 Development of stress-strain-temperature relations from creep curves. (*Based on McVetty, Ref. 62.*)

applies more to short-time than to long-time tests, as any eccentricity will diminish with time.

In the design of members subjected to high temperatures, the allowable stress for a given operating temperature is defined by a permissible plastic strain over a specified time of service. Since these service periods may run into several years, it is usually necessary to extrapolate data obtained from relatively short-time tests. In some cases creep strains have been estimated on the assumption that creep proceeds at a constant rate for periods many times greater than the actual periods of observation; the creep is calculated by extending a tangent to the end of the creep curve at some point in the second stage (as stylized in Fig. 15.1). Thus for a greater time t than is covered by the test, the total creep or plastic strain ϵ_p can be determined by the equation $\epsilon_p = \epsilon_o + v_o t$.

An example of the reduction of typical creep data is shown in Fig. 15.7. Figure 15.7a shows creep curves plotted directly from experimental data for a temperature of 425°C. Extrapolated creep curves for various stresses are shown in Fig. 15.7b. Inasmuch as this scheme of evaluation is invalid for strains extending beyond the inflection toward the third phase of creep, it is recommended that extrapolations based on it be carried no farther than the equivalent of 1 percent elongation. To obtain an allowable

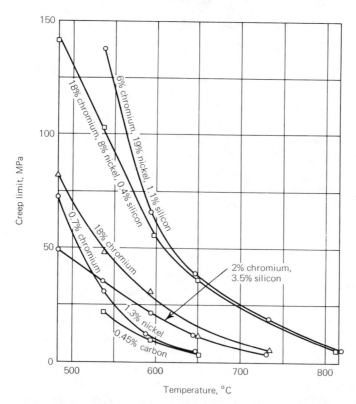

Figure 15.8 Creep limits (stresses producing 1% creep in 100 000 h) for various steels. (*From Norton, Ref. 72.*)

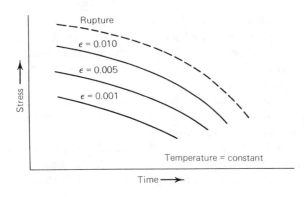

Rupture

$\epsilon = 0.010$

$\epsilon = 0.005$

$\epsilon = 0.001$

Stress

Temperature = constant

Time ⟶

Figure 15.9 Constant-strain and rupture curves for a specific test temperature.

stress (a stress producing a specified creep in a stated time) a graph is first made between stress and percentage of creep for various periods of time (Fig. 15.7c); such relations are obtained directly from Fig. 15.7b. Thus, for a creep of 1 percent in 10 years, the allowable stress for a temperature of 425°C is 90 MPa. For a temperature of 425°C these stresses are then plotted as ordinates in Fig. 15.7d. The stresses for other temperatures are also determined and the relations between stress and temperature for various creep values plotted as shown in Fig. 15.7d.

The limiting creep stresses for various types of steel at various operating temperatures for a period of 100 000 h with 1 percent elongation are shown in Fig. 15.8. A similar diagram could be constructed for rupture rather than limiting strain.

Instead of summarizing the data by drawing constant-strain curves for a particular time as in Fig. 15.7d, the data may be presented for a particular temperature, which is illustrated in Fig. 15.9. If the specimens are tested to rupture, a rupture curve can be added to the diagram, as shown. Because the design temperature is usually given, such a diagram is a convenient aid in that it readily shows under what combinations of stress and duration of stress certain distortions or rupture are to be expected.

Occasionally the times on these diagrams are plotted to a logarithmic scale, particularly for some nonmetallic materials such as plastics.

15.5 EFFECTS OF VARIABLES

The influence of various factors on the creep of steel can be summarized as follows [13, 82]:

The desired operating temperature determines the most suitable composition of the metal. At temperatures below the lowest temperature of recrystallization the creep resistance may be increased either by certain elements that largely enter into solid solution in the ferrite, such as nickel, cobalt, and manganese, or by the carbide-forming elements, such as chromium, molybdenum, tungsten, and vanadium. At temperatures above the lowest temperature of recrystallization, however, the carbide-forming elements are the most effective in increasing the creep strength. Small additions of titanium and columbium to chromium-nickel stainless steels appreciably reduce their creep characteristics over a considerable range of high temperatures.

Creep resistance is influenced by heat-treatment. At temperatures of 550°C or greater, the maximum resistance is usually produced by normalizing, provided the drawing temperature is about 100°C above the test temperature; the lowest resistance is produced by quenching and drawing. The creep resistance of steel in an annealed condition is between the resistances for the other two treatments.

Creep resistance is influenced by the manufacturing process. Results available indicate electric-arc-furnace steel to be superior to the open-hearth product, whereas induction-furnace material is superior to electric-arc-furnace steel.

Creep resistance is also influenced by the grain size of the steel. At temperatures below the lowest temperature of recrystallization a fine-grained steel possesses the greater resistance, whereas at temperatures above that point, a coarse-grained structure is superior.

PROBLEM 15 Creep Test of Lead

Object: To determine a creep curve for lead at room temperature.
Preparatory reading: ASTM E 139 and E 150.
Specimens: Two specimens cut from lead sheet.
Special apparatus: Low-capacity tension machine.
Procedure:
1. Perform a static tension test to determine the ultimate strength of the material.
2. Carefully measure the creep specimen and insert it into the tension machine. Make certain that the machine is free from vibration and properly leveled and that the specimen is well aligned.
3. Attach and zero a dial gage or other deformeter.
4. Apply a load causing a stress equal to about two-thirds of the strength of the specimen. Do this rapidly and smoothly, start a timer, and take an initial reading.
5. Take readings at about 1-min intervals, perhaps starting with shorter intervals and increasing them later. Continue until the specimen ruptures.
Report: Report the temperature and the stress level, as well as all circumstances surrounding the test and all known properties of the material and specimen. Present the results in the form of a creep curve, plotting strain vs. time (min), both to an arithmetic scale. Try to distinguish the three stages of creep. Inspect the broken specimens with a magnifying glass, make appropriate sketches, and comment on any differences between the static- and creep-test specimens.

DISCUSSION

1. What happens to sails after they have been used for a few seasons? Can you think of any similar examples?
2. What are the four basic variables involved in creep tests? What are the possible ways of relating them graphically?
3. If you had an appropriate full creep curve available for a material, what might be a good limit for the service life of a part not sensitive to distortions of a few percent?
4. How is prestressed concrete affected by creep?
5. Explain how you could obtain constant-creep curves for a certain temperature from a family of creep curves.
6. How could the creep of adhesives be tested? Are there ASTM standards for this application?
7. We did not say much about specimens in this chapter. What are some ASTM requirements for metal and plastic specimens?

8. Is creep of the steel girders in an office building a serious problem? What about creep of the cylinders in an internal-combustion engine?

9. Creep is almost automatically associated with high temperatures. Why aren't there more structural materials that creep under normal conditions?

10. Devise a creep demonstration using materials and implements you have at home.

READING

Conway, J. B.: *Numerical Methods for Creep and Rupture Analysis*, Gordon and Breach, New York, 1967.

Finnie, I., and W. R. Heller: *Creep of Engineering Materials*, McGraw-Hill, New York, 1959.

Garofalo, Frank: *Fundamentals of Creep and Creep-Rupture in Metals*, Macmillan, New York, 1965.

Hult, A. H.: *Creep in Engineering Structures*, Blaisdell, Waltham, Mass. 1966.

Kennedy, A. J.: *Processes of Creep and Fatigue in Metals*, Wiley, New York, 1963.

Penny, R. K., and D. L. Marriott: *Design for Creep*, McGraw-Hill, London, 1971.

Pomeroy, C. D.: *Creep of Engineering Materials*, Mechanical Engineering Publications, London, 1978.

SIXTEEN

NONDESTRUCTIVE TESTS

16.1 NONDESTRUCTIVE TESTS

Often it is desirable to know the characteristic properties of a product without subjecting it to destructive tests. With the exception of some hardness tests and proof loadings, the methods of testing discussed in the previous chapters will not permit the attainment of this objective, since most of the procedures, instead of using a finished product, use specially prepared specimens and test them to either partial or complete destruction.

Nondestructive tests may be divided into two general groups. The first group consists of tests used to *locate defects*. In it are various simple methods of examination such as visual inspection of the surface as well as the interior by the use of drilled holes, tests involving the application of penetrants to locate surface cracks, the examination of welded joints by use of a stethoscope to detect changes in sounds caused by hidden flaws, and highly technical methods involving radiographic, magnetic, electrical, and ultrasonic techniques. Radiographic tests include those that make use of the radiation of short electromagnetic waves such as X-rays and gamma rays. With the aid of these test methods it is possible to inspect the interior of opaque objects of appreciable thickness to ascertain their general homogeneity. Tests that use the magnetic characteristics of the materials inspected include *electromagnetic analysis*, which depends on changes in the electromagnetic characteristics of the material from point to point to detect any corresponding change in its mechanical or structural characteristics; and *magnetic-particle tests*, which are used to locate defects by noting irregularities in the magnetic flux of the object. The class of tests that depend on observations of *electrical characteristics* includes the Sperry detector for locating flaws in rails that are in place and in tube and bar stock. *Ultrasonic tests* depend on reflected sound effects to locate internal defects.

The second group of nondestructive tests consists of those used for determining *dimensional*, *physical*, or *mechanical characteristics* of a material or part. In this group are tests for the thickness of paint or nickel coatings on metallic bases, the thickness of materials from only one surface, the determination of moisture content of wood by electrical means, certain hardness tests, proof tests of various kinds, surface-roughness tests, and methods employing forced mechanical vibrations to determine the changes in natural frequency of the system due to changes in the properties of the material. One type of vibration test uses the sonic analyzer for determining the natural frequency, from which the modulus of elasticity can be computed. The simple determination of the latter property is of interest, since for some materials it is generally related to the quality of the material.

A faint and rather blurred line distinguishes nondestructive testing from *experimental stress analysis*. Obviously both kinds of testing concern stresses and both rely on experimentation. Materials testing, however, both destructive and nondestructive, emphasizes the determination of the properties (including defects) of *materials*, whereas the main objective of experimental stress analysis is to investigate the distribution of magnitudes of stresses in *structures* in certain configurations, made of certain materials, and subjected to certain loads. Although these purposes differ, the tools and methods are largely the same.

16.2 RADIOGRAPHIC METHODS

X-rays, beta rays, and the still shorter gamma rays are electromagnetic waves used in industry to penetrate opaque materials and give a permanent record of the result on sensitized film. These short wavelengths of radiant energy have become a useful tool for the inspection of the interior of metals and other materials. When the rays pass through material of nonuniform structure containing defects such as cavities, cracks, or portions of variable density, the rays passing through the less dense parts of the object are absorbed to a smaller extent than the rays passing through the adjacent sound material. On development of a light-sensitive film placed at the far side of the object exposed to short-wave radiation, a picture of light and dark areas results, the dark areas representing parts of the material having a lower density. This film is called an exograph when produced by X-rays and a gammagraph when produced by gamma rays; both types of film are termed *radiographs*. To be successful, a radiograph must show the size and shape of any significant defect or nonhomogeneity. The general principle involved in the production of a radiograph using X-rays and gamma rays as sources of radiant energy is illustrated in Fig. 16.1. The recommended practice for radiographic testing is covered in ASTM E 94, and related terms are defined by ASTM E 586. Proper safety precautions are essential in all such testing.

The two most common applications of industrial radiography are in the examination of welded products and castings; installations that test the former are the most numerous. Installations in foundries are generally used to develop a proper foundry technique and less frequently for routine inspection. X-ray and beta-ray instruments are also used to measure the thickness of sheet stock moving as fast as 10 m/s.

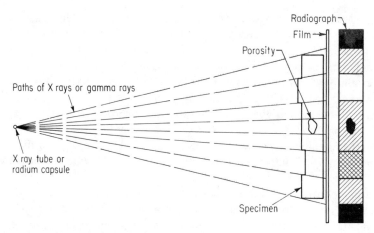

Figure 16.1 Production of a radiograph.

The X-ray method is much more rapid than the gamma-ray method, requiring seconds or minutes instead of hours. In the case of annular objects this difference may be partly offset when using gamma rays by placing films around the circumference, with the radium in the center, and exposing the whole circumference at one time.

Equipment X-ray tubes in common use are usually the Coolidge type, as shown in Fig. 16.2. The X radiation is produced when electrons traveling at very high velocities are suddenly stopped by impact with the tungsten anode. The intensity or quality of the radiation depends on the rate of liberation of electrons from the cathode, and this determines the exposure time necessary to produce a satisfactory radiograph. The release of electrons is governed by the temperature of the incandescent filament and can be regulated by a rheostat in a separate low-voltage circuit. The intensity is directly

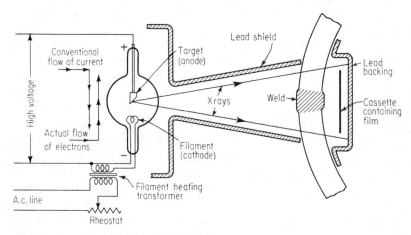

Figure 16.2 Arrangement for radiographing a welded joint. *(From St. John and Isenburger, Ref. 85.)*

proportional to the amperage of the high-voltage current that flows through the tube. The wavelength or quality of the radiation determines its penetrating power. This property depends on the velocity with which the electrons are driven from the cathode to the target and is controlled by an adjustment of the high voltage supplied to the tube; the higher the voltage, the shorter the wavelength of the radiation and the greater the penetrating power. For industrial radiography the current supplied to the X-ray tube may be about 1 to 10 mA, and the voltage may range from 50 to 2000 kV.

Xeroradiography is an advanced X-ray technique employing a reusable dry plate to record the image. Xeroradiography uses static electricity, much as a comb does in picking up a piece of paper. The static electricity arranges a fine powder on a specially coated aluminum plate and produces an image similar to that on conventional X-ray film. Using this process, an X-ray image is available for viewing within 45 s, whereas conventional X-ray film requires 1 h to develop and dry for viewing.

Xeroradiography uses a conventional X-ray tube and requires no new X-ray equipment. The need for a darkroom is eliminated, and a permanent record of the image on the plate may be made in a few seconds by conventional photographic methods.

Radium and its salts decompose at a constant rate, giving off gamma rays that are of much shorter wavelength and more penetrating than ordinary X-rays. It is estimated that the wavelengths of gamma rays (about 2 pm) correspond to the average for X-rays excited by about 750 kV.

Radium sulfate, sealed in small silver capsules and enclosed in an outer duralumin container to facilitate ease in handling and protection against loss, is the form commonly used. For commercial radiography, 100 to 300 mg is usually placed in a unit, the intensity of the radiant energy varying directly with the amount used.

The apparatus necessary for gamma-ray radiography is very simple. The container of radioactive material is supported rigidly in front of the specimen to be inspected, and films in suitable holders are fastened to the back of the specimen. Cobalt 60, an isotope produced by neutron irradiation, can be used in place of radium, and since it is much cheaper it may replace the latter.

The radiograph A radiograph is simply a record of the differences in intensity of the radiation that penetrates through the various parts of the test object to produce on sensitized film a shadow picture called the image. The intensity of the emergent radiation and of the resultant image is a function of the distance between the source of the ray and the film, being inversely proportional to the square of this distance. It also depends on the amount of radiation absorbed, which varies with the thickness and the material of the object examined; the absorption increases very rapidly with the thickness according to an exponential equation and increases with the density of the material. The absorption is also a function of the wavelength of the radiation: the longer the wavelength, the greater the absorption. Films have various emulsion properties.

Several of these variable factors can be adjusted by the operator so that the radiation image has sufficient intensity to be registered adequately within a reasonable time on the film; this may take from a few seconds to several hours. To assist the operator in selecting the proper conditions, exposure charts are constructed for specific conditions of use.

For a radiograph to have value, it must show all *significant* defects present and their size. This requires that small differences in density or effective thickness of the object produce sufficiently large differences in intensity of the emergent radiation to be clearly recorded on the film. Furthermore, there must not be any disturbing influence that distorts or fogs the shadow picture produced.

To interpret a radiograph accurately, it is necessary to know the size of the smallest defect that it can be depended on to show. The size of the smallest discontinuity detectable on the radiograph depends on the definition or sharpness of the record and the contrast between adjoining areas.

The size of the ray source is important in determining *sharpness*, because, as shown in Fig. 16.3, the smaller the source, the larger the intense umbral shadow u and the narrower the penumbral ring (p minus u). Also, it is evident from the figure that the closer the flaw is to the film, the sharper the shadow. Both of these conditions tend to facilitate the detection of the flaw. The same result is obtained by using a large distance D between the ray source and the flaw; the distance should be made as large as possible without unduly increasing the time of exposure. In general, the ratio D/t should never be less than 7; higher values up to about 30 are preferable. Modern X-ray tubes are made so that the focal spot is small enough to produce sharp images at reasonable distances D, say, 75 cm. To secure clear images in gamma-ray work the distance D should be governed by the amount of radium used.

Low voltage enhances *contrast*. The darker portions of the radiograph (negative) indicate the less dense parts of the object, whereas the lighter portions indicate the more opaque parts; the reverse is true for a print from the negative. This nonuniformity may be caused by (1) differences in thickness due to the form of the object, (2) differences in thickness due to voids, or (3) differences in density of the component

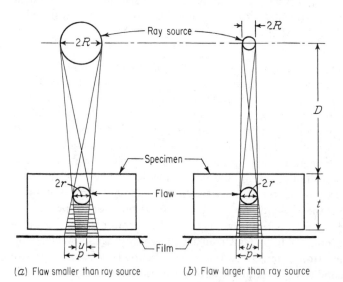

(*a*) Flaw smaller than ray source   (*b*) Flaw larger than ray source

Figure 16.3 Influence of size of source on size and sharpness of image. *(From Lester, Ref. 49.)*

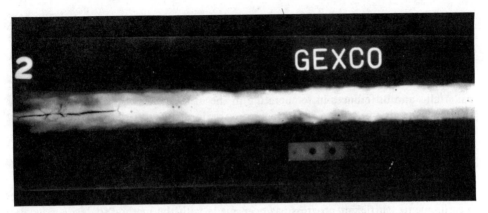

Figure 16.4 Radiograph of a 20-mm weld. (*Courtesy of General Electric X-Ray Corp.*)

parts, for example when metal parts are embedded in a material such as a plastic or when slag inclusions occur in welds.

Common defects and their characteristic appearance on the negatives of castings are as follows [64]:

1. Gas cavities and blowholes are indicated by well-defined circular dark areas.
2. Shrinkage porosity appears as a fibrous irregular dark region having an indistinct outline.
3. Cracks appear as darkened areas of variable width.
4. Sand inclusions are represented by gray or black spots of an uneven or granular texture with indistinct boundaries.
5. Inclusions in steel castings appear as dark areas of definite outline. In light alloys the inclusions may be more dense than the base metal and thus cause light areas.

In steel welds the most common defects are slag inclusions, porosity, cracks, and incomplete fusion. Since these flaws are less dense than steel, they produce darkened areas on the negative. Incomplete fusion produces a dark line parallel to the joint; the other imperfections appear as described for castings. Figure 16.4 shows a radiograph of a welded joint, which reveals a large crack at the left and a profusion of gas pockets through the centerline. In the lower right-hand part of the photograph one may notice the image of a thin plate with holes of various sizes. This is a calibration device called a penetrameter, which helps in determining the smallest detectable defect.

16.3 ELECTROMAGNETIC METHODS

The magnetic characteristics of a material are related to its composition or structure and its mechanical properties. It appears possible, at least theoretically, to locate differences and discontinuities of structure and variations in the dimensions and various properties of a given material by making magnetic measurements on the material and comparing them with corresponding measurements on some standard or reference material. Certain practical limitations are involved, however, owing to the fact that the magnetic properties may be subject to irregular changes that do not correlate well with changes that occur in the mechanical properties of the material. In addition to variables that influence the mechanical properties, a magnetic test is sensitive to internal strains and differences in temperature in the material, which exert little effect on the mechanical properties. The effect of variations due to internal strains is most noticeable when high frequencies or low magnetizing forces are employed. These conditions make it most difficult to determine definite quantitative relations and to make satisfactory practical applications of the principles.

While numerous factors such as variations in chemical composition, structure, internal strains, temperature, and dimensions are known to influence the magnetic measurements, sufficient progress has been made with the method to devise magnetic tests for a variety of commercial applications that are in daily use. In these applications the difference in magnetic properties between the object tested and a piece known to have the requisite mechanical properties and freedom from defects is determined. The magnitude of the difference that can be tolerated is determined by experiment. In these industrial applications, only one magnetic property is ordinarily used in a given type of test for evaluating the comparative quality of a material. This property may be simply the magnetic permeability, some quantity such as residual induction or coercive force derived from the hysteresis loop, or the waveform of an alternating induced voltage as determined by the shape of the hysteresis loop.

While in general dc methods have considerable value for certain commercial applications, it has been found that ac methods have many advantages. For example, observations that depend on the shape of the hysteresis loop developed by the use of alternating current are often found to be a better indication of structural characteristics than simple permeability as determined by the use of direct current. In the ac methods the sample under observation forms the core of a transformer. The various characteristics of the core cause shifts in the phase angle between the current and the voltage of the induced current and give rise to harmonics in the ac wave. By measuring the shifts or the harmonics, differences in the magnetic characteristics of the cores (samples) can be distinguished. Alternating-current methods are most suitable for material of uniform cross section.

Equipment is available that permits continuous magnetic analysis of bar stock and pipe 6 to 90 mm in outside diameter [107]. It includes a test-coil assembly, an indicator with control cabinet, and a feeding mechanism. The test-coil assembly consists of a heavy primary coil and a secondary coil. The material to be inspected is passed through the secondary coil, located coaxially within the primary coil, and is energized by alternating current. The secondary coil contains a number of windings connected to the test circuits, and thus it acts as a pickup or detector.

One of three available circuits is used as a flaw detector. Its use is based on the magnetic-flux leakage method and depends on the action of special detector coils in connection with a compensating network and high-gain amplifier. The output of the amplifier is measured by means of a microammeter and controls the flashing of a light. If an average speed of about 0.6 m/s is used, short defects may not cause impulses sufficient for a noticeable deflection of the microammeter, although they are strong enough to flash the light. Since this method is based on a comparison of one section of the test bar with another section of the same bar, it is limited to indicating the beginning and end of a defect only. For this reason, continuous defects (seams) even of considerable depth are sometimes not detectable.

The other two circuits are of the wave-analyzer type. They permit the investigation of variations of amplitude at certain arbitrarily fixed phase points of the curves representing the two electromotive forces induced in the two separate detector windings. These circuits are used for the detection of flaws as well as for the investigation of composition variations. Although less sensitive than the main flaw-detector circuit, they outline the full length of detectable defects. Because the circuits use alternating current the depth of penetration of this method is limited, precluding the possibility of locating internal defects and restricting magnetic inspection to the detection of surface flaws.

In commercial work one bar is selected as the standard of comparison, preferably after a thorough check to ascertain correct composition, dimension, heat-treatment, and freedom from flaws. As this bar is passed through the coil, the test circuits are individually and successively compensated to show zero deflection for the desired degree of sensitivity and selectivity. The operator then passes the bars in succession through the coil, watching the resultant meter deflections as well as the light flashes. The types of product particularly suited to magnetic inspection are hot-rolled and cold-drawn straight bar stock and several types of pipe and tubing.

Sucker rods used in drilling oil wells have been inspected by magnetic methods for the presence of fatigue cracks. In this application the sucker rod is passed through a suitably energized test coil as it is withdrawn from the oil well. A pickup coil is connected to a recorder to make a graph of the magnetic characteristics. An analysis of these graphs permits the detection of parts of rods having abnormal magnetic characteristics. Past records appear to show that such characteristics indicate rods that are susceptible to fatigue failures.

Magnetic methods are used for measuring the *thickness* of enamel, paint, and nickel coatings on metallic bases. One application is the portable electric enamel-thickness gage developed by the General Electric Company, which is used to measure the thickness of enamel or paint on a flat steel surface [23]. The reluctance of the magnetic circuit of the sensitive gage head when placed on a coated steel surface varies with the thickness of the coating. The indicator unit connected to the gage head is calibrated to read thickness directly in thousandths of an inch.

Another type of instrument for measuring the thickness of coatings on metal has been developed at NBS [10]. The instrument employs a portable spring balance for measuring the force required to detach a permanent magnet from the surface under test. The thickness of nickel coatings on nonmagnetic base metals is determined by one form of the instrument while a second, employing a smaller magnet and a stiffer spring, is used to measure the thickness of nonmagnetic coatings on ferrous metals [9, 10]. The greater the thickness of a nickel coating, the greater the force required to detach the magnet from the coating, whereas for nonmagnetic coatings on iron or steel the force decreases with the thickness. Both instruments must be calibrated.

Standard terminology and symbols relating to magnetic testing are defined by ASTM A 340 and E 268.

16.4 THE MAGNETIC-PARTICLE METHOD

The magnetic-particle method[†] of inspection is a procedure used to determine the presence of defects at or near the surface of ferromagnetic objects. It is based on the principle that, if an object is magnetized, irregularities in the material, such as cracks or nonmetallic inclusions, which are at an angle to the magnetic lines of force, cause an abrupt change in the path of a magnetic flux flowing through the piece normal to the irregularity, resulting in a local flux leakage field and interference with the magnetic lines of force. This interference is detected by the application of a fine powder of magnetic material, which tends to pile up and bridge over such discontinuities. Under favorable conditions, a surface crack is indicated by a line of the fine particles following the outline of the crack, and a subsurface defect by a fuzzy collection of the fine particles on the surface near the defect [18].

Various types of defects are detectable by the magnetic-particle method. A few of the more common ones are cracks caused by quenching, fatigue, and embrittlement, seams, and subsurface flaws. An example of fatigue cracks in an airplane gear that were invisible to the eye but detected by the magnetic-particle method is shown in Fig. 16.5. Subsurface defects can be located only when they are relatively close to the surface, but usually the desirability of locating them increases as they approach the surface. The materials to be examined must necessarily be capable of being magnetized to an appreciable degree.

[†]Sometimes called magnaflux inspection, a term that derives from a trade name of the Magnaflux Corp.

Figure 16.5 Fatigue cracks in airplane gear detected by the magnetic-particle method. (*From Doane, Ref. 18.*)

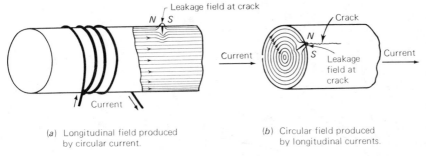

(a) Longitudinal field produced
    by circular current.

(b) Circular field produced
    by longitudinal currents.

Figure 16.6 Orientation of magnetic fields. *(From McCune, Ref. 61.)*

Various magnetizing methods may be used for practically any steel part, some more satisfactory than others. One basis of classification divides them into *residual* and *continuous* methods. In the first method the residual magnetism in the part is relied on when the magnetic powder is applied. In the continuous method the current inducing the magnetic flux in the part to be inspected is allowed to flow while the powder is applied.

Another basis of classifying the methods of magnetizing objects for this type of inspection depends on the character of the induced magnetic field; circular fields produce circular magnetization, and solenoid fields produce longitudinal magnetization. These two types of magnetization are illustrated in Fig. 16.6.

A third basis of classifying the methods used in magnetizing depends on the kind of current, direct or alternating, that is used. Direct current is often employed, since it appears to permit the detection of defects lying more deeply in the section, but in many cases either alternating or direct current can be used, so that the character of the available power supply may be the deciding factor.

The magnetic powder, the base of which is generally iron or black magnetic iron oxide, is ground to pass a 149-μm sieve.[†] The individual particles are elongated rather than globular to obtain a better polarization, and those of metallic iron are coated to prevent oxidation and sticking. The powder may be applied *dry* or *wet* and is available in both black and red. The color is selected that shows up to best advantage on the parts inspected. Some magnetic particles are prepared with a fluorescent coating, inspection being carried out under ultraviolet light so that every crack is marked by a glowing indication. The magnetic-particle method is described in ASTM E 709.

For the dry method the powder is applied in the form of a cloud or spray. For the wet method the powder is suspended in a low-viscosity noncorrosive fluid such as kerosene. The magnetized part is either immersed in the liquid or the liquid is flowed or sprayed over the part.

An advantage of the dry powders is that they are not so messy to work with as is oil. Where large areas are to be inspected and recovery of the oil-paste suspension would be difficult, the dry method is used extensively. From a magnetic standpoint an advantage of the dry powder over the wet is that it is better for locating near-surface

[†]No. 100 sieve.

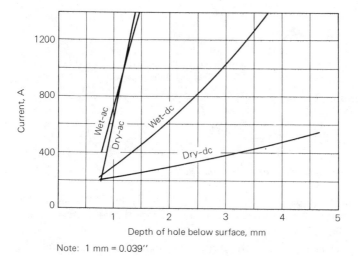

Note: 1 mm = 0.039''

Figure 16.7 Threshold indications of near-surface cavities. *(From Doane, Ref. 18.)*

defects, as shown in Fig. 16.7. The wet method is often considered superior for locating minute surface defects, because in general the available paste is finer than the available powder. The ability to reach all surfaces of the part being inspected, including vertical surfaces and the underside of horizontal surfaces, by hosing or by immersion, is a considerable advantage in favor of the wet method.

The results of an experiment made to show the effect of various test procedures on the indications obtained are shown in Fig. 16.7. The specimen was a 40-mm length of 100-mm-diameter SAE 1020 seamless tubing with a 20-mm wall. Nine holes made with a 0.7-mm-diameter drill were spaced at intervals around the ring at various distances from the surface. The "threshold" current is the lowest current that would produce a distinctly noticeable powder pattern on the ring. Referring to the figure, it will be noticed that as the depth of the defect increases, the amount of alternating current required for threshold indications increases considerably faster than does the amount of direct current needed. In addition, if the curves are extended below the point where they intersect, it can be concluded that for defects at or extremely near the surface alternating current at a given amperage may reveal more than direct current. The curves also show that there is very little difference between the sensitivity of the wet and dry method when alternating current is used, whereas when direct current is used, the dry method gives satisfactory indications at considerably lower current values than those required for the wet method.

Reasonably smooth surfaces of uniform color offer the most favorable conditions for formation and examination of the powder pattern. When the dry method is used, grease and soft loose rust should be removed. Paint and metallic plating have the effect of converting surface flaws into subsurface flaws. In considering whether or not paint is to be removed, the relative thickness of the paint layer and the size of the smallest flaws that are being sought by the inspection must be considered. Rough surfaces, for example that of a rough weld, interfere with the powder pattern; such surfaces should be ground reasonably smooth.

The placing of a magnetized part in service after a magnetic-particle inspection may result in the attraction of steel particles to areas of local polarity and cause wear. If for this or other reasons demagnetization is necessary, the most practical and convenient method for accomplishing it is to subject the piece to the action of a magnetic field that is continually reversing in direction and at the same time gradually decreasing in strength. Alternating current at the available frequency is employed for demagnetizing to a greater extent than any other current, usually in connection with an open solenoid. The part to be demagnetized is placed in the solenoid and then slowly withdrawn while the current is flowing.

16.5 ELECTRICAL METHODS

Electrical analysis methods are applicable to any material that can conduct an electric current, regardless of whether the material is magnetic or nonmagnetic. Thus electrical methods overcome one of the principal limitations of magnetic methods.

Flaws in rails The Sperry apparatus for the detection of flaws in railway rails is installed in a self-propelled car, which is run over the track to be tested at a speed of about 10 km/h. Several thousand amperes of low-voltage direct current from a generator are fed into the track by the main brushes, which are mounted between the wheels of the rear truck. Auxiliary brushes mounted on the forward truck are used to preenergize the track so that the indications obtained will be more reliable.

A flaw of sufficient magnitude to cause a deviation in the direction of the current in the rail will produce a corresponding deviation in the direction of the magnetic field around the rail, which is determined by detector coils mounted on the rear truck between the main brushes and located just above the rail, one ahead of the other. The currents induced in these coils are opposed to one another and are normally in balance. When unbalanced by the influence of any defect passing under one or the other of the two coils, the current generated is amplified sufficiently to operate a recording device. A continuous record is made on a moving tape of all deviations in the magnetic field and of the location of the rail causing such deviations. The detector coils also control a paint gun that makes a spot of paint on the rail at the locations of any appreciable magnetic deviations.

When the record tape shows that a flaw has been located, the car is stopped and the magnitude of the flaw is determined by a drop-of-potential method using portable equipment. Transverse fissures, compound fissures, horizontally split heads, and vertically split heads are the defects most commonly located.

Flaws in tubes The Sperry electrical detection of flaws in cylindrical metal tubes is accomplished by a different method from that used for rails. It consists in causing a controlled electric current to flow through the tube in a direction transverse to the direction of a possible flaw. Any flaw in the path of the current will increase the resistance of the circuit and produce a difference in the value of the current, which can be measured with suitable instruments.

Most flaws in tubing extend in a longitudinal direction, so the induced current is

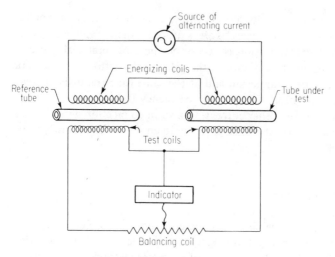

Figure 16.8 Simplified diagram of Sperry electric detector for flaws in tubes. *(From Knerr, Ref. 45.)*

made to flow circumferentially in the tube. This is done by passing an alternating current through a set of energizing coils surrounding the tube, inducing the desired currents.

A simplified diagram showing a typical application of the method is presented in Fig. 16.8. The two tube lengths shown, one sufficiently perfect to be taken as a standard or reference tube and the other the tube length under test, are surrounded by identical energizing coils carrying the same alternating current. Two identical test coils also surround the tubes and are in electrical balance when no defect is present. A flaw in the tube under test upsets the balance of the test circuit as shown by the indicator.

In commercial installations the tubes under test are passed through the equipment at a rate of approximately 0.25 m/s, and the machine is arranged so that the presence of a defect will stop automatic-feed rolls. The defective tube is then marked.

Machines of this general type, with certain refinements in the electric test circuits, are said to be capable of detecting flaws 3 mm long, extending halfway through the wall of a tube of ordinary low-carbon steel that has a diameter as small as 15 mm and a wall thickness of 1.0 mm.

Metal parts and stocks that appear alike but differ in their composition, heat-treatment, and so on can be separated quickly and easily with a somewhat similar electric instrument called a metals comparator, which is effective for sorting either magnetic or nonmagnetic materials.

Surface roughness Direct numerical measurement of surface roughness is possible with electronic instruments that measure the average height of the surface irregularities. Such instruments include the Profilometer (Physics Research Co.) shown in Fig. 16.9 and the Surface Analyzer (Brush Instruments). The tracer unit that is moved over the surface contains a diamond-tipped stylus. In the Profilometer the stylus is spring loaded and moves up and down with the irregularities as the tracer head is moved.

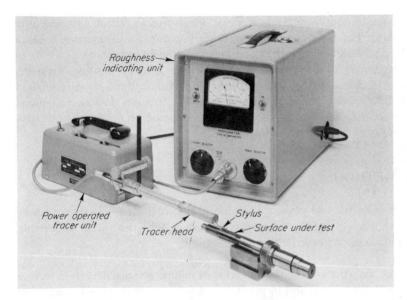

Figure 16.9 Electronic device for measuring surface roughness. *(Courtesy of Physics Research Co.)*

These mechanical fluctuations are converted into corresponding electric fluctuations as a tube connected to the stylus moves up and down within the coil of an electric system. The electric fluctuations are amplified to operate the instrument that shows the average height of surface roughness of the part.

Moisture in wood Although the moisture content of wood can be determined by cutting out a small sample and drying it in an oven, frequently it is desired to determine the moisture content without cutting the wood. This can be done within a few seconds using an electric moisture meter. Such equipment is suitable for sorting lumber on the basis of moisture content and can test finished woodwork in place without serious damage to the wood. Two types of meters are available, one that determines the moisture content by measuring the electric resistance of the wood and another that measures the electric capacity of the wood.

Below the fiber-saturation point (a moisture content; see Chap. 20), the *electric resistance* of wood increases greatly as the moisture content decreases. Although the resistance is affected by temperature and the species of wood, corrections supplied with the instrument are easily made.

The electrical contact points are needles mounted so that they can be easily driven into the wood and then withdrawn after testing. For wood that is not of a uniform moisture condition, the meter indicates the moisture content at or near the points of the needles, since the wood becomes a better conductor at greater depths from the surface, where the moisture content is higher. Thus it is possible to determine the moisture content at any distance from the surface by driving the electrodes to that depth. In general, the moisture content at a depth of one-fifth the thickness of the material is at about the average value for the wood. For thin material use is made of surface-plate electrodes (instead of needle electrodes), which make contact with opposite surfaces of a board. For both types of electrodes the actual measurement is made by balancing the resistance between the electrodes with known resistances.

The *electric capacity* of wood varies directly with the moisture content. Temperature effects are negligible, as is the species of the wood. Although the capacity method is excellent for determining the mass of water in wood, it is not possible to convert this to a percentage without knowing the density of the wood. In practice, the density value is usually taken to be the average for the species.

16.6 SONIC METHODS

The response of materials and objects to induced high-frequency vibrations can be used both to detect defects and to determine properties. Vibrations with frequencies beyond the audible range are called *ultrasonic* vibrations.

Defects Ultrasonic vibrations allow the detection of minute internal defects in ferrous and nonferrous metallic objects and in plastics, ceramics, and so on. The frequency of the vibrations used is in the range of 100 kHz to 20 MHz, whereas the audible or sonic range is only 20 Hz to 20 kHz. Both sonic and ultrasonic vibrations or waves are transmitted through solid materials much more readily than through air; in fact, the waves initiated at one face of solid objects are reflected back at any air gap in the material or at the opposite face, shown in Fig. 16.10. The test method for locating defects makes use of this phenomenon by electronically determining the relative times for the ultrasonic waves to be reflected back from the defect and from the opposite face.

Ultrasonic waves are usually produced by the piezoelectric effect, which is the mechanical deformation in certain crystals, such as quartz, when placed in electric

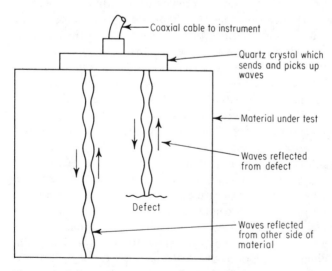

Figure 16.10 Detection of defects by ultrasonic waves.

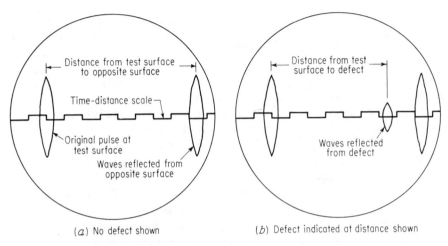

(*a*) No defect shown   (*b*) Defect indicated at distance shown

Figure 16.11 Oscilloscope screen of ultrasonic tester.

fields. An alternating voltage produces mechanical oscillations of the crystal at the same frequency. The probe containing the crystal is placed against the test piece, which is then subjected to these oscillations. The waves, like light, tend to travel in beams with a divergence angle determined by the ratio of the wavelength to the diameter of the source. In steel a sound at 5-MHz frequency has a wavelength of only 1.25 mm, so that for a crystal less than 25 mm in size the divergence of the beams is relatively small.

Usually one crystal probe both sends and receives sound. It is placed against the test piece using an oil film for better transmission of the sound waves. One cycle of the continuous operation of the equipment is about as follows: if a 5-MHz frequency is used, electric oscillations of this frequency would be applied to the crystal for, say, 1 μs so that five waves of the 5-MHz frequency would be sent through the surface of the object. The crystal then stops sending and is ready to receive any reflected waves after they have traveled through the steel at about 6.225 km/s.[†] The reflected waves vibrate the crystal, producing electric impulses that are fed into a cathode-ray oscilloscope. This cycle is repeated 60 times per second. The oscilloscope is timed so that the beam sweeps to the right 60 times per second in coordination with the sending cycle. The original vibrations striking the sending surface makes a sharp peak (or *pip*) at the left side of the oscilloscope screen, and the reflected waves are indicated by a pip toward the right at a distance dependent on the thickness traveled and the time required. While the crystal is receiving for $\frac{1}{60}$ of a second there is time for a wave to travel 50 m and return before the next cycle begins. If the piece of steel has a defect, most of the beam striking the defect will be reflected back to the crystal before the beam that reaches the opposite surface is reflected back to the crystal. The oscilloscope would then show a pip not only for the entering waves and for those reflected back from the opposite surface but also one in between for the waves reflected from the defect, as shown in Fig. 16.11. A time (or distance) scale in the form of a square

[†]In air, the velocity is only 344 m/s (at 20°C).

wave is constantly shown on the oscilloscope; this indicates the distance from the surface where the crystal is applied to the reflecting surface, whether the latter is the defect or the opposite side of the object. The distance scale may be varied so that one cycle of the wave may indicate, say, 10 or 100 mm.

Since the beam from the crystal is very narrow, it is necessary that every bit of the surface be covered by a progressive movement of the crystal. Furthermore, since some cracks may be parallel to the waves and therefore reflect very little of the beam, it is necessary to conduct two series of tests in which the ultrasonic beams are normal to each other. For checking the performance of ultrasonic testing equipment it is customary to use standard reference blocks having various sizes of holes drilled in them at various distances from the test surface, as specified in ASTM E 127 and E 428. Other ASTM specifications cover the use of ultrasonic equipment for locating defects.

Large forgings are inspected for internal soundness before expensive machining operations are carried out, moving strip or plate is inspected for laminations and thickness, routine inspections of locomotive axles and wheel pins for fatigue cracks are performed in place, and rails are inspected for bolt-hole breaks and other failures without the necessity of dismantling rail-end assemblies. Many industrial applications of ultrasonic inspection are of large products that could not be examined satisfactorily by any other known method.

Thickness The thickness of plates, tube walls, and the like can be determined from one side only by a procedure somewhat similar to the ultrasonic reflection method for locating defects in material. The Sonizon equipment shown in Fig. 16.12 uses the resonance principle and is suitable for thicknesses of any solid material up to 100 mm.

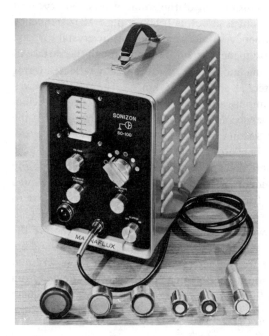

Figure 16.12. Ultrasonic tester for measuring thickness from one side only. (*Courtesy of Magnaflux Corp.*)

In this equipment a continuously varying frequency is fed to a quartz crystal held against the test piece, the mechanical resonant frequency of which varies with its thickness. When the varying frequency of the crystal hits this resonance, there is an electric change within the crystal causing a pip on the calibrated thickness scale of the oscilloscope screen. Six different crystal probe units are shown in the figure.

Stiffness When a solid homogeneous object is struck with a hammer, it emits a clear ringing sound, whereas a defective (cracked) object has a well-known dull sound. This ancient technique, simple but only qualitative, may be extended to give a quantitative measure of some mechanical property, especially stiffness.

If the induced or forced vibrations caused by a blow are transverse to the longitudinal axis of the object, the modulus of elasticity of the material can be computed by use of an appropriate flexure formula if the distance between nodes and the frequency of vibration are known. Within the audible range, comparison with sounds produced by vibrations of known frequencies is one way to make quite accurate estimates of frequency, provided the operator has good sound perception. Within or beyond the audible range, some tuned radio-frequency circuit operating an oscillograph, together with electrically produced vibrations, may be used to determine accurately the natural frequency of an object being tested. A so-called "sonic" or "dynamic modulus of elasticity" is determined by this method, which has been applied to such materials as concrete; dynamic values of the modulus correlate with those obtained by the slower static-loading method. However, the latter method is often inapplicable, since values of the modulus of some materials are influenced by inherent plastic flow.

Changes in the modulus of elasticity can be used as a criterion for estimating the effects of certain test conditions imposed on a specimen of a material such as concrete, for example continued exposure to the deteriorating influence of alternating cycles of freezing and thawing [38]. A progressive measure of deterioration can then be obtained by periodic determinations of the modulus of elasticity, since the test is nondestructive.

The sonic method also has been proposed for determining the modulus of elasticity of structural members or parts under load. An ultrasonic pulse applied at one face of a concrete member is received by a sensitive unit at the opposite face. The time taken by the pulse to pass through the concrete, as measured by an electronic timing circuit, is used to compute the modulus of elasticity, which is also used to estimate the compressive strength [42, 73].

Terms relating to ultrasonic testing are defined by ASTM E 500.

16.7 MECHANICAL METHODS

Not all mechanical testing methods are destructive. In fact, types of nondestructive mechanical tests too numerous to mention are routinely used both to determine properties and to control quality.

Figure 16.13 Calibrating a Schmidt concrete test hammer. (*Courtesy of Pacific Coast Aggregates.*)

Concrete test hammer A nondestructive impact test for determining the hardness and the probable compressive strength of concrete in a structure is made using the Schmidt concrete test hammer. The tubular unit is pressed against the surface of the concrete to be tested, as shown in Fig. 16.13, causing a spring-loaded hammer inside the tube automatically to strike the concrete. The rebound of the hammer after impact is indicated on a scale. The rebound number thus obtained, in combination with a calibration curve, can be used to give a fairly good value of the compressive strength of the concrete. The result serves as a convenient substitute for test cylinders cured at the site for evaluating the compressive strength of the structure at early ages, or for cores cut from the hardened concrete. During calibration, the test cylinder must be supported solidly to prevent any movement while the hammer is used; the cylinder's compressive strength is then determined for correlation with the hammer reading.

Hardness tests Hardness tests have been discussed in Chap. 12 with destructive tests, but in many cases they are, in effect, nondestructive. Owing to the general relation that exists between hardness and the physical properties of materials, hardness tests are commonly made on parts to be used in service to control heat-treating operations and to ensure satisfactory physical characteristics in materials. The test may be made on surfaces located so that the indentation and all accompanying strain effects will be removed by later machining. If this is not possible, care should be exercised to see that the size, location, and nature of the indentation will not be objectionable or result in later damage.

Proof tests Many types of so-called proof tests are used. Perhaps the most common examples are the proof tests of chain, eyebolts, wire ropes, crane hooks, and other parts used for lifting. The applied proof load is usually well above the allowable service load. The proof load applied to welded wrought-iron chain is commonly high enough to deform the links. For eyebars that are to be proof-loaded, the eyes are usually made undersize so that after loading the effect of any distortion is cut away when the eyes are enlarged to correct size.

The specifications of Lloyd's Register and the American Bureau of Shipping require that steel castings for ship parts and anchors shall be subjected to a drop test onto a hard surface and then be suspended clear of the ground and well hammered all over with a heavy sledge hammer to test the soundness of the material. The value of such a test is questionable, since it probably does not subject the casting even to normal loads. A proof tension test for anchors is often specified and appears to have considerably more merit than the dropping and hammering test.

Pressure tests of various types of tanks or pipe systems constitute a type of proof test of the material or the structure itself. Such tests reveal not only defects but also regions of low strength, for these regions undergo excessive deformations. The pressure applied is commonly some multiple of the design pressure, such as 1.5 times the service pressure. Water is much safer than air as the pressure medium. For high-temperature tests, steam or hot oil may be employed, but their use is rather dangerous. While under load the container or pipe being tested may be struck with a specified weight of hammer.

16.8 VISUAL METHODS

Visual inspection of the object should never be omitted whenever it is necessary to detect the presence of possible surface defects. Although it may appear unnecessary to list this as a test method, some tend to overlook the advantages to be gained by a careful visual inspection using low-power magnifying glasses as well as microscopes, if necessary. Microscopes equipped with photographic attachments are often used to get permanent records of defects, questionable areas, and variations in structure. The eye may be aided by penetrants, colored liquids that emphasize defects.

Surface roughness In various industrial operations it is necessary to produce smooth metallic srufaces for good appearance, low friction in bearings, and so on. There are several crude methods of determining roughness. *Touch inspection* involves moving one's fingernail along the surface at a speed of about 2 to 3 cm/s. Irregularities as small as a few micrometers can be felt by this method. *Visual inspection* with the aid of illuminated magnifiers is limited to rougher surfaces, and the results vary with the observer. *Microscopic inspection* involves a comparison of the test surface with a standard finish. Since only a small portion of the surface is visible at one time, the determination of a reliable average requires many observations.

Penetrant tests All liquid-penetrant processes are nondestructive testing methods for detecting discontinuities that are open to the surface. They can be effectively used for metal products and for nonporous, nonmetallic materials such as ceramics, plastics, and glass. Surface discontinuities such as cracks, seams, laps, laminations, or lack of bond are indicated by these methods. They are applicable to in-process, final, and maintenance inspection. The various methods of liquid-penetrant inspection are covered in ASTM E 165.

The liquids used enter small openings such as cracks or porosities by capillary action. The rate and extent of this action are dependent on such properties as surface tension, cohesion, adhesion, and viscosity. They are also influenced by factors such as the condition of the surface of the material and the interior of the discontinuity.

For the liquid to penetrate effectively, the surface of the material must be thoroughly cleaned of all material that would obstruct the entrance of the liquid into the defect. After cleaning, the liquid penetrant is applied evenly over the surface and allowed to remain long enough to permit penetration into possible discontinuities. The liquid is then completely removed from the surface and either a wet or a dry developer applied. The liquid that has penetrated the discontinuities will then bleed out onto the surface, and the developer will help delineate them, showing the location, general nature, and magnitude of any discontinuities present. To hasten this action the part may be struck sharply to produce vibrations to force the liquid out of the defect.

The oil-whiting test is one of the older and cruder penetrant tests used for the detection of cracks too small to be noticed in a visual inspection. In this method the piece is covered with a penetrating oil, such as kerosene, then rubbed dry and coated with dry whiting. In a short time the oil that has seeped into any cracks will be partially absorbed by the whiting, producing plainly visible discolored streaks delineating the cracks.

Fluorescent-penetrant inspection makes use of a penetrant that fluoresces brilliantly under so-called black light having a wavelength of 365 nm, which is between the visible and ultraviolet in the spectrum. The coating of developer used with it is not fluorescent but is dark when viewed under black light. It acts to subdue background fluorescence on parts and causes any defects to show up more distinctly under black light. Pores show as glowing spots, cracks as fluorescent lines, and where a large discontinuity has trapped a quantity of penetrant, the indications spread on the surface.

Some nonfluorescent penetrants can be easily seen in daylight or with visible light. They are usually deep red in color so that indications at defects produce a definite red color as contrasted to the white background of the developer. Depth of surface discontinuities may be correlated with the richness of color and speed of bleed out.

"Filtered-particle" inspection depends on the unequal absorption into a porous surface of a liquid containing fine particles in suspension. This preferential absorption causes the fine particles in the solution to be filtered out and concentrated directly over the crack, producing a visual indication.

In a test method for locating cracks in nonconducting materials such as plastics, ceramics, and glass, the object is first covered with a special penetrant that conditions the defects, the surface is dried, and a cloud of fine electrically charged particles is blown over the surface, causing a buildup of powder at the defect. The penetrant is not required when the nonconductor is backed by a conductor, as in enamelware.

16.9 PHOTOELASTICITY

One problem that arises in the design of solid structural members and machine elements is the magnitude and distribution of stresses throughout the part. Of particular interest are the zones of stress concentration and magnitudes of the maximum principal and shear stresses. Stress concentrations may occur due to the way the external forces are applied or due to the shape of the part.

For conditions of plane stress in an isotropic material, a means of experimentally determining the stress distribution or stress concentration within the range of elastic behavior is by photoelastic analysis.[†] In this technique a model of the part is made of a suitable transparent material, such as transparent annealed Bakelite or a plastic designated CR-39. By passing polarized light through the loaded model it becomes birefringent (i.e., doubly refractive), and a pattern of colored or light and dark bands or *fringes* can be observed, the pattern being a function of the stress distribution in the part. These observations, together with other supplemental measurements, enable the state of stress to be determined at points of interest or throughout the part.

A simplified arrangement of apparatus for making a photoelastic analysis is shown schematically in Fig. 16.14. A variety of arrangements of optical systems and of component parts are in use. The model in which the stress distribution is to be studied, shown at M, is in a beam of parallel polarized light. The light from source S has been polarized by the polarizer P. The polarizing device may be a Nicol or similar prism, but a sheet of material called *Polaroid* is often used. The rays through the polarizer and the model are made parallel by a lens system L_1. After leaving the model, the light passes through another polarizing unit at A, called the analyzer; the plane of polarization of the analyzer is set at right angles to the plane of polarization of the polarizer. Another lens system at L_2 serves to project the emerging light pattern on a viewing screen V. With the polarizer and analyzer "crossed" at right angles, if there is no model in the polariscope, and if the polarizing units are fully efficient, the beam of light is extinguished and no light reaches the screen.

[†]Some procedures for three-dimensional photoelastic analysis have been developed [34].

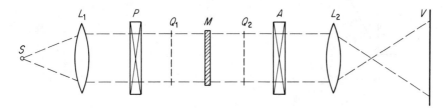

Figure 16.14 Schematic arrangement of a polariscope.

Plane polarization The method depends, essentially, on two experimentally demonstrated facts. When polarized light of some given wavelength is passed through a strained transparent plastic body (1) the incident ray of polarized light at each point is broken up into two component rays of light vibrating in mutually perpendicular planes that are parallel to the directions of the principal stresses and (2) each component ray is transmitted through the transparent solid material with a velocity that is a linear function of the principal stress with which it is associated.

The process of composition and resolution of the rays of light passing through the system is diagrammed in Fig. 16.15. The two component rays emerge from the stressed material out of phase and vibrate in planes at right angles to each other.

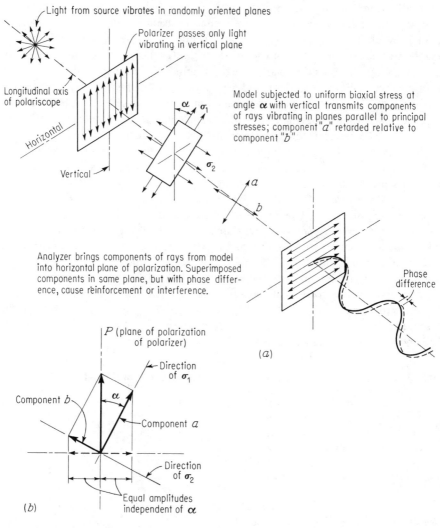

Figure 16.15 Principle of simple polariscope.

When these component rays are passed through the second polarizing device (the analyzer), whose plane of polarization is at right angles to the initial incident polarized ray, the component vibrating light waves are again made to vibrate in the same plane (the plane of polarization of the analyzer). However, although the emerging components vibrate in the same plane and the waves now have the same amplitude, they are still out of phase by an amount proportional to the difference in the principal stresses at the point through which the ray passed in the stressed solid.

When the components are out of phase by one-half wavelength ($\lambda/2$) or an odd integer multiple ($3\lambda/2$, $5\lambda/2$, and so on) the ray will be extinguished. For other phase differences the resultant ray will be partially or completely transmitted.

If the entire beam of monochromatic light (all rays) passing through the stressed solid is now projected on some viewing screen, a pattern of light and dark bands may be observed; each dark band, called a *fringe*, is the locus of points having the same difference in principal stress (or constant maximum shear stress). If white light is used, each band is made up of a sequence of colors of the spectrum; lines of the same color (representing the locus of points having the same principal stress difference) are called *isochromatic* lines.

If the stressed solid is subjected to a uniform state of stress throughout, the image on the screen will be all dark, all light, or all the same color, depending on the magnitude of the difference in the principal stresses and the kind of light used.

Successively larger differences between the two principal stresses produce correspondingly larger phase differences in the final superimposed beam. Corresponding to the odd-integer multiples of $\lambda/2$ are a series of fringes (dark bands, or spectral bands) of successively *higher order*. For a given material and thickness, the stress difference for one fringe order can be determined by stressing a sample of the material of which the model is made under uniform uniaxial stress.

Then, beginning at a corner where both principal stresses are zero, the number of fringes to a given point can be counted and a measure of the principal stress *difference* at that point determined. If there is no corner at which to begin, first locate an *isotropic* point (i.e., where both principal stresses are equal). This point will be where the black isoclinic lines (described below) are the same for both white and monochromatic light. The point is also one where the difference of the principal stresses is zero. Therefore, it can be used as a starting point to count the number of fringes to any other point. Alternatively, an experimenter can load the specimen gradually, counting the fringes as they pass. With either method, the principal stress differences (which equal twice the maximum shear stresses) can be mapped.

One way to find the *sum* of the principal stresses is to measure the change in thickness (Δt) of the model under stress. The change is a function of the sum of the principal stresses (σ_1 and σ_2), the modulus of elasticity E, and Poisson's ratio μ. Thickness measurements are usually made by very precise calipering, although light interference caused by reflection of light from the front and rear surfaces of the model can be used. In the direction normal to the plane of the model the stress (σ_3) is zero and the strain therefore is

$$\frac{\Delta t}{t} = 0 - \mu \frac{\sigma_1}{E} - \mu \frac{\sigma_2}{E} = -\frac{\mu}{E}(\sigma_1 + \sigma_2)$$

A calibration factor C can be determined by stressing the model material in uniform tension and used to find the sum of the stresses from Δt:

$$\sigma_1 + \sigma_2 = \left(\frac{E}{\mu t}\right)\Delta t = C \Delta t$$

Once the sums *and* the differences of the principal stresses are known, the magnitude of each of the principal stresses may be computed for each point of interest in the model by simple algebra, that is

$$\sigma_1 = \frac{(\sigma_1 + \sigma_2) + (\sigma_1 - \sigma_2)}{2}$$

$$\sigma_2 = \frac{(\sigma_1 + \sigma_2) - (\sigma_1 - \sigma_2)}{2}$$

Once the principal stresses are known, all stresses can be obtained.

To find the stress sums is, to be sure, experimentally difficult. But the stress differences alone are often of value, as at many *boundaries* the minor principal stress is zero. The major principal stress then is known and quite often critical.

When the *directions* of the principal stresses are parallel to the planes of polarization of the polarizer and analyzer, the light is extinguished. In an image of a stressed model, the locus of points having the direction of the principal stresses parallel to the planes of polarization appears as a black line or zone; such lines of constant direction or inclination of the principal stresses' directions are called *isoclinic* lines. By rotating both the polarizer and analyzer, the direction of the principal stresses throughout the stressed models may be determined. When white light is used, the isoclinic lines are more readily distinguishable from the stress fringes.

The directions of the principal stresses are inclined at 45° to the directions of maximum shear.

Circular polarization In the plane polariscope the light can vanish under two conditions:

1. The phase differences are odd multiples of half a wavelength, giving fringes (with monochromatic light).
2. The directions of the principal stresses coincide with the polarizer and analyzer axes, yielding isoclinic lines (with any light).

To avoid confusion of the isoclinic lines with the stress fringes, a quarter-wave plate is inserted in the light beam on each side of the model (at Q_1 and Q_2 in Fig. 16.14) to eliminate the isoclinics.

The quarter-wave plates transform the plane polarized beam into two mutually perpendicular polarized components, one being $\lambda/4$ out of phase with the other. As these two waves pass an observer, their resultant continuously sweeps through a circle. Hence this is called *circularly polarized light*.

In a circular polariscope, the first quarter-wave plate causes the plane polarized light to rotate at constant speed and amplitude as it passes through the model. This does not affect the retardation causing the fringes, but the directional effect is eliminated (just as an airplane's propeller casts a shadow when still but not when rotating). The second quarter-wave plate, crossed at right angles with the first, restores the resultant light to oscillation in a single plane. The result of the entire procedure is that nothing is changed once the light passes the analyzer, but the quarter-wave plates have one important effect: the isoclinics are eliminated.

Models If the models are small, they must be carefully shaped and annealed. They should have uniform thickness and flat surfaces.

Good quantitative stress determination by photoelastic analysis requires careful technique, precise work, and good equipment. However, inexpensive apparatus and fairly crude procedures can often give helpful qualitative pictures of the nature of stress concentrations.

Figure 16.16 shows a fringe photograph of a model of a simple-span beam subjected to two concentrated loads. The closeness of the fringes at a support and at the two load points indicates that high stress concentrations exist there.

PhotoStress A special photoelastic method of determining stresses in actual structures or components of any material while under load, on flat or curved surfaces, is called *PhotoStress*. It acts as an infinite number of strain gages of practically zero gage length, uniformly distributed over the surface to be studied.

In this method, liquid PhotoStress plastic is applied and allowed to polymerize directly on the part, or flat or contoured sheets of transparent PhotoStress material are bonded to the part, which has been previously made reflective by spraying the surface with aluminum paint.

When a load is applied to the object, strains developed in the bonded plastic coating cause it to become birefringent, as noted earlier. When illuminated with polarized white light, the birefringent plastic coating displays a pattern of colored fringes, which is a function of the strain distribution of the entire surface and from which peaks of strain can be determined in terms of their location, direction, sign, and magnitude.

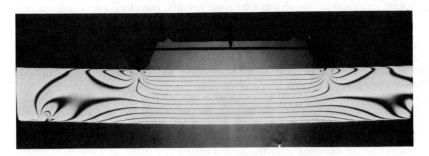

Figure 16.16 Fringe pattern for a beam subjected to two concentrated loads.

Two sets of bands are visible in the strained plastic coating when observed through a *reflection polariscope*. Black bands, or isoclinics, give the directions of the principal strains. The colored fringes, or isochromatics, give the magnitude of maximum shear strain when only one reading is made of fringes with light going under normal incidence through the plastic, and give separate values of principal strains when two readings are made of fringes, one with light going under normal incidence through the plastic and the other under oblique incidence.

Because the birefringence in the plastic is proportional to the thickness as well as to the strain, the thickness must be accurately known. When liquid plastic is used, its thickness can be measured with commercial gages or with the proper PhotoStress instrument, but when sheets are used their thickness is known. The determination of the directions of principal strains (isoclinics) is independent of the thickness of the coating.

PROBLEM 16 Photoelastic Analysis

Object: To determine lines of constant maximum shear and lines of direction of maximum shear.
Preparatory reading: ASTM E 671 and reference on beam stresses and Mohr's circle.
Specimens: Photoelastic tension and bending specimens, as assigned. Material properties and thickness should be identical.
Special apparatus: Polariscope.
Procedure:
1. Measure both specimens carefully and calculate the appropriate section properties.
2. Arrange the tension specimen in a tension tester and tighten it with a very small load.
3. Place the polariscope into position, with the planes of polarization of the polarizer and analyzer normal to each other. Using white light, rotate both plates together so that a gray band is visible along the centerline of the specimen. If quarter-wave plates are used, the directions should be mutually perpendicular and inclined at 45° to the planes of polarization of the polarizer and analyzer; in that case, repeat with monochromatic light.
4. Gradually increase the load, observing the development of isochromatic fringes. Record the load whenever a reddish-blue line passes, continuing for four or five cycles.
5. After removing the tension specimen arrange and load the flexure specimen. At a suitable load level sketch (and perhaps photograph) the fringe pattern.
6. Remove the quarter-wave plates, if present. Trace the isoclinics by rotating the polarizer and analyzer gradually, starting from the vertical/horizontal position, and observing the moving black lines.
Report:
1. Analyze the calibration by graphing the tensile stress vs. the fringe number.
2. On a picture of the flexure specimen draw lines of constant shear (the fringe pattern), and assign values to the maximum shear by using the results from the calibration and counting fringes from locations of known zero shear.
3. Prepare a similar sketch showing the directions of maximum shear by observing that the isoclinics, that is, the directions of the principal stresses, are inclined by 45° to the maximum shear directions.
4. Analyze the given beam by simple structural theory, prepare sketches similar to the experimental results, and compare the analytical and experimental results.
5. Comment on the reasons for discrepancies you discovered during the comparisons.

DISCUSSION

1. Most people do some simple proof testing in everyday life. Can you think of any examples?

2. A 100-mm cast iron pipe is suspected of severe corrosion of the inside. How could its condition be checked without disconnecting or damaging it?

3. What method could be used to locate the reinforcing bars in an existing concrete wall?

4. Sonar is a method of locating submerged submarines. From your knowledge of nondestructive testing, can you speculate how it works?

5. At the pool, a high-diver walks out to the end of the 3-m board and jumps up and down a few times before preparing to dive. Explain this action in materials-testing terms.

6. You have a 3-m-long pipe fully coated with a thin opaque film. List some tests that, without damaging the pipe in any way, might help you identify the material.

7. There are battery-operated instruments that tell if a potted plant needs water. Explain the principle and name an engineering application.

8. Discuss the advantages and disadvantages of using wet powder in the magnetic-particle method.

9. You see some sunglasses in a store and wonder if they are of the polarized type. Explain a quick test that will give you the answer.

10. Devise a proof test to ensure that the heads of sledge hammers are firmly attached to the handles.

READING

Betz, C. E.: *Principles of Penetrants*, Magnaflux Corp. Chicago, 1963.

Dally, J. W., and W. F. Riley: *Experimental Stress Analysis*, 2d ed., McGraw-Hill, New York, 1978.

Durelli, A. J., and W. F. Riley: *Introduction to Photomechanics*, Prentice-Hall, Englewood Cliffs, N. J. 1965.

Hendry, A. W.: *Elements of Experimental Stress Analysis,* SI ed., Pergamon, Oxford, 1977.

Hughes, T. P.: *Elmer Sperry, Inventor and Engineer*, John Hopkins, Baltimore, 1971.

Kuske, A., and G. Robertson: *Photoelastic Stress Analysis*, Wiley, London, 1974.

Libby, H. L.: *Introduction to Electromagnetic Nondestructive Test Methods*, Interscience-Wiley, New York, 1971.

McGonnagle, W. J.: *Nondestructive Testing,* Gordon and Breach, New York, 1961.

Zandman, F.: "Photostress, Principles and Applications," *Handbook*, Society of Non-Destructive Testing, 1959.

THREE

PROPERTIES OF MATERIALS

SEVENTEEN

METAL

17.1 COMPOSITION AND CHARACTERISTICS

The elements called metals together surely constitute a most important engineering material. They can be combined with one another and with some nonmetals to form alloys that have characteristics superior to any pure metal for most purposes. Alloys with a great many combinations of properties can be made.

The reference to pure metals has to be taken, almost literally, with a grain of salt. Even when absolute purity is desirable, which is the case quite rarely, it cannot be fully attained. Whenever we speak of "pure" metals, therefore, we mean metals with less than 1 percent impurities.

Alloys are essentially solutions of one or more metals, called *solutes*, in another metal, the *solvent*. These solid solutions are quite similar to liquid ones and in fact are usually obtained by cooling a liquid solution. To have a solution in any state, the atoms or molecules of the solute must be uniformly and randomly distributed throughout the solvent. At a given temperature there is a limit to the amount of salt that can be dissolved in water. With metals similar limits exist, with but a few exceptions, so that for most combinations only a certain amount of a metal can be dissolved in another.

Metals are thought of as being divided into two broad groups, *ferrous* and *nonferrous* metals, depending on whether the major constituent is iron or not. The word "ferrous" is derived from the Latin noun *ferrum*, iron.

Ferrous metals Ferrous metals are principally iron-carbon alloys containing small amounts of sulfur, phosphorus, silicon, and manganese. Some are alloyed with copper, nickel, chromium, molybdenum, vanadium, or other elements to alter their physical and mechanical properties.

305

The three common forms of ferrous metal are steel, wrought iron, and cast iron. *Steel* is essentially a solid solution of carbon in iron. Since iron at room temperature will not hold in solution more than about 1.7 percent carbon by mass, this becomes the upper theoretical limit of carbon in steel; however, commercial steels rarely contain more than 1.2 percent carbon.

Wrought iron is practically a low-carbon steel except that it contains a small amount of slag, usually less than 3 percent. The carbon content is generally less than 0.10 percent, and the small particles of slag that are thoroughly distributed throughout the metal appear as long fibrous elements owing to the rolling operations employed in its manufacture; these slag fibers serve to distinguish it from steel having the same carbon content.

Ordinary *cast irons* contain 2.2 to 4.5 percent carbon. Slow cooling of cast iron results in the separation of as much as 0.9 of the carbon in the form of graphite, the remainder being chemically combined with iron to form iron carbide or cementite. Ordinary cast iron is called "gray" cast iron because of the color of its fracture. Rapid cooling of molten cast iron does not permit the separation of the graphite, and most of the carbon remains combined with the iron. Such material is called "white" cast iron, since a fractured surface has a characteristic silvery white metallic color.

The approximate carbon content for the principal classes of ferrous metals, designated by commonly used names, is presented in Table 17.1. The properties of ferrous metals can be improved by heat-treatment or cold-working.

Nonferrous metals Nonferrous metals and their alloys are also an important group of engineering materials. Some have a high strength-mass ratio, whereas others have good antifriction qualities and resistance to corrosion, and still others are suitable for die-casting and extrusion. Heat-treatment does not generally improve their properties to the same extent as it does those of steel alloys. Cold-working, however, quite effectively increases the yield strength of most nonferrous metals.

Annealed *copper* is relatively weak, but many of its properties such as strength and fatigue resistance may be improved by the addition of alloying elements and subsequent heat-treatment. It is one of the best electrical conductors besides being highly resistant to atmospheric corrosion.

The principal use of *nickel* is as an alloy for the purpose of improving the properties of copper, iron, and steel. Its mechanical properties approximate those of medium-carbon steel, and it is much stronger, tougher, and stiffer than any other commercial nonferrous metal. It is highly resistant to corrosion, retains its strength at elevated

Table 17.1 Carbon content of principal ferrous metals

Material	Carbon, percent by mass	Material	Carbon, percent by mass
Wrought iron	Trace to 0.09	Tool steel	0.61 to 1.20
Boiler steel	0.10 to 0.15	Special cast iron	1.21 to 2.20
Structural steel	0.16 to 0.30	Cast iron	2.21 to 4.50
Machine steel	0.31 to 0.60		

temperatures, and can develop these properties in metals with which it is alloyed. As pure nickel, however, it is too costly for general use.

The ordinary mechanical properties of *zinc* do not have much significance because it tends to creep at low stresses and temperatures. It has good resistance to corrosion, and when used as a protective coating on iron and steel plate, the product is known as galvanized iron.

Aluminum is one of the relatively strong, lightweight metals of construction, and only gold is more malleable. Its value as an electrical conductor, even though its conductivity is twice as good as copper of the same length and mass, is greatly reduced by its low resistance to fatigue stresses. Pure aluminum has good resistance to atmospheric corrosion but is readily attacked by strong alkalies and certain acids. Since its mechanical properties can be considerably improved by alloying it with small amounts of other metals, most aluminum is used in the alloyed state.

Magnesium is the lightest metal available in sufficient quantities for use in engineering construction. It is not used unalloyed. Humid air produces a self-healing film of magnesium hydroxide on its surface and thus protects the underlying metal from further oxidation.

Beryllium has about the same density as magnesium, but it has a much higher strength-weight ratio, which together with a high tensile modulus of elasticity gives it great possibilities in lightweight construction. In fact is has one of the highest known stiffness-density ratios, but high cost and poor workability limit its present use. Its principal value lies in its use as an alloy, especially with copper.

Lead is the densest of the common metals and therefore is often used for radiation shielding. Lead is also highly resistant to corrosion. Its very poor mechanical properties can be improved somewhat by alloying, mainly with antimony and tin. One application of lead alloys is for bearings, as the alloys used are able to retain a film of lubricant under adverse conditions.

The outstanding property of *zirconium* is its low neutron absorption. This makes it valuable as a structural material in nuclear reactors, where it may be used to make the cladding rods that house the fuel pellets.

17.2 PRODUCTION METHODS

Metals are found in nature in the form of mineral-bearing rocks called ores. Minerals are chemical compounds containing one or more metals and other elements, commonly either oxygen or sulfur. Examples are magnetite (Fe_3O_4), alumina (Al_2O_3), and galena (PbS). The extraction of metals from ores is divided into two phases, namely mineral dressing and process metallurgy.

Mineral dressing consists of mechanical operations during which the desired minerals are separated from the waste material, known as gangue or tailings. The rocks are first crushed and ground to a suitable size. Subsequently the minerals can be separated from the tailings, or concentrated, by gravity, flotation, magnetism, or other techniques.

Process metallurgy reduces the impure minerals chemically to refined metals with the desirable properties. These processes may use heat (pyrometallurgy), electricity (electrometallurgy), or water (hydrometallurgy). Ferrous metals are processed by pyrometallurgy, aluminum by electrometallurgy; depending on the ores used, copper may be produced by hydrometallurgical or pyrometallurgical methods, and is then often refined electrolytically.

After the desired chemical composition of a metal (or alloy) is achieved, the metal must be processed by physical means. *Metal processing* serves the dual purpose of shaping the material and of altering its structure and properties. It includes operations such as hot- and cold-working (rolling, drawing, extruding, spinning), casting, surface treatments (hot dipping, cementing), and heat-treatment (annealing, hardening, tempering).

Because they are so widely used, we will take a closer look at the processing of steel and aluminum. This will also illustrate the preceding very brief discussion of process metallurgy and metal processing.

Iron and steel production Iron ores, consisting chiefly of iron oxide, are deoxidized and partially purified in a blast furnace to produce pig iron. In the operation coke is used both as a reducing agent and a fuel, and limestone is used as a fluxing agent. Some of the pig iron from the blast furnace is remelted and used for making iron castings, but most of it is made into steel.

The refinement of pig iron to produce steel is usually accomplished in a Bessemer converter, an open-hearth furnace, or an electric furnace. The open-hearth process is still used more than any other, but the electric furnace is being used in many installations, particularly for the production of high-quality steel. These two latter methods have an advantage over the Bessemer method, since they can be used with any grade of pig iron or mixture of pig iron and scrap steel, whereas the Bessemer method is suitable only with special grades of pig iron. Some high-grade steels are made from steel scrap or high-grade iron by the crucible process. In all methods of making steel the principal object is the decrease in the amount of the impurities to specified limits. Excess carbon is removed as carbon dioxide gas; the oxides of other impurities form a slag on top of the molten steel. The Bessemer converter uses no additional heat as do the open-hearth and electric furnaces; instead, the air blown through the steel oxidizes the impurities and thus generates enough heat to keep the charge liquid.

When molten steel is cast directly into its final form, the resulting product is coarsely crystalline and relatively brittle and weak. However, this condition can be improved by suitable heat-treatment of the product. Casting the steel into ingots that are subsequently hot-worked to the desired shape greatly improves the properties of the finished steel.

Most steel is hot-rolled to its final shape, but many shapes that cannot be made by rolling are formed by a forging process. This operation produces a denser mass by closing minute cavities and reducing the grain size. Industrial forming operations on cold metal are known as cold-working. In general, cold work might be said to be any plastic deformation carried on below the lowest critical temperature. Such cold-forming operations as rolling, cupping, and drawing produce strain hardening, which is accompanied by an appreciable increase in tensile strength and a marked decrease in ductility of the metal, as shown in Fig. 17.1.

The *heat-treatment* of steel is a heating and subsequent cooling procedure for the

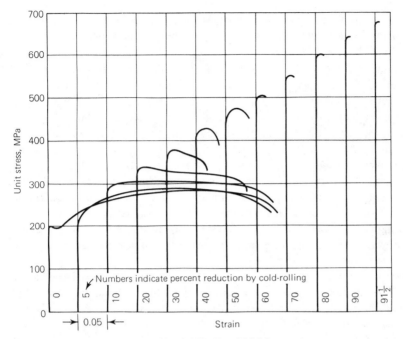

Figure 17.1 Effect of cold-rolling on stress-strain diagrams for ingot iron. (*From* Metals Handbook—1948, *American Society for Metals, Cleveland, Ohio, 1948.*)

purpose of improving certain desirable properties of the metal. The three principal operations classified as heat-treatment are annealing, hardening, and tempering. *Annealing* involves heating the steel to about 800°C[†] and maintaining that temperature long enough for sufficient refinement of grain size to occur to produce softness, to relieve any internal strains caused by cold work or rapid cooling, or to eliminate any coarse crystalline condition due to cooling from excessive temperatures. The heating is followed by slow cooling, which leaves the steel in a strain-free condition and with the fine grain size attained during the heating period. The annealed steel is weaker but more ductile and tougher than the unannealed steel. When steel is heated as in annealing but cooled rather rapidly, usually in still air, it is said to be "normalized." Such steel also has a refined structure, free of the principal internal strains.

Hardening of steel is accomplished by using the same heating process as for annealing and then cooling by sudden immersion in some *quenching* medium such as oil, water, or brine. In the quenched condition, the hardened steel is generally so brittle that its practical usefulness is greatly restricted. To correct this condition the steel is usually *tempered* by reheating to some temperature between 200 and 550°C, which is well below that of the preliminary heating, followed by either rapid or slow cooling. This operation is sometimes called drawing. It produces a steel that is somewhat softer but much tougher than the steel quenched at the higher temperature only. Tempered

[†]800°C = 1472°F.

steel is generally superior to hardened steel, since ordinarily it can be made hard enough for many uses yet tough enough so that it will not shatter in service.

Aluminum production Unlike iron and most other metals of interest to engineers, aluminum is normally produced by an electrolytic process rather than by directly smelting its ores. The mineral used is alumina, which is refined from very impure but also extremely abundant ores called "bauxites" because of their initial discovery near Les Baux in France. After oxygen and silicon, aluminum is the third most abundant element on earth.

Alumina is usually produced from bauxite by the Bayer process. The procedure involves four steps: (1) digestion, during which the bauxite is ground, mixed with a solution of sodium hydroxide, and dissolved under heat and pressure; (2) clarification, during which the resulting saturated solution of sodium aluminate is filtered; (3) precipitation, during which aluminum hydrate crystals are precipitated and cleansed; and (4) calcination, during which the water is driven off at temperatures of about 1000°C. The result is pure alumina or aluminum oxide, a very stable compound resembling sugar. It is the principal raw material for making aluminum but also has other uses.

Almost all aluminum is currently produced by reducing alumina in the Hall-Heroult process. Steel cells lined with carbon—often arranged in "potlines" of a hundred or more cells—are filled with an electrolyte consisting mainly of the mineral cryolite (Na_3AlF_6). From carbon electrodes a current (about 5 V, 100 kA) passes through the electrolyte to the bottom of the cell. A crust forms on top, onto which alumina is placed to be preheated and dried. At intervals the crust is broken and the alumina stirred into the bath. The alumina is dissolved and electrolytically decomposed (we are not sure exactly how), and alumina is deposited on the bottom and can be siphoned off. The oxygen part of the alumina combines with carbon and escapes as carbon dioxide. About 1000 kg of metal can be produced per day in one cell. Each kilogram requires about 50 to 60 MJ of electric energy, a prodigious amount.

The highly pure molten aluminum is siphoned into crucibles, from which it can be cast into molds to become "foundry ingots" or first alloyed to form "fabricating ingots." The major alloying elements are copper, manganese, silicon, magnesium, and zinc.

Certain alloys are best suited for casting; others can be used as wrought aluminum. In general, wrought aluminums do not allow as much alloying material as do cast aluminums, because alloying tends to destroy ductility while increasing strength and hardness. Annealing, however, will improve the ductility of wrought alloys.

Aluminum can be formed and worked by all methods. It can be cast, hot- and cold-rolled, extruded, drawn, and forged. Plates and shapes are made by hot-rolling, and the plates can then be made into very thin sheets by cold-rolling. In extruding and drawing, the metal is pressed through an orifice whose shape it retains; for extrusions the metal is heated, for drawing it is not. In forging, a preheated ingot is subjected either to hammer blows or to pressure in a press. Cold-working hardens and strengthens the metal.

Heat-treatments improve the final characteristics of both cast and wrought alloys. Ingots are usually preheated before cold- or hot-working to improve workability. Annealing is similar, but times are shorter and temperatures lower (typically 3 h at

425°C), and cooling is carefully controlled. To ensure that the alloying elements are properly dissolved, the alloy is heated for up to 24 hours and then quenched. Heating may also be used to relieve residual stresses caused by fabricating effects.

17.3 PROPERTIES OF CAST IRON

The properties of cast iron are mainly determined by the amount and form of the carbon present, by alloying elements, and by annealing.

Carbon has a very important effect on cast iron, since not only the amount but also the form in which the carbon exists affect the strength, hardness, brittleness, and stiffness of the metal. In gray cast iron, some of it is in the graphitic form as free carbon and the metal is weak and soft, whereas in white cast iron the carbon is in the form of cementite, which makes the iron strong and so hard that it can be finished only by grinding. The differences between several properties of these two types of cast iron are shown in Table 17.2. The strength and hardness of white cast iron may be three to four times that for gray iron, but the modulus of elasticity in tension is usually less than twice as great.

Nickel is often added to improve the properties of cast iron. Nickel cast iron may be divided into three classes. In the first, normal nickel cast iron, the nickel, which is in amounts of 0.5 to 3.0 percent, decreases the presence of hard spots in the metal and gives uniform machinability. In the second class, commonly known as chilled cast iron, the nickel content ranges from 3 to 5 percent. Usually some chromium and manganese are used together with the nickel, and with or without heat-treatment they produce a metal having a surface with a high resistance to abrasion. Its Brinell hardness number ranges from 600 to 750. When the nickel content is high, say, 12 to 20 percent, the metal is called corrosion-resistant cast iron because of its ability to give good service under conditions causing corrosion.

Chromium when added to cast iron is essentially a carbide former. It increases the "chill" (the hard outer shell) hardness and tensile strength at elevated temperatures up to certain limits. So-called chrome-nickel alloys of cast iron are quite common. *Molybdenum* has somewhat the same effect on cast iron as chromium, although it is less drastic in its action.

The *annealing* of white cast iron changes the combined carbon to a finely divided, free, amorphous form composed of small rounded particles called temper carbon. The resulting metal is malleable cast iron. It is weaker, softer, and much more ductile than white cast iron.

The strength parameters most often specified for cast iron, for example by ASTM A 48, are the minimum tensile strength and the breaking load applied at the center of various standard flexural specimens. The tensile strength of gray cast iron varies from about 140 MPa to 400 MPa. Table 17.2 shows typical values of mechanical properties of irons. Irons have densities of about 7.65 Mg/m^3 and coefficients of linear thermal expansion of about 10 to 12 MK^{-1}.[†] The melting temperature of iron is 1539°C.

[†]1 $MK^{-1} = 10^{-6} K^{-1} = (\mu m/m)/K$. This means a change of 1 μm for each 1 m of length per 1 K of temperature change. 1 K = 1°C = 1.8°F (temperature *differential*). See also App. A.

Table 17.2 Mechanical properties of iron†

Material	Strength in tension, MPa‡		Compressive yield strength, MPa‡ (0.2% set)	Strength in torsional shear, MPa‡		Modulus of elasticity, GPa		Elongation in 50 mm, percent	Brinell hardness No.	Modulus of toughness, MJ/m³ (MPa)‡	Fatigue limit, reversed bending, MPa‡
	Yield strength (0.2% set)	Ultimate		Yield strength (0.2% set)	Ultimate	Tension	Shear				
Gray cast iron	---	140	240	---	255	105	40	1	130	0.5	75
White cast iron	---	415	690	---	415	140	55	---	400	---	---
Nickel cast iron, 1.5% nickel	---	310	415	---	---	140	55	1	200	---	---
Malleable iron	230	345	230	130	330	170	70	14	120	---	180
Ingot iron, annealed, 0.02% carbon	165	290	145	105	205	205	85	45	70	---	180
Wrought iron, 0.10% carbon	205	345	205	125	240	185	70	30	100	95	170

†Based on F. B. Seely, *Resistance of Materials*, Wiley, New York, 1947; and *Metals Handbook–1948*, American Society for Metals, Cleveland, Ohio, 1948.

‡1 MPa = 145 psi.

17.4 PROPERTIES OF STEEL

The quality and mechanical properties of steel are determined by (1) the method of production, (2) the composition, (3) the mechanical work, and (4) the heat-treatment.

Carbon content Carbon is employed as the controlling constituent in regulating the properties of both common steels (so-called plain carbon steels) and alloy steels. Carbon's most important influence is on the strength, hardness, and ductility of the metal. Beginning with pure iron, which is soft and ductile, additions of carbon to normally cooled steel increase the hardness and strength and decrease the ductility. For each 0.1 percent of carbon added until "eutectoid" composition (0.84 percent carbon) is reached, the proportional limit is raised nearly 40 MPa, the tensile strength is increased about 70 MPa, and the elongation is reduced by about 5 percent, as shown in Fig. 17.2. Above the eutectoid point further additions of carbon result in increasing hardness and strength but more rapidly increasing brittleness.

The yield point, as shown in Fig. 17.3, and yield strength, as measured by the offset method, increase with an increase in carbon content. However, only steels with carbon contents below 0.40 percent have well-defined yield points.

Carbon has no appreciable effect on the stiffness of steel. The modulus of elasticity is practically the same (200 GPa) for all grades of steel, as shown in Fig. 17.3, but is slightly lower for wrought iron (185 GPa).

Alloying *Chromium* in steel forms a stable carbide. It produces stainless and heat-resisting iron and steel and makes steel hard and strong. Chromium is principally used with other alloying elements such as nickel and vanadium.

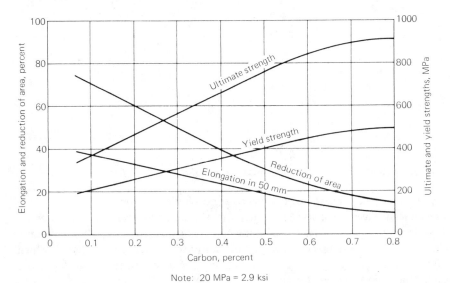

Note: 20 MPa = 2.9 ksi

Figure 17.2 Effect of carbon on tensile properties of hot-worked carbon steels. (*From Sisco, Ref. 83.*)

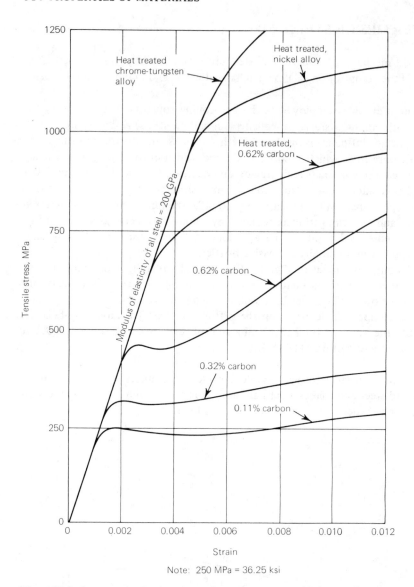

Figure 17.3 Stress-strain diagrams for various steels. (*From Mulenbruch, Ref. 71.*)

Copper increases the resistance of steel to atmospheric corrosion.

Manganese in steel has a strong affinity for oxygen and sulfur; it tends to eliminate these harmful components by withdrawing them into the slag. The manganese left in the steel after the removal of oxygen and sulfur forms the carbide, Mn_3C, which causes it to roll and forge better and also slightly increases the tensile strength and hardness. Manganese gives steel a fine-grained structure, and when present in 0.8 to 1.5 percent carbon steels, in amounts from 1.0 to 1.4 percent, it produces extreme hardness and resistance to abrasion.

Molybdenum is, next to carbon, the most effective hardening agent used in the manufacture of steel. It facilitates effective heat-treatment and causes the alloy to retain its hardness and cutting edge at high temperatures.

Nickel produces a fine-grained crystalline structure when added to steel, makes quenching effective to greater depths, and adds strength with little loss of ductility. It is useful in many types of stainless and heat-resisting steels. Low-nickel steels (2 to 4 percent) are used without heat-treatment for structural purposes and machine parts. The best qualities of such steels, however, are made available only by proper heat-treatment.

Oxygen causes steel to be both "hot-short" (brittle at red heat) and "cold-short" (brittle at ordinary temperatures). Overblown Bessemer steel without deoxidation contains only 0.15 percent oxygen, yet this renders the metal unfit for use. Oxygen combines with iron, forming an iron oxide, which appears as microscopic inclusions. These inclusions cause the metal to be known as "dirty" steel, which is weak, especially under fatigue and impact stresses.

Phosphorus causes steel to be cold-short, although some evidence has been given to prove that up to 0.1 percent it does not produce harmful brittleness. Most steel specifications limit phosphorus to about 0.04 percent.

Aluminum is a powerful deoxidizer, for it facilitates the escape of gases from molten steel. It is commonly added to plain carbon steel as well as to various alloy steels.

Silicon is a deoxidizing agent when added to steel and tends to diminish blowholes in ingots and castings. It is desirable because it increases both ultimate strength and elastic limit without decreasing ductility. In common steels each 0.1 percent of silicon increases the Brinell hardness number by about 6; this means that silicon is about one-third as effective as carbon in increasing hardness. Because silicon has a marked tendency to prevent the solution of carbon in iron, care must be exercised in treating steels having high carbon and silicon contents to avoid prolonged heating at high temperatures, since otherwise graphite may be formed. Typical steels include about 0.4 percent silicon.

Sulfur in the form of iron sulfide causes steel to be hot-short, but it can be partially neutralized by manganese so that it is comparatively harmless. However, most specifications commonly limit the sulfur content to about 0.05 percent. Since a high sulfur content produces a steel that is more easily machined, rods for making screws commonly have a high sulfur content.

Tungsten produces increased strength of steel, but this is accompanied by brittleness, which unfits it for structural uses. It makes steel suitable for permanent magnets, self-hardening tool steels, and high-speed steels, which retain their hardness and cutting edge at high temperatures.

Vanadium gives steel high elastic and tensile strengths. It produces a fine-grained "clean" metal.

Perhaps the highest tensile strength obtainable with present commercial alloy steel rods is about 2000 MPa,[†] using heat-treated silicon-molybdenum steel, which in its normalized condition has a tensile strength of about 850 MPa. However, the tensile

[†]1 MPa = 145 psi.

strength of plain carbon steel containing 0.80 percent carbon, when drawn into fine wires, approaches 3500 MPa. This may well be the maximum tensile strength of commercial steel available.

Processing In general, the superior qualities of a high-carbon steel or an alloy steel are not obtained by simple additions of carbon or metals to the steel. It is only by proper heat-treatment that the desired potential qualities are brought out.

Hot-working processes leave the metal in a comparatively strain-free condition and with a fine-grained structure, producing a steel of higher yield strength and tensile strength and of considerably greater ductility than existed before the hot-working operation.

Cold-working produces a higher yield strength, tensile strength, and indentation hardness (not scratch hardness) than hot-working, but the plasticity and formability decrease owing to the lowered ductility.

Various standards, such as those of ASTM, specify minimums and/or maximums for mechanical properties. These values, like the chemical requirements, generally differ within a particular standard depending on the section (shapes, plates, or bars), the size or thickness, and in some cases the particular grade. To obtain precise values the standards should be consulted.

Standards Common ASTM designations for structural steels include the following:

A 36   The basic steel for construction, with a yield point of about 250 MPa.

A 283   Carbon steels for plates, with relatively low strengths and cost.

A 441   High-strength low-alloy steels containing vanadium for strength and weldability and copper for corrosion resistance.

A 242   Weathering high-strength low-alloy steels that are self-protecting in clean atmospheres; welding them is difficult.

A 514   Steel types with yield points up to 700 MPa used for plates; the strengths are achieved by heat-treatment.

Another widely used specification is that of the Society of Automotive Engineers (SAE) in which various carbon and alloy steels are classified and the permissible ranges of the principal constituents are stated for each class. In this system the first digit indicates the type to which the steel belongs; thus 1 indicates a carbon steel; 2, a nickel steel; and 3, a nickel-chromium steel. In the case of the simple alloy steels the second digit generally indicates the approximate percentage of the predominant alloying element. Usually the last two digits indicate the carbon content in "points" or hundredths of a percent. For example, an SAE 1020 steel is a carbon steel that contains 0.20 percent of carbon and corresponds to a structural grade of steel.

Table 17.3 shows representative mechanical properties of some steels and indicates the effects of alloying and of processing. All steels have a density of approximately 7.85 Mg/m^3[†] and a coefficient of linear thermal expansion of about 11.7 MK^{-1}.

[†]$1 \ Mg/m^3 = 0.036 \ lbm/in^3$.

Table 17.3 Mechanical properties of steel[†]

Material	Strength in tension, MPa[‡]		Compressive yield strength, MPa[‡] (0.2% set)	Strength in torsional shear, MPa[‡]		Modulus of elasticity, GPa		Elongation in 50 mm, percent	Brinell hardness No.	Modulus of toughness, MJ/m³ (MPa)[‡]	Fatigue limit, reversed bending, MPa[‡]
	Yield strength (0.2% set)	Ultimate		Yield strength (0.2% set)	Ultimate	Tension	Shear				
Wrought iron, 0.10% carbon	205	345	205	125	240	185	70	30	100	95	170
Steel, 0.20% carbon											
Hot-rolled	275	415	275	165	310	200	85	35	120	115	215
Cold-rolled	415	550	415	250	415	200	85	15	160	85	275
Annealed castings	240	415	240	145	310	200	85	25	130		
Steel, 0.40% carbon											
Hot-rolled	290	485	290	170	380	200	85	25	135		
Heat-treated for fine grain	415	620	415	250	515	200	85	25	190		
Annealed castings	240	450	240	145	380	200	85	15	130		
Steel, 0.60% carbon											
Hot-rolled	435	690	435	255	550	200	85	15	200	85	345
Heat-treated for fine grain	540	825	540	325	690	200	85	15	235	105	380
Steel, 0.80% carbon											
Hot-rolled	505	825	505	305	725	200	85	10	240		
Oil-quenched, not drawn	860	1240	860	515	1035	200	85	2	360		
Steel, 1.00% carbon											
Hot-rolled	570	930	570	345	795	200	85	10	260	75	415
Oil-quenched, not drawn	965	1515	965	580	1275	200	85	1	430	15	690
Nickel steel, 3.5% nickel, 0.40% carbon, max. hardness for machinability	1035	1170	1035	620	965	200	85	12	350	95	525
Silicomanganese steel, 1.95% Si, 0.70% Mn, spring tempered	895	1200	895	540	795	200	85	1	380	145	

Note: Most steels depend on heat-treatment as well as on composition to develop particular mechanical properties.

[†] Based on F. B. Seely, *Resistance of Materials*, Wiley, New York, 1947; and *Metals Handbook–1948*, American Society for Metals, Cleveland, Ohio, 1948.

[‡] 1 MPa = 145 psi.

17.5 PROPERTIES OF NONFERROUS ALLOYS

With rare exceptions, nonferrous metals are used in alloyed form. Alloys exhibit properties superior to those of pure metals, and they can be heat-treated. Some common nonferrous alloys are briefly described in this article.

Copper Copper is used as the principal metal, that is, as the solvent, in a variety of nonferrous alloys. *Brass* made of 60 to 90 percent copper and 10 to 40 percent zinc is the one most used. For resistance to corrosion by water, the red brasses having a high copper content are preferred. For the highest strength a combination of approximately two parts copper to one part zinc, termed "standard" or "high brass," is best. Brasses suitable for hot-working, by forging, rolling, or extruding, usually contain only about 60 percent copper.

Ordinary *bronze* consists of copper with less than 20 percent tin. The tin hardens the copper but does not appreciably increase its tensile strength. With more than 25 percent tin, the alloy is weak and either brittle or soft. Very small percentages of phosphorus added to bronze produce a phosphor bronze that has good resistance to corrosion, excellent strength, and fair ductility.

A number of so-called bronzes contain very little or no tin; they are not true bronzes. Manganese bronze is essentially brass with a little tin and manganese. It has good mechanical properties and high resistance to corrosion. Aluminum bronze is copper with 5 to 10 percent aluminum. It also has good mechanical properties and high resistance to corrosion and is a good antifriction metal.

The most effective alloy of copper contains about 2.5 percent beryllium; besides retaining many of the desirable properties of copper, the alloy attains great strength. In the annealed condition it is as strong as mild steel, and proper heat-treatment will increase its tensile strength to nearly 1400 MPa. Its high fatigue limit, 240 to 300 MPa, is maintained even under corrosive conditions.

Aluminum The addition of small amounts of copper to aluminum serves to harden the aluminum and to increase its strength considerably. One such alloy is known as duralumin. By suitable heat-treatment and aging, its mechanical properties can be made to approach those of mild steel. The addition of small amounts of magnesium to aluminum also increases the strength. Such alloys may not be as strong as the aluminum-copper ones, especially when hard-rolled, but they are slightly less dense.

There are a great many aluminum alloys commercially available, and a standard classification code has been devised by the Aluminum Association. The designations are four-digit numbers for wrought alloys and three-digit numbers for castings. In each case the first digit indicates the major alloying element: 1 means pure aluminum; 2 designates copper; 3, manganese; 4, silicon; 5, magnesium; 6, magnesium and silicon; 7, zinc; and 8, any other element. The remaining digits designate the specific alloy. The number is followed by a hyphen and a code consisting of a letter and up to two digits, which indicates the temper. A typical designation, for example, is 6061-T6 for a magnesium and silicon wrought alloy. Other common types are 2014, 5083, 5086, 5454, 5456, and 6062.

Other alloys *Monel metal* is an alloy made of about 66 percent nickel, 28 percent copper, and small amounts of iron, manganese, and sometimes aluminum. It is highly resistant to corrosive liquids. It is very ductile, tough, and strong and retains its strength at high temperatures better than most other commercial ferrous and nonferrous alloys.

When *magnesium* is alloyed with 5 to 10 percent aluminum and small amounts of manganese, it produces an exceptionally lightweight metal that possesses good mechanical properties. The use of magnesium alloys in the construction of certain airplane parts, where high strength is not essential, is advantageous.

Alloys of zinc are most commonly used for *die-casting*, but alloys of aluminum, magnesium, and copper are also suitable. The zinc alloys have the highest strength and ductility.

Nonferrous alloys are generally classified as heavy or light alloys, depending on the density of the solvent. Tables 17.4 and 17.5 show the important mechanical properties of these two types of alloys, respectively. The coefficient of linear thermal expansion tends to be slightly above 23 MK^{-1} for aluminum alloys and slightly above 25 MK^{-1} for magnesium alloys. The melting temperatures of nonferrous metals differ considerably: aluminum, 660°C; copper, 1084°C; magnesium, 650°C; lead, 327°C; and zinc, 419°C.

17.6 CORROSION

When metals are exposed to the environment, chemical reactions often occur at the surface between the metal and the environmental chemicals. The chemicals may include acids, bases, brines, and the like, but they may be as seemingly innocuous as oxygen and water. The environment may be air, water, another gas or liquid, or soil; it may be natural or polluted.

This ubiquitous phenomenon is known as "corrosion." It is benign in some cases, but injurious in most. As metals corrode, they lose strength and other desirable properties. To compound the detriment, the corrosion layer may add significantly to the load on a structure, as is the case with a rusted steel bridge.

Direct chemical reaction Corrosion may take the form of a direct chemical attack, which is unaccompanied by any macroscopic flow of electric current and happens quite uniformly over a surface. The deposit may adhere to the surface or scale off. The thickness of adherent layers is approximately proportional to the square root of the time of exposure. The tarnishing of silver is a well-known example.

Corrosion of *iron* and *steel* is commonly due to a direct chemical reaction with the oxygen and water in moist air: rusting. Iron hydroxide is formed and then usually decomposes in turn into iron oxide and water:

$$Fe + \tfrac{1}{2}O_2 + H_2O \longrightarrow Fe(OH)_2 \longrightarrow FeO + H_2O$$

This process occurs at normal temperatures, provided that both oxygen and water are available.

Copper corrodes in much the same way. If polluted moist air contains carbon dioxide, copper carbonate is formed on the surface, lending the characteristic green

Table 17.4 Mechanical properties of heavy nonferrous alloys[†]

Metal or alloy	Approximate composition, percent	Tensile yield strength (0.2% set) MPa[‡]	Tensile strength, MPa[‡]	Tensile modulus of elasticity GPa	Elongation in 50 mm, percent	Shear strength, MPa[‡]	Rockwell hardness no.	Density, Mg/m³
Copper	Cu							
Annealed		33	209	125	60	---	---	8.85
Cold drawn		333	344	112	14	---	B37	8.85
Free-cutting brass	Cu 61.5, Zn 35.5, Pb 3							
Annealed		125	340	85	53	205	F68	8.3
Quarter hard, 15% reduction		310	385	85	20	230	B62	8.3
Half hard, 25% reduction		360	470	95	18	260	B80	8.3
High-leaded brass (1 mm thick)	Cu 65, Zn 33, Pb 2							
Annealed, 0.050-mm grain		105	325	85	55	230	F66	8.3
Extra hard		425	585	105	5	310	B87	8.3
Red brass (1 mm thick)	Cu 85, Zn 15							
Annealed, 0.070-mm grain		70	270	85	48	215	F66	8.6
Extra hard		420	540	105	4	305	B83	8.6
Aluminum bronze	Cu 89, Al 8, Fe 3							
Sand cast		195	515	---	40	---	---	8.3
Extruded		260	565	125	25	---	---	8.3
Beryllium copper	Cu 97.9, Be 1.9, Ni 0.2							
A (solution annealed)		---	500	125	35	---	B60	8.9
HT (hardened)		1035	1380	125	2	---	C42	8.9
Manganese bronze (A)	Cu 58.5, Zn 39, Fe 1.4, Sn 1, Mn 0.1							
Soft annealed		205	450	90	35	290	B65	8.3
Hard, 15% reduction		415	565	105	25	325	B90	8.3
Phosphor bronze, 5% (A)	Cu 95, Sn 5							
Annealed, 0.035-mm grain		150	340	90	57	---	B33	8.9
Extra hard, 0.015-mm grain		635	650	115	5	---	B94	8.9
Cupronickel, 30%	Cu 70, Ni 30							
Annealed at 760°C		140	380	150	45	---	B37	8.9
Cold drawn, 50% reduction		540	585	150	15	---	B81	8.9

[†]Based on *Metals Handbook–1979*, Ref. 64.
[‡]1 MPa = 145 psi.

Table 17.5 Mechanical properties of light nonferrous alloys[†]

Metal or alloy	Approximate composition, percent	Tensile yield strength (0.2% set) MPa[‡]	Tensile strength, MPa[‡]	Tensile modulus of elasticity, GPa	Elongation in 50 mm percent	Shear strength, MPa[‡]	Rockwell hardness no.	Fatigue limit for reversed bending, MPa[‡]	Density, Mg/m³
Aluminum	Al								
Sand cast, 1100-F		40	75	60	22	---	---	---	2.7
Annealed sheet, 1100-O		35	90	70	35	---	---	---	2.7
Hard sheet, 1100-H18		145	165	70	5	---	---	---	2.7
Aluminum alloy 2024	Al 93, Cu 4.5, Mg 1.5, Mn 0.6								
Temper O		75	185	73	20	125	H90	90	2.8
Temper T36		395	495	73	13	290	B80	125	2.8
Aluminum alloy 2014	Al 93, Cu 4.4, Si 0.8, Mn 0.8, Mg 0.4								
Temper O		95	185	73	18	125	H92	90	2.8
Temper T6		415	485	73	13	290	B83	125	2.8
Aluminum alloy 5052	Al 97, Mg 2.5, Cr 0.25								
Temper O		90	195	69	30	125	H82	110	2.7
Temper H38		255	290	69	8	165	E85	140	2.7
Aluminum alloy 5456	Al 94, Mg 5.0, Mn 0.7, Cu 0.15, Cr 0.15								
Temper O		160	310	---	24	195	---	---	2.5
Temper H321		255	350	---	16	205	---	---	2.5
Aluminum alloy 7075	Al 90, Zn 5.5, Cu 1.5, Mg 2.5, Cr 0.3								
Temper O		105	230	---	17	150	E65	---	---
Temper T6		505	570	---	11	330	B90	160	---
Magnesium	Mg								
Cast		21	90	40	2-6	---	E16	---	1.74
Extruded		69-105	195	40	5-8	---	E26	---	1.74
Rolled		115-140	200	40	2-10	---	E51	---	1.74
Magnesium alloy AM100A	Mg 90, Al 10, Mn 0.1								
Cast, condition F		85	150	45	2	125	E64	70	1.8
Cast, condition T61		150	275	45	1	145	E80	70	1.8
Magnesium alloy AZ63A	Mg 91, Al 6, Zn 3, Mn 0.2								
Cast, condition F		95	200	45	6	125	E59	75	1.8
Cast, condition T6		130	275	45	5	140	E83	75	1.8

[†]Based on *Metals Handbook–1979*, Ref. 64.
[‡]1 MPA = 145 psi.

color to the metal. In salty air near the ocean, copper reacts with sodium chloride to form copper hydroxide and eventually copper chloride.

High-temperature oxidation of iron is another type of direct chemical attack and takes place without water. The hot iron combines directly with oxygen to form an inner layer of FeO, an intermediate layer of Fe_2O_4, and an outer layer of Fe_2O_3. This coating is known as mill scale when it forms during manufacture.

Electrolytic corrosion Sometimes corrosion is caused by ion movement from anode to cathode areas on metal surfaces within an electrolyte, causing the anodes to corrode. This phenomenon of electrochemical corrosion is localized in small areas and does not attack the metal uniformly. The process may involve concentration cells when one metal is exposed to different concentrations of an electrolyte, or it may involve galvanic cells when two different metals are exposed to the same electrolyte.

An example of such electrochemical corrosion due to *concentration cells* is the pitting of many metals when there are surface irregularities or loose dirt. Such irregularities may seem perfectly innocent, like lap joints in common structural joints with their unavoidable crevices. The action occurs because the concentration of the electrolyte will vary.

Corrosion caused by *galvanic cells* is also common. For example, when a cast iron pipe connected to copper tubing carries an electrolytic solution, even a weak one such as drinking water, a galvanic process is set up. The iron pipe corrodes in this case, as iron is a more "active" metal than copper (that is, it has a higher single-electrode potential) and becomes the anode. Impurities present in almost all metals may cause a galvanic form of pitting. Table 17.6 shows several metals arranged in order of potential in an "electromotive force series." If the pipes were made of iron and zinc, it would be the zinc pipe that would corrode. The area ratio of the cathode to the anode is important. The higher the ratio, the more concentrated and severe the corrosion.

Corrosion protection In some cases metal will corrode in such a manner as to form a coating that adheres closely and acts as a protective shield against further corrosion. This process is called *passivation*.

Table 17.6 Electromotive force series

Potential	Metal	Ion
Most active	Magnesium	Mg^{++}
	Aluminum	Al^{+++}
	Zinc	Zn^{++}
	Chromium	Cr^{++}
Decreasing potential	Iron	Fe^{++}
	Cadmium	Cd^{++}
	Nickel	Ni^{++}
	Tin	Sn^{++}
	Lead	Pb^{++}
	Copper	Cu^{++}
	Silver	Ag^{+}
	Platinum	Pt^{++}
Least active	Gold	Au^{++}

The best-known example of passivation occurs on the surface of aluminum and aluminum alloys. Under normal atmospheric conditions a dense coating of aluminum oxide forms, protecting the metal from further corrosion. Rust would do the same for iron and steel, but it is not dense enough and does not adhere well enough. However, certain steels are alloyed to provide the same type of protection. This may be achieved by the addition of copper (ASTM A 441), which results in a steel twice as resistant as normal carbon steel. Weathering steels have even greater resistance, and stainless steels, which are alloys containing chromium, are afforded greater protection yet. Chromium forms a coating of Cr_2O_3 that also protects against high-temperature oxidation of ferrous alloys.

A guide to the selection of materials that are to be used together consist of "galvanic series" for particular electrolytes. In such a series various metals and alloys are arranged in much the same way as in the electromotive force series, but the passivation of the materials is taken into account. It is best to choose materials near one another in the galvanic series, which may be found in reference books on corrosion.

In some cases it may be possible to treat not the metal but the electrolyte by neutralizing the latter with the aid of inhibitors, such as the soluble oils added to the cooling water in a car's radiator. In other cases it may be feasible to remove or alter the corrosive medium, perhaps by desiccating the air in a space containing machinery susceptible to oxidation.

Finally, galvanic action may be countered by adding, and sacrificing, a piece of active metal to the environment; for example, a bar of zinc suspended in a steel storage tank will itself corrode but will save the tank. Alternatively, a current may be artificially introduced so as to make cathodic a part that would normally act as an anode.

17.7 WIRE ROPE

Wire rope or cable finds many engineering applications: suspension bridges and roof systems, cranes and hoists, ships' rigging, cable cars and ski lifts, and the like. It was invented in the 1830s and first used in mining.

A wire rope is a cord made of strands of wire of any ductile metal twisted together. The nominal diameter is that of a circle that will just enclose the rope. Wire rope, which may have a fiber core, resembles rope made solely of natural or synthetic fibers but is considerably stronger and stiffer. Some typical comparative values for 15-mm ropes are

Material	Tensile strength, kN	Unit mass, kg/m	Maximum strain	Strength-to-mass ratio, kN · m/kg
Natural (3-strand manila)	16	0.11	0.13	145
Synthetic (3-strand nylon)	40	0.10	0.45	400
Steel (6-strand, corrosion resisting, fiber core)	140	0.60	0.005	233

Most often, wire rope is made of steel wire, which may be protected by coating with zinc. Some special ropes that must resist corrosion are made of brass, bronze, or stainless steel. For steel wires, the tensile strength of the rope generally increases with higher carbon content of the steel. The common names for some steels and their ranges of carbon content are as follows:

Grade of wire	Carbon, percent
Iron	0.05–0.15
Traction steel	0.20–0.50
Mild plow steel	0.40–0.70
Plow steel	0.65–0.80
Improved plow steel	0.70–0.85

Rope construction Ordinary rope is composed of a number of strands, each containing a group of wires, usually laid around a hemp center. The pattern or arrangement and number of wires in the strand, the fundamental rope unit, is influenced by the use to be made of the rope, which in turn determines the properties the rope must possess. The seven-wire strand pattern is a single central wire surrounded by a layer of six other wires, all of the same diameter. Nominal 19- and 37-wire strands use different diameters of wire; there may also be some thin filler wires not included in the nominal count. Instead of the central wire, natural or synthetic fibers may be used.

When the rope is to resist long-continued static loading in such applications as guying for derricks, ship rigging, stacks, and small bridge cable, the only bending stresses will be those incidental to the anchorages and splices. For such applications flexibility is not required, and hence, ropes having few wires and a long pitch are the least costly for the required strength. The standard construction for this kind of service is 6 by 7 (six strands of seven wires each) and is listed variously as coarse lay, haulage, transmission, and standing rope.

If wire ropes are subjected to bending, as when they are passed over sheaves, flexibility is an important consideration. Flexibility is obtained by making ropes out of numerous small wires. In addition to bending stresses, wire ropes often carry dynamic loads such as in mine hoists where starting, acceleration, and vibration produce additional stresses so that strength with moderate flexibility, since the sheaves are usually large, is important. When sheaves are small in diameter and the loads relatively light, flexibility becomes more important, and the classification in most catalogs is hoisting rope. The standard hoisting rope is 6 by 19. Extra flexible types are 6 by 37, and 8 by 19 ropes.

The hemp center, when used, is designed to support and lubricate the wire strands. Obviously it is impossible to put enough lubricant into a rope during construction to last for the life of the rope. Internal friction will be reduced and resistance to corrosion increased if the rope is periodically treated with a lubricant that will adhere to the metal and penetrate to the hemp core. Such treatment is especially necessary if the rope is subjected to acid fumes or acid or salty water.

Most rope is made *right lay*, in which the strands are laid around the center like right-hand screw threads. However, it is supplied left lay also.

The terms *regular lay* and *lang lay* refer to types of rope construction. For regular lay (either left or right lay) the wires in the strands and the strands in the rope are laid in opposite directions, thus making the wires on the outside of the rope lie parallel with the axis of the rope. This is standard construction. In lang-lay construction both wires and strands are laid in the same direction, with the result that the wires, which now lie at an angle with the axis of the rope, have a greater exposed length and hence the wearing qualities of the rope are increased. The lang-lay rope, although it has a high resistance to abrasion if not damaged, is not generally used because it is more likely to curl and kink and is difficult to handle, often being called a cranky line.

Manufacture Rope is made in two stages, the first of which is the forming of the strands. Briefly, the individual wires, which have been previously wound on bobbins, are twisted together in a predetermined pattern in a machine to form a strand. The wires pass to the front end of the stranding machine, where they go through holes in a twister-head plate. They are then brought together and pulled through a round hole in a hardened steel die. The effect of revolving the cage carrying the bobbins and twister-head plate in the stranding machine and pulling the strand out at a uniform rate is that one or more layers of wires are laid in helical paths around a center or core. The strands are wound on haul-off drums. In the second stage, strands instead of wires are wound on larger bobbins that are mounted in a rope-closing machine, and the finished rope is taken off as the strand was. Thus it is seen that the wires and the strands are merely sprung into position and that the finished rope is under initial stress.

Another process, known as *preforming*, is accomplished by properly bending the wires and the strands in the closing operations. That the usual torsional forces exerted in the ordinary wire rope are removed can be easily demonstrated by cutting a finished preformed rope, which maintains its shape without the use of the usual seizings. The preforming is accomplished by passing the wires over rollers (placed between the twister-head plate and the closing die in the stranding machine) of such diameter and position relative to one another that the wires assume the helical shape they will occupy in the finished strand. In a similar operation the strand is made to assume the exact helical shape it will occupy in the finished rope. Preformed wire ropes are said to have much longer life owing to the reduction of internal friction, and also the worn ropes are more easily handled since the broken wires do not stick out as they do in the ordinary worn rope. In fact it is very difficult to detect broken wires in a preformed rope, and hence it is likely that such a rope might be kept in service long after it should have been retired since the number of broken wires is the principal criterion for the determination of the time of replacement.

Occasionally wire strands find other uses than in ropes. Prestressing systems for concrete are the major example. In these systems strands of seven wires with a strength of 1800 MPa are typically used. The strands have nominal diameters ranging from about 10 to 15 mm.

Mechanical properties Typical properties, including the fatigue limit, of six types of wires are presented in Table 17.7.

The efficiency of a wire rope is computed by dividing its actual tensile strength by the sum of the strengths of the individual wires; this value is usually about 80 to 85 percent. Three factors influence the efficiency:

Table 17.7 Typical properties of wire[†, ‡]

Type	Carbon, percent	Tensile strength, MPa[ƒ]	Twists in length of 100 diam	90°C reversed bends	Fatigue limit MPa[ƒ]	Fatigue ratio
Iron	0.06	585	94	86	370	63
Toughened steel	0.36	1090	34	86	490	45
Cast steel	0.41	1385	44	88	625	45
Mild plow steel	0.59	1480	33	52	505	34
Plow steel	0.72	1895	38	64	725	38
Improved plow steel . . .	0.71	1960	34	84	615	31

[†]Based on A. V. de Forest, and L. W. Hopkins, "The Testing of Rope Wire and Wire Rope," *Proc. ASTM*, vol. 32, pt. II, 1932.
[‡]Diameter of wire 0.043 in (1.1 mm).
[ƒ]1 MPa = 145 psi.

1. The longer the pitch of the wires and strands and the smaller the rope, the more nearly parallel are the wires to the axis of the rope and the greater the efficiency.
2. The more uniform the quality of the steel wires, the higher the efficiency because then they tend to fail simultaneously rather than progressively.
3. A uniform distribution of load to the several wires has the same effect as uniform quality in developing a high efficiency. To attain this condition all wires should be of the same length, be under the same initial stress, and have the same general position with respect to the axis of the rope. This last condition is not realized in ropes with steel centers. In short test specimens the wires may not be of the same length, thus tending to cause one strand to fail before the other strands are carrying an equal load; the computed efficiency in such a case would be low.

The modulus of elasticity of wire rope is difficult to determine because the rope tends to twist as it is loaded. This difficulty can be overcome by the use of a special extensometer arranged to compensate for the torsional displacement of the two attachment collars. Care must be taken to keep the collars securely clamped to the rope because the diameter of the rope becomes considerably smaller as load is applied. The modulus of a new steel rope may vary from 30 to 55 GPa.[†] It becomes successively greater with each cycle of loading until the strands settle themselves in the hemp center and with respect to one another. The modulus of elasticity may ultimately reach a value of 100 GPa. This tendency for the modulus to increase is less pronounced in a rope having a steel center. The modulus of a rope having a steel center is somewhat greater than for one of the same diameter having a hemp center.

Resistance to abrasion is proportional to the tensile strength and size of the outer wires in the strand. Radial crushing strength is inversely proportional to the number of strands and the number of wires in the strand. It also depends on the type of center.

In determining the required strength of a wire rope, the factors to be considered

[†]1 GPa = 145 ksi.

include the nominal load, the speed at which the load is to be applied, the rate of acceleration and deceleration, the vibration, the additional load due to friction, and the decrease in strength due to bending around sheaves or drums. A factor of safety is finally applied; this varies from 4 to 10, depending on the relative necessity for safety.

Durability Long life for a wire rope depends on the type of wire, the suitability of the construction of the rope for the given use, avoidance of high bending stresses by use of large-diameter sheaves, avoidance of wear and distortion of the rope by use of the proper size grooves in sheaves, avoidance of overloading or kinking, avoidance of lateral crushing on drums or sheaves, adequate use of correct lubricants, and prevention of corrosion and abrasion.

The diameter of sheave used in wire rope installations has an important influence on the life of the rope. Therefore it is customary to use minimum sheave diameters of 18 times the rope diameter for 6-by-37 ropes and 42 times the rope diameter for 6-by-7 ropes. However, whenever possible, it is desirable to use sheaves having diameters at least 50 percent larger than these minimum sizes.

Specifications Commonly used specifications include those of the American Petroleum Institute (API), which were developed to cover the application and use of wire rope in the oil industry. There are also numerous and varied ASTM specifications for wire rope; ASTM A 370 describes test procedures.

General notes and tolerances (similar to the API specifications) can be stated as follows:

1. The pitch or lay of a 6-by-7 rope should not exceed 8.0 times its nominal diameter.
2. The pitch or lay of a 6-by-19, 6-by-37, or 8-by-19 wire rope should not be more than 7.25 times its nominal diameter.
3. In like-positioned wires total variations of wire diameter should not exceed these values:

Wire diameter, mm	Total variation, μm
0.45-0.75	37.5
0.75-1.50	50.0
1.50-2.50	62.5
2.50-3.30	75.0

4. The total number of wires selected and tested should be equal to the number of wires in any one strand. They should be selected from all strands of the rope so as to use, as nearly as possible, an equal number from each strand. Thus the specimens selected will constitute a complete composite strand exactly like a regular strand in the rope.
5. Specimens for tension tests of wires should be at least 450 mm[†] long, and the distance between the grips of the testing machine should be at least 300 mm. The speed of the testing machine should not exceed 25 mm/min. Any specimen breaking within 6 mm from the jaws should be disregarded and a retest made.

[†]1 mm = 0.039 in.

Table 17.8 API strength requirements for 6-by-7 bright wire ropes

Nominal diameter, mm	Breaking strength, kN[†]			
	Plow steel		Improved plow steel	
	Minimum	Nominal	Minimum	Nominal
10	53	55	62	64
15	107	110	124	128
20	191	195	218	223
25	293	300	338	347

[†]1 kN = 225 lbf.

6. In making torsion tests the distance between the jaws of the testing machine should be at least 200 mm when the wire is straight, and the wire should be twisted at a uniform speed not exceeding 60 full twists per minute. Tests in which breakage occurs within 3 mm of the jaw should be disregarded.
7. The finished rope specimen for tension tests should be at least 1.0 m long between sockets for ropes up to 25 mm in diameter, at least 1.5 m long between sockets for larger ropes, and the sockets should be sized for a rope 5 mm larger in diameter than the one under test.

The API specifications list the minimum acceptable tensile loads and the minimum number of twists for a wide range of wire sizes for both plow steel and improved plow steel. The specifications also list the required breaking strengths for many sizes in each of several types of rope constructions. Table 17.8 shows values for 6-by-7 bright wire ropes having a fiber core. For galvanized ropes the breaking strengths are somewhat lower.

These specifications are not severe, especially the requirement for the tensile strength of the finished rope, but the individual wires must meet tensile strength and twisting requirements, so that undesirable material is easily eliminated. It is important that wire ropes also be subjected to some tests for fatigue strength, which generally depends on the sizes and arrangement of the wires as well as on the materials and the fabricating processes.

PROBLEM 17 Heat-Treatment and Tests of Steel

Object: To study the effects of annealing and quenching on the tensile strength, ductility, toughness, and hardness of a steel.
Preparatory reading: Chaps. 8 and 12.
Specimens: Three tension specimens of carbon steel.
Special apparatus: High-temperature oven, testing machine with automatic recording equipment.
Procedure:
1. Preheat the oven to 800°C and, after obtaining the dimensions of all three specimens, place two specimens into the oven.
2. Subject the third specimen to a normal tension test to rupture. From the stress-strain curve obtain the ultimate tensile strength, the total elongation, and the modulus of toughness. Subject the specimen to a Rockwell hardness test, obtaining the mean value from several trials.

3. After at least one hour of heating, remove the other two specimens from the oven. Put one aside to cool in the air and quench the other in a bucket of water or oil.

4. Subject each of the heat-treated specimens to the same tests as the untreated one.

Report:

1. Report the actual treatment procedures used, including temperatures, durations, type of quenching medium, and so on.

2. Report the types of tests used, including specimen dimensions, type of steel, apparatus, load rates, type of fracture, and so on.

3. Plot the stress-strain curves for the three specimens on one sheet of paper. Also demonstrate the differences in hardness by suitable graphics.

4. Tabulate the major strength parameters obtained for the various specimens, and comment on the effects of the treatment methods used.

DISCUSSION

1. What is the effect of carbon content on the properties of iron?

2. Name some alloys that have special names.

3. What metals might be suitable to make skis?

4. Steel is sometimes galvanized to protect it from corrosion. Why and how does this work?

5. Select a steel beam and determine how large a uniform load it can support in addition to its own weight. If the outside millimeter of the steel rusted, by how much would the load-carrying ability be reduced? (The atomic mass of oxygen is 16.00, of iron 55.85.)

6. How can electrolytic corrosion be prevented? Describe several methods.

7. Why, do you suppose, is aluminum more expensive than steel?

8. What are weathering steel and stainless steel?

9. Look up and summarize the provisions of ASTM A 36.

10. Compare the various methods of making steel.

READING

Allen, D. K.: *Metallurgy Theory and Practice*, American Technical Society, Chicago, 1969.

Bosich, J. F.: *Corrosion Prevention for Practicing Engineers*, Barnes & Noble, New York, 1970.

Davies, R., and E. R. Austin: *Developments in High-Speed Metal Forming*, The Machinery Publishing Co., Brighton, England, 1970.

Forsyth, P. J. E: *The Physical Basis of Metal Fatigue*, Blackie, London, 1969.

Hume-Rothery, W., R. E. Smallman, and C. W. Haworth: *The Structure of Metals and Alloys*, 5th ed., The Metals and Metallurgy Trust of the Institute of Metals and the Institution of Metallurgists, London, 1969.

The Making, Shaping, and Treating of Steel, 8th ed., United States Steel Corporation, Pittsburgh, 1964.

Metals Handbook, American Society for Metals, Metals Park, Ohio, various volumes published at different times.

Street, A., and W. Alexander: *Metals in the Service of Man*, 5th ed., Penguin, Harmondsworth, U. K., 1972.

Uhlig, H. H.: *Corrosion and Corrosion Control, An Introduction to Corrosion Science and Engineering*, 2d ed., Wiley, New York, 1971.

Van Horn, K. R., ed.: *Aluminum*, American Society for Metals, Metals Park, Ohio, 1967; Vol. 1, *Properties, Physical Metallurgy and Phase Diagrams;* Vol. 2, *Design and Application;* Vol. 3, *Fabrication and Finishing*.

Woldman, S. E.: *Engineering Alloys*, 4th ed., Reinhold, New York, 1962.

EIGHTEEN

CONCRETE

18.1 COMPOSITION

For good reasons, portland cement concrete—so named because the hardened cement resembles the gray limestone near Portland, England—is a widely used building material. Because this concrete can readily be formed into many shapes, architectural distinction of buildings, bridges, and other structures made with it is possible. Its resistance to corrosion in the presence of most soils and waters makes it especially useful for pavements and foundations. Portland cement exhibits low tensile strength, but this disadvantage can be overcome by reinforcing or prestressing it, normally with steel.

Concrete may be thought of as a mass of inert filler material (aggregate) held together by a matrix of binder (cement paste). The properties of the concrete, both in the freshly mixed and in the hardened state, are intimately associated with the characteristics and relative proportions of these components.

The solid portion of the hardened concrete is composed of the aggregate and a new product that is the result of a chemical combination of cement with water. The remaining portion of the space occupied by a given volume of concrete is composed of free water and air voids. The latter are comparatively small, usually not over 1 or 2 percent in ordinary freshly made concrete. After any considerable period, the amount of free water depends on the extent of combination that has taken place with the cement and the loss due to evaporation.

The binder material, the cement-water paste, is the active component and has three main functions: (1) to provide lubrication of the fresh plastic mass, (2) to fill the voids between the particles of the inert aggregates and thus to produce watertightness in the hardened product, and (3) to give strength to the concrete in its hardened state. The properties of the paste depend on the characteristics of the cement, the relative proportions of cement and water, and the completeness of the chemical reaction between the cement and water, which is termed *hydration*. Extensive hydration of the cement requires time, favorable temperatures, and the continued presence of moisture. The period during which concrete is definitely subjected to these conditions is known as the "curing" period. On construction work, this period may vary from four to 10 days; in the laboratory, a common curing period is 28 days. Good curing is essential for the production of quality concrete.

The aggregate has three principal functions: (1) to provide a relatively inexpensive filler for the cementing material, (2) to provide a mass of particles that are suitable for resisting the action of applied loads and abrasion, and (3) to reduce the volume changes caused by the setting and hardening process and the drying of the cement-water paste. The properties of concretes resulting from the use of particular aggregates depend on (1) the mineral character of the aggregate particles, especially as related to strength, elasticity, moisture volume changes, and durability; (2) the surface characteristics of the particles, especially as related to the workability of the fresh concrete and bond within the hardened mass; (3) the grading of the aggregates, especially as related to the workability, density, and economy of the mix; and (4) the amount of aggregate in a unit volume of concrete, especially as related to strength and elasticity, to volume changes due to drying, and to cost.

It should be noted that the word "cement" without qualification may designate any of various binding agents or adhesives, including asphalt, epoxy, and the like. Similarly, the term "concrete" in general may be used to describe various mixtures of a cement and a filler material or aggregate, notably asphalt concrete. Whenever the danger of confusion exists, the descriptive term "portland" should be included, although it often is omitted when the context is clear. Under no circumstances may the word "cement" be used to mean "concrete."

18.2 MATERIALS FOR CONCRETE MAKING

Unlike standard metal alloys, concrete mixes are designed for a particular job using ingredients that are largely obtained locally and combined on or near the job. This article introduces those materials: portland cement, admixtures, aggregates, and water.

Cement Portland cement is produced by burning in a rotary kiln, almost to the point of fusion, a properly proportioned mixture, the essential components of which are lime and clay. The kiln product, known as "clinker," is subsequently ground, with a small amount of gypsum, to the desired degree of fineness. The function of the gypsum is to retard the time of setting, as without gypsum the cement would develop a flash set. For an average cement an oxide analysis shows that the CaO content is about 64 percent, the SiO_2 content is about 22 percent, and the Al_2O_3 content is about 5 percent. The raw materials react during burning to form four complex compounds. These compounds, with their common abbreviated designations and their approximate percentages in a normal (Type I) cement, are tricalcium silicate (C_3S), 45 percent; dicalcium silicate (C_2S), 27 percent; tricalcium aluminate (C_3A), 11½ percent; and tetracalcium aluminoferrite (C_4AF), 8 percent.

Altogether there are five types of portland cement; they differ primarily in compound composition but also in fineness. These types and their uses are as follows:

Type I   For use in general concrete construction when the particular properties specified for the four other types are not required

Type II   For use in general concrete construction exposed to moderate sulfate action, or where moderate heat of hydration is required

Type III   For use when high early strength is required

Type IV   For use when a low heat of hydration is required

Type V   For use when a high sulfate resistance is required

Four physical tests are usually prescribed for the acceptance of a cement. These are (1) fineness, (2) soundness, (3) time of set, and (4) strength of a mortar made with the cement. Pertinent ASTM specifications are C 109, C 115, C 151, C 188, C 190, C 191, C 359, and C 451.

Fineness is expressed in terms of "specific surface," that is, the surface area of the particles, in square meters, contained in a kilogram of cement. This value for fineness can be determined by measuring the air permeability of a specially compacted bed of cement. Other methods involve (1) a sedimentation procedure employing Stokes' law regarding the settlement of fine particles or (2) a determination of the turbidity (using a photoelectric cell) of a specified cement-kerosene mixture. An average commercial cement has a specific surface of about 280 to 320 m^2/kg[†] when it is measured with the air permeability method. Since greater fineness induces more rapid hydration, many of the commercial high-early-strength cements are simply produced by fine grinding.

The *soundness* or volume constancy of a cement is determined by measuring the expansion of a cement bar after it has been cured for five hours in an autoclave at a maximum steam pressure of 2.0 MPa. For the cement to be acceptable the expansion must not exceed 0.50 percent.

To ensure sufficient time to place concrete while it remains plastic, a minimum limit is imposed on the *time of "initial" set*, which may be taken as the condition of the mass when it begins to stiffen appreciably. ASTM specifications require that initial set should not take place within one hour. Depending on the test used to determine it, initial set usually takes place within two to four hours. To ensure that a cement will harden for use, a maximum limit is imposed on the *time of "final" set*. ASTM specifications require that final set occur within 10 hours. With many commercial cements final set occurs within five to eight hours. The condition of initial and final set is determined by the penetration of standard needles or rods into a "neat" (straight cement) paste of specified consistency.

[†]1 m^2/kg = 10 cm^2/g = 4.89 ft^2/1bm.

Strength tests are made on tension briquets (sometimes on 50-mm compression cubes) made with Ottawa standard sand. The mortar is made up of one part cement to three parts standard sand by mass, and the tests are done at specified ages. Standard curing conditions prescribe moist air for one day followed by water storage thereafter, all at a temperature of 21°C. For a Type I cement to be acceptable under the ASTM specifications, briquets made and stored under standard curing conditions must possess tensile strengths exceeding 1.9 MPa at seven days and 2.4 MPa at 28 days. For compression cubes made of a 1:2.75 graded standard sand mortar the corresponding required compressive strengths are 14.5 and 24.0 MPa.

Water Water containing acids, alkalies, salts, silt, and organic matter should be avoided. As a rule, any water that is potable is satisfactory for concrete mixes. Minor impurities might be tolerated, however, provided that the resultant loss in strength of the concrete does not exceed about 10 percent. Under that criterion some seawaters may be acceptable.

Admixtures Small amounts of various admixtures can advantageously be added to concrete mixes to modify some of their characteristics in both the fresh and hardened states. Certain admixtures act as wetting agents to increase the fluidity of the mix or to permit a reduction in the water content for a given fluidity. Other admixtures tend to entrain minute globules of air, which serve (1) to lubricate the mass in the same way as wetting agents and (2) to increase markedly the resistance of the concrete to freezing and thawing.

Aggregates Aggregates may be classified according to source, mode of preparation, and mineralogical composition.

With reference to source, aggregates may be natural or artificial. Natural mineral aggregates may be the result of weathering and the action of running water, producing sands and gravels, or may have been prepared by crushing, yielding "stone" sands and crushed stone. Natural aggregates result from attrition or disintegration of any or all of the rock types: igneous, sedimentary, and metamorphic. Artificial aggregates are usually lightweight materials produced by burning special clays or shales in a rotary kiln.

Material passing a 4.76-mm[†] screen is classified as sand or fine aggregate, although some fine aggregates may be considerably finer than this size. Usually two or three gradations of fine aggregate and several sizes (size groups) of coarse aggregates are available for concrete work.

To qualify for use in concrete, aggregates must be clean, free of organic matter, hard, tough, strong, durable, of the proper gradation, and nonreactive with cement. Fortunately, there is usually little difficulty in securing materials that meet these requirements. The question of obtaining satisfactory materials is largely a matter of cost. Pertinent ASTM specifications are C 33 and C 330.

[†]Tyler No. 4. Standard screen or sieve sizes are listed in App. B.

Properties of an aggregate that should be known to proportion a concrete mix and calculate batch quantities are the true density of the mineral particles, moisture content, gradation, and bulk density.

The *true density* is determined by displacement methods and is based on the "solid" volume of the aggregate particles, including impermeable voids within the particles. For typical sands and gravels the true density averages about 2.65 Mg/m^3.

The *moisture content* of an aggregate is based on its oven-dry mass, although it is the free or surface moisture in excess of the absorption, causing the saturated, surface-dry condition, that usually is the inportant value. The various moisture conditions are (1) oven dry—all moisture, external and internal, driven off by heating at 100°C; (2) air dry—no surface moisture on the particles, some internal moisture, but not saturated; (3) saturated surface-dry—no free or surface moisture on the particles, but all voids within the particles filled with water; and (4) damp, or wet—saturated and with free or surface moisture on the particles. The excess of condition (3) over condition (1) is the absorption capacity, or total possible internal moisture, of the aggregate.

In a concrete mix, the free or surface moisture becomes a part of the mixing water and must be taken into account in determining the quality of water to be added to a batch. Also, a dry aggregate absorbs some of the mixing water. Hence, in the proportioning and batching of concrete mixes, all calculations are based on the saturated surface-dry condition. The excess or deficiency of moisture that produces this condition may be determined in either of two ways: (1) by completely drying out the aggregate and applying the absorption capacity as a correction or (2) by displacement methods employing the value for the density of the aggregate based on a saturated surface-dry condition. In the second method either a specially constructed flask or a pycnometer is employed. Approximate absorption capacities and surface-moisture contents of common aggregates are given in Tables 18.1 and 18.2. The finer the aggregate, the more free water it can and usually does carry.

The film of surface moisture on the particles of a fine aggregate holds them apart so that the gross or bulk volume is increased. This phenomenon is called "bulking," and when batching quantities for a concrete mix by volumetric methods, it must be taken into account if the proper proportions of ingredients are to be obtained. For ordinary concrete sands, maximum bulking occurs when the moisture content is about 4 to 7 percent by mass, and the increase in volume may be as much as 25 to 40 percent. Larger percentages of water tend to decrease the bulking until, when the moisture is roughly 12 to 20 percent by mass, the sand is completely submerged or

Table 18.1 Approximate absorption capacities of various aggregates

Aggregate	Absorption capacity, percentage by mass
Average concrete sand	0 to 2
Granitic-type rocks	0 to $\frac{1}{2}$
Average gravel and crushed limestone	$\frac{1}{2}$ to 1
Sandstone	2 to 7
Lightweight porous materials	Up to 25

Table 18.2 Approximate surface moisture on ordinary aggregates

Aggregate	Approximate surface moisture	
	Percentage by mass	Liters per cubic meter
Moist gravel and crushed stone	$\frac{1}{2}$–2	8–30
Moist sand	1–3	15–45
Moderately wet sand	3–5	45–65
Very wet sand	5–10	65–130

inundated and occupies about the same space it did in the dry, loose condition. The bulking of coarse aggregates is negligible.

Gradation of particles is determined by a "sieve analysis," for which a series of sieves or screens is used. Standard mesh openings are given in ASTM E 11 (and App. B). A suitable gradation of the combined aggregate in a concrete mix is necessary in order to obtain proper workability and density. Harsh or otherwise deficient gradings usually result in concretes of inferior quality or in poor economy. For much routine work, if properly sized commercial materials are available, the aggregates may be combined in arbitrary proportions, based on experience. On important jobs, however, it is clearly advantageous to analyze the materials for their particle-size distribution (by means of a series of sieves; see App. B) and proportion the various sizes so as to secure a satisfactory gradation of materials from fine to coarse. A cumulative frequency analysis done as explained in Chap. 3 may be required. Fortunately, provided proper adjustments are made in the mix, a fairly wide latitude in grading may be tolerated without seriously affecting the properties of the resulting concrete. The approximate proportions of fine to coarse aggregate for concretes used in building construction are given in Table 18.3.

The *bulk density* of aggregates is used for computing quantities in batching by volume and in estimating quantities of materials. It is determined by weighing the aggregate required to fill a container of known volume under specified conditions of compaction or moisture content: (1) loose or compact, and (2) dry or damp. Like

Table 18.3 Percentage of sand in total aggregate for various maximum sizes of aggregate

Maximum size of aggregate, mm	Percentage of sand in total aggregate, by mass[†, ‡]	
	Range for various job materials	Recommended maximum
10	55–75	60
20	45–60	55
30	40–55	50
40	35–50	45
50	. . .	40

[†] For aggregates having densities within the range of about 2.5 to 2.8 Mg/m^3.
[‡] The finer the sand or the richer the mix, the lower the percentage required.

true density, bulk density is given in Mg/m^3,[†] but it differs from the former in that the air spaces among the particles are counted in the volume. Bulk densities normally range from about 1.5 Mg/m^3 to 1.8 Mg/m^3 for sand and are about 1.7 Mg/m^3 for gravel.

Reactive aggregates contain certain forms of silica that combine with the alkali in the cement and cause differential expansions of the concrete. This leads to crazing, cracking, and other forms of disintegration of the concrete. Petrographic examinations and long-time expansion tests can be used to evaluate a new source of aggregates. If any reactive silica is present in an aggregate, only a low-alkali cement should be used with it. However, appreciably reactive aggregates should not be used under any circumstances. Some tests for such reactivity are described by ASTM C 289 and C 441.

Other aspects of aggregates are discussed in Chap. 19 in connection with bituminous mixes.

18.3 MANUFACTURE OF CONCRETE

As is the case for the production of other materials, the object of the manufacture of concrete is to obtain a satisfactory product at the least cost, using the materials that are available. The term "satisfactory" concrete here implies concrete that has the necessary and desired properties, such as workability and homogeneity in the fresh mixture and strength, watertightness, durability, and volume constancy in the hardened product. The securing of the desired properties, as well as the economy of the work, involves the selection of a suitable combination of materials and the control of the process of manufacture until the concrete is the proper age. ASTM specifications pertinent to quality control of concrete mixes include C 31, C 39, C 143.

Design of the concrete mix The first step in the manufacture of concrete after materials have been selected is the design of the mix. This usually resolves itself into proportioning the ingredients (including the water) for a given strength, within general limits imposed by the cement content and workability, which in turn are dictated by general experiences with regard to design requirements and to weathering. Strength has been found to be a fair index of the most desirable properties of concrete; impermeability or watertightness is the next most inportant criterion since, other factors being the same, the greater the impermeability of a concrete, the greater its durability. A mix containing much cement relative to the aggregates is called "rich" and one containing little is called "lean." The principal methods of proportioning are:

1. *Using arbitrary proportions of cement, sand, and coarse aggregate, without any preliminary tests.* The proportions of a concrete mix are usually stated by mass, but sometimes by volume; for example, a $1:2:3\frac{1}{2}$ mix means 1 part cement, 2 parts sand, and $3\frac{1}{2}$ parts coarse aggregate. This method is now used only on relatively

[†]1 Mg/m^3 = 62.4 lbm/ft.^{3}

small jobs or where the character and gradation of the materials vary within rather narrow limits.

2. *Using a fixed cement content per cubic meter of concrete, a specified grading of combined aggregates, and a specified consistency of fresh concrete.* The specifications are the result of experience with concrete for a particular class of work.

3. *Making trial mixes, using a specified strength (or water-cement ratio) and specified consistency.* This method is the most flexible, and maximum economy may be derived from its use, although it requires facilities for testing and some experience with concreting methods. Briefly, the trial-mix method of proportioning involves making a series of trial batches and tests from which a water-cement ratio vs. strength relation and the relation between proportions and consistency are established for the given materials. From these data, the quantities for the desired mix may be computed. Water-cement ratio vs. strength diagrams for both normal cement concrete and high-early-strength concrete are given in Fig. 18.1, and recommended consistency, cement content, and aggregate size for various classes of work are given in Table 18.5 (Art. 18.4).

4. *Following the American Concrete Institute (ACI) calculation method.* This procedure is based on data tabulated from observations of a large number of trial mixes.

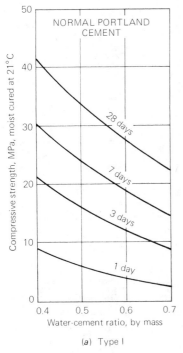
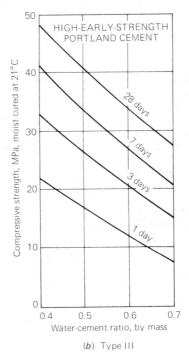

Note: 10 MPa = 1450 psi

Figure 18.1 Water-cement ratio vs. compressive strength. (*From Portland Cement Ass'n., Ref. 17.*)

On almost all work of any importance, the aggregates and cement are in bulk and are batched by mass. On small jobs, scales may not be at hand, in which case measurements are made by volume. The cement is supplied in 94-lb (42.75-kg) sacks,[†] and the batches are such as to use integral multiples of one sack. The quantity of water should be, and usually is, carefully controlled. Correction is made at the mixer for the amount of free water contained in wet aggregates or absorbed by dry aggregates. On the small job, if batching is done by volume, the bulking of the aggregates, due to moisture, must be taken into account.

Hence batching calculations are best carried out by considering mass, although for small jobs the results may have to be converted to volumes. In that case the bulk densities of the aggregates must be determined or estimated. Batching concrete is like cooking: recipes stated in terms of mass result in greater consistency, but some cooks find it more convenient to measure volumes.

For example, if specified proportions by mass (water:cement:sand:gravel) are given as 0.5:1.0:2.5:4.0, the mass of the mix will be 8.0 times the mass of the cement. This ratio is the "yield" of the mix. The amounts of ingredients can be determined for particular requirements. If a certain volume of mix is needed (concrete is often sold by volume), its density may be assumed to be about 2.4 Mg/m^3 or it can be calculated more accurately from the proportions and relative densities of the ingredients. Here are a few cases to illustrate:

Requirement	Water	Cement	Sand	Gravel	Total
True density (Mg/m^3)	1.00	3.10	2.65	2.65	2.4
Bulk density (Mg/m^3)	1.00	1.51	1.70	1.55	
Mix proportion	0.5	1.0	2.5	4.0	8.0
Use 1 Mg cement	0.5 Mg	1.0 Mg	2.5 Mg	4.0 Mg	8.0 Mg
Make 10 t mix	0.625 Mg	1.250 Mg	3.125 Mg	5.000 Mg	10.0 Mg
Make 1 m^3 (2.4 Mg) mix	0.15 Mg	0.30 Mg	0.75 Mg	1.20 Mg	2.4 Mg
Use 1 sack (42.75 kg) cement	21.4 kg	42.75 kg	107 kg	171 kg	342 kg
Use 1 sack (0.028 m^3) cement	0.021 m^3	0.028 m^3	0.063 m^3	0.110 m^3	0.222 m^3

The bulk densities of the aggregates were, of course, assumed here for illustration. Also, the aggregates were assumed to be saturated surface-dry, or their surface moisture would have to be subtracted from the water requirement and added to the aggregate requirements.

Suppose that the aggregates in the example are wet, the sand having a surface moisture content of 4 percent and the gravel of 2 percent. Including the surface moisture the proportion of sand would be (2.5)(1.04) = 2.60 and of gravel (4.0)(1.02) = 4.08, while the proportion of water would be reduced to 0.5 − (2.5)(0.04) − (4.0)(0.02) =

[†]In past practice in the United States; this amount (94 lbm) corresponds to one cubic foot (0.0283 m^3); a bulk density of 1.51 Mg/m^3 is thus implied.

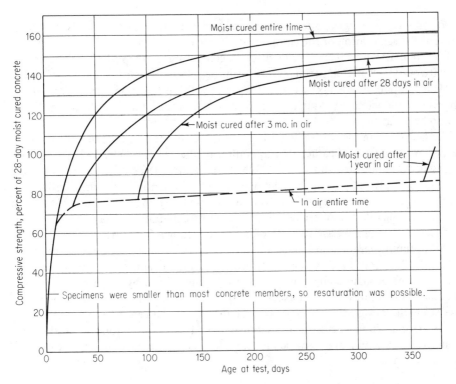

Figure 18.2 Effect of curing conditions on compressive strength. (*From Portland Cement Ass'n., Ref. 17.*)

0.32, so that corrected proportions of 0.32 : 1.00 : 2.60 : 4.08 could be used. Correcting the amount of water is significant and important, whereas adjustments in the aggregate amounts are less essential. If the proportion of water in the example had not been changed, the water-cement ratio would have been changed from 0.5 to 0.68. An inspection of Fig. 18.1a shows that for Type I cement the 28-day compressive strength would have been reduced by about a third.

If the aggregates were drier than saturated surface-dry, corresponding adjustment would be made and the amount of water increased. In contrast to the earlier case, however, omission of this adjustment would result in a stronger concrete.

If air were entrained, the volume of the mix would, of course, be increased by the volume of the entrained air (usually stated as a percentage), while the mass would remained unchanged.

Mixing, placing, and curing Machine mixing is always done in batch mixers, usually of the revolving-drum type. Concrete is often batched at a central plant and mixed while en route to the job in special truck mixers. Although mixers are available that can handle small batches satisfactorily, hand mixing is sometimes resorted to in the laboratory.

Methods of transporting concrete from the mixer to the forms vary to suit the job. Wheelbarrows, buggies, buckets, and pumps are used. Precautions must be taken to prevent the segregation of the component materials of the mix.

The compaction of concrete in place is commonly accomplished by the use of power-driven vibrators. Superior results can be obtained with vibratory tamping over hand tamping if the mix has been designed for this purpose and proper technique in the use of the vibratory equipment is observed. Impact forces applied to the forms may also be used.

The last step, and an exceedingly important one in the manufacture of concrete, is curing. We pointed out earlier that hydration of the cement will take place only in the presence of moisture and at favorable temperatures. Depending on the type of construction, moist conditions may be maintained by retention of forms, sprinkling, ponding, or moist coverings such as wetted burlap and straw. These curing methods retain in the concrete the water required for complete hydration. The effect of curing conditions on the compressive strength of concrete is shown in Fig. 18.2.

The optimum temperature range for the curing of ordinary concretes appears to be between 20° and 40°C. Concrete is usually seriously damaged by temperatures below freezing, the hardening process is considerably slowed by temperatures below 10°C, and temperatures much in excess of 50°C have sometimes resulted in concretes whose strength retrogressed at later ages. Hydration of the cement is an exothermic reaction, so that cooling is generally required; this is another reason for wetting the hardening concrete with water, which is then allowed to evaporate, drawing heat from the concrete mass.

18.4 PROPERTIES OF FRESH CONCRETE

Although the main purpose of mix design is to achieve a satisfactory final product, the properties of the fresh concrete are also important, even if indirectly. If the wet concrete cannot be placed properly, difficulties and delays in construction occur and the end result is a poorer, perhaps inadequate, more expensive structure.

Consistency Consistency, one property of the fresh concrete, is an important consideration in the securing of a workable concrete that can be properly compacted in the forms. Workability is a relative term referring to the comparative ease with which concrete can be placed on a given type of work; for example, concrete considered workable for pavement construction may be entirely unworkable for a thin wall or a narrow beam. The term *consistency* relates to the state of fluidity of the mix and embraces the range of fluidity from the driest to the wettest mixtures. The common tests to determine consistency are the "slump" test and the "ball penetration" test. These tests give only rough measures of consistency, but they do give satisfactory measures of this property for most practical work.

The *slump* test is made by measuring the subsidence of a pile of concrete 300 mm (12 in) high, formed in a mold that has the shape of the frustum of a cone (ASTM C 143).

The *ball penetration* test is made by measuring the settlement of a 150-mm steel ball (weighing 13.6 kg with its handle) into the surface of the concrete (ASTM C 360).

Tests have shown that approximately two times the settlement equals the slump for the same concrete.

For convenience various degrees of wetness of a mix may be roughly classified as dry, stiff, medium, wet, or sloppy. A concrete is said to have medium or plastic consistency when it is just wet enough to flow sluggishly—not so dry that is crumbles or so wet that water or paste runs from the mass. The range in values of slump corresponding to these arbitrary degrees of consistency is indicated in Table 18.4. The values for each class are not to be taken as absolute. Consistencies employed on various types of concrete construction are shown in Table 18.5.

The principal factors affecting consistency are

1. *The relative proportions of cement to aggregate.* For given cement-water pastes the more aggregate that is crowded into the paste, the stiffer is the resulting concrete.
2. *The water content of the mix.* For fixed proportions of cement and aggregate, the more water, the more fluid the resulting mix tends to be. There is a limit to the amount of water that a mix will hold without serious segregation of the aggregate from the cement.
3. *The size of the aggregate.* For a fixed water-cement ratio and a fixed aggregate-cement ratio, the finer the aggregate, the stiffer the mix. In a given volume of aggregate, the more finely divided the particles are, the greater the surface area of the particles; hence the more paste required to coat them to produce a given consistency.
4. *The shape and surface characteristics of the aggregate particles.* Angular particles or those with rough surfaces require a greater amount of paste for the same mobility of mass than is necessary for smooth well-rounded particles.

Table 18.4 Approximate range in slump of concrete for various degrees of consistency

Consistency	Slump,[†] mm[‡]	Remarks
Dry	0–25	Crumbles and falls apart under ordinary handling; can be compacted into rigid mass under vigorous ramming, heavy pressure, or vibration, but unless care is used exhibits voids or honeycomb
Stiff	15–65	Tends to stand as a pile; holds together fairly well but crumbles if chuted; with care and effort can be tamped into solid dense mass; satisfactory for vibratory compaction
Medium	50–140	Alternate terms: plastic, mushy, quaking. Easily molded, although some care required to secure complete compaction
Wet	125–200	Pile flattens readily when dumped; can be poured into place
Sloppy	175–250	Grout or mortar tends to run out of pile, leaving coarser material behind

[†]Ball penetration is approximately half the slump.
[‡]1 mm = 0.039 in.

Table 18.5 Consistency, cement content, and aggregate size employed on various types of concrete construction

Type of construction	Typical structures	Consistency	Cement content, kg/m^3[†]	Max. size aggregate, mm[‡]
Massive	Dams, heavy piers, large open foundations	Stiff	140–280	75–150
Semimassive	Piers, heavy walls, foundations, heavy arches, girders	Stiff; medium	225–340	50–100
Pavement	Road surfaces, heavy slabs, moderately heavy footings	Stiff; medium	250–340	40–65
Heavy building . . .	Large structural members, small piers, medium footings. Wide to moderately wide spacing of reinforcement	Medium wet	280–390	25–50
Light	Small structural members, thin slabs, small columns, heavily reinforced sections, closely spaced reinforcement	Wet	310–390	10–25

[†] $1 kg/m^3 = 0.062 lbm/ft^3$.
[‡] $1 mm = 0.039 in.$

5. *The fineness and type of cement and the kind and amount of admixture.* These factors may affect the fluidity of the paste and thus the consistency of the concrete.

Bleeding The tendency for water to rise to the surface of freshly placed concrete is known as "water gain," or "bleeding." It results from the inability of the constituent materials to hold all the mixing water. As a result of water gain the top portion of a lift becomes overly wet, and water films accumulate under the particles of coarse aggregate and under horizontal reinforcement bars. Concrete subject to water gain is not as strong, durable, or impervious as properly designed concrete. Water gain can be at least partially controlled by making a workable concrete mix with a minimum amount of water, higher cement content, and natural sands having an adequate percentage of fines.

Air entrainment The entrainment of air is beneficial to fresh concrete as well as to the finished product. At a given water content it makes the mix more workable, which, of course, provides the option of reducing the water content and therefore the cement content. It also reduces bleeding and segregation of aggregates.

18.5 STRENGTH CHARACTERISTICS

Because portland cement concrete is mainly used as a structural material, its behavior under stress is its most important characteristic. The determination of this behavior is greatly influenced by the age of the hardened concrete at the time of testing, as concrete gains strength almost indefinitely, if at an increasingly slow rate. Review Fig. 18.1. An arbitrary age of 28 days is considered rather standard for testing, but in any case the age of testing must be specified.

Also important are the size and shape of the specimen. In the United States 150-by-300-mm (or 6-by-12-in) cylinders are frequently specified for compression tests, while in European countries 200-mm cubes are most often used. The cylinders yield about 80 percent of the apparent strength of the cubes.

Strength The compressive strength of concrete is taken as an important index of its general quality. Strength tests are relatively simple to make, and since strength is a first requisite for the structural designer, this property is the one most frequently determined. The compression test of a 150-by-300-mm cylinder at age 28 days, after moist storage at a temperature of 23°C, is a standard test (ASTM C 31, C 39). The compressive strength of concrete, made and tested under standard conditions, ordinarily varies from 15 to 40 MPa or higher.

The tensile strength of concrete is roughly 10 percent of the compressive strength and is often neglected by structural engineers. The flexural strength of plain concrete, as measured by the modulus of rupture, is about 15 to 20 percent of the compressive strength.

The principal factors affecting strength are

1. *Water-cement ratio*. The results of various investigations have shown that the water-cement ratio may be taken as the most important factor in controlling strength. The effect of the water-cement ratio on the strength of *workable* mixtures, for standard curing conditions, is shown in Fig. 18.1. In practice, the quality of concrete is usually rated by its compressive strength at the age of 28 days, the relation of the water-cement ratio to strength ordinarily being established by trial for each particular set of job conditions.
2. *Age*. The strength of moist concrete generally increases with age. For normal cements, stored under standard conditions, an empirical relation is

$$S_{28} = S_7 + 2.5 \sqrt{S_7}$$

where S_{28} is the strength at 28 days and S_7 is the strength at seven days in MPa. Compressive strength at various ages is shown in Fig. 18.1.
3. *Character of the cement*. Both the fineness of grinding and the chemical composition of the cement affect the strength of concrete, particularly at early ages. Fine cement and high proportions of tricalcium silicate (C_3S) and tricalcium aluminate (C_3A) promote high strength at early ages. The two groups of curves shown in Fig. 18.1 show the influence of the character of the cement on strength.
4. *Curing conditions* (*moisture and temperature*). The greater the period of moist storage (as shown in Fig. 18.2) and the higher the temperature (for a temperature range of 5° to 40°C), the greater the strength at any age.
5. *Moisture content of the concrete at the time of test*. The higher the moisture content, the lower the strength.
6. *Richness* (*high cement content*) *or leanness of the mix and character of the aggre-*

gate. These factors largely affect the strength through their influence on the water-cement ratio required to produce the desired consistency.

Stiffness The stress-strain diagram for concrete, as determined from an ordinary compression test, is a curved line. The slower the rate of loading, the sharper the curvature due to plastic deformation, which takes place with time. The modulus of elasticity may therefore be expected to exhibit considerable variability. Furthermore, concrete is not, strictly speaking, truly elastic, so that successive loadings may be expected to result in different values of the modulus.

It was explained in Chap. 2 that the secant modulus of elasticity is determined from the slope of a straight line drawn through the origin of the stress-strain curve and some point of interest on the curve, or at some percentage of the strength, as shown in Fig. 18.3. The secant modulus is the one most commonly used when no qualification is given. The slope of a line tangent to the stress-strain diagram at the origin is sometimes used; this is called the "initial tangent" modulus. Other possibilities are the

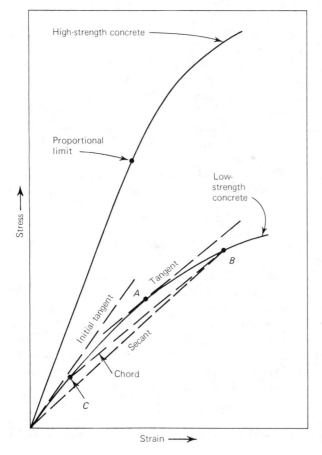

Figure 18.3 Typical stress-strain curves.

"tangent" and the "chord" moduli, which are also shown in Fig. 18.3. The secant modulus for concrete ranges from about 15 to 40 GPa.[†]

The factors that influence the strength of concrete similarly influence the modulus of elasticity. The following observations indicate the general influence of certain variables on the modulus of elasticity: (1) The stronger the concrete, the higher the modulus. (2) The use of lightweight aggregates reduces the modulus. (3) The modulus increases with age, sometimes to a very marked degree. (4) The drier the concrete at the time of test, the lower the modulus; wet concrete is stiffer although often weaker. (5) For the same consistency and water-cement ratio, the larger the maximum size of aggregate and the coarser the grading, the higher the modulus.

The ACI Building Code [112] assumes that the modulus of elasticity of concrete for use in reinforced concrete design is given closely enough by the relation

$$E_c = w^{1.5} \; 1.36 \; \sqrt{f_c'}$$

where E_c = modulus of elasticity, GPa

f_c' = compressive strength of concrete, MPa

w = relative density of concrete, for the range of 1.5 to 2.5

For ordinary concrete, a rough estimate of $E_c = 1000 \, f_c'$ is sometimes used.

Figure 18.4 shows some typical stress-strain curves from 28-day short-time compression tests on standard cylinders.

Creep Concrete is susceptible to creep, which is sustained yielding under stress. In contrast to many other materials, in concrete creep occurs at normal temperatures. Its effects are significant in various situations. In a column, for example, creep of the concrete will cause an increasingly large portion of the load to be supported by the longitudinal reinforcing bars, as steel does not creep at normal temperatures. Similarly, a prestressed beam will gradually lose its prestressing force as the tendons shorten.

[†]1 GPa = 145 ksi.

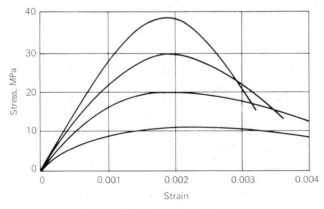

Note: 10 MPa = 1450 psi

Figure 18.4 Moduli of elasticity.

The seepage of absorbed colloidal water into internal voids and to the exterior is believed to be the primary cause of creep. Creep is more pronounced when the water content is greater and when the hydration periods are shorter before loading. It obviously depends on stress levels and their duration, but it also depends on various other factors: the aggregate, cement type, temperature, humidity, and size of the mass.

The total creep strain per megapascal of compression for different concretes ranges from about 30 to 300 μm/m, 150 μm/m being a usual value. This amounts to several times the maximum elastic deformation. Creep is more pronounced when the load is applied to relatively new concrete. After application of the stress, the creep is approximately proportional to the natural logarithm of the time since application, i.e., it increases rapidly at first and insignificantly after a few years.

18.6 OTHER PROPERTIES

Some properties of hardened concrete are not directly related to strength but nevertheless are significant. How durable or how watertight concrete is may be of interest, or what volume changes can be expected. This article describes some of these properties but leaves others, such as wear resistance, extensibility, and thermal conductivity, for future reading.

Density The density of concrete made with common mineral aggregates is 2.4 Mg/m^3 or can be rounded to this value. By using light aggregates, the density can be reduced to 1.6 Mg/m^3 and even to 0.5 Mg/m^3 if substantial strength is sacrificed. By using heavy aggregates, such as metallic ores, densities above 4.0 Mg/m^3 can be achieved to provide nuclear shielding, for example, since the effectiveness of shielding is a function of the mass.

Durability and impermeability A factor that has been too little considered in the design of concrete mixtures is durability. The action of weather in the deterioration of concrete structures is due in part to the cyclic expansion and contraction under changing moisture and temperature conditions, in part to the expansive force of ice crystals as they are formed in the pores of the concrete, and in part to the leaching of soluble compounds from the mass by water. It follows, therefore, that a relatively watertight concrete is a more weather-resistant concrete.

Although the principal factors that affect strength also affect watertightness, important variables are richness of mix, gradation of aggregate (particularly in the fine particles), compaction of the concrete in place, and curing.

The tests for watertightness and durability do not yield results as definite as do strength tests, but they serve for comparison of different concretes on the same basis. The permeability test is made by subjecting one surface of a slab or block of concrete to a head of water and measuring the amount that passes into or through the concrete. An accelerated durability test is made by subjecting concrete test specimens to cycles of freezing, thawing, drying, and wetting and then noting the extent of deterioration by (1) testing the specimen in compression and comparing the results with those from companion specimens that have been stored under standard conditions, (2) measuring the progressive expansion of the concrete resulting from its disintegration, or (3) noting reductions in the sonic modulus of the concrete [38].

Many tests have shown the marked improvement in the resistance of concrete to freezing and thawing when it contains about 4 percent of air entrained in the form of minute globules. These air cells serve to relieve the internal pressures exerted by the ice, which tends to form in the pores of the concrete, and thereby prevent deterioration of the concrete. Because the reduction in compressive strength may become appreciable when the air content exceeds about 4 percent, success in the use of entrained air requires careful measurement of the ingredients and control of the concrete-making process to regulate the amount of entrained air.

Various methods for measurement of the volume of entrained air have been developed, and three have been standardized by ASTM. Two of the latter methods involve a comparison of the volume of a batch of air-entrained concrete with the solid volume of the same batch as computed from the summation of the solid volumes of the several components of the mix (ASTM C 138) or as determined by displacement of

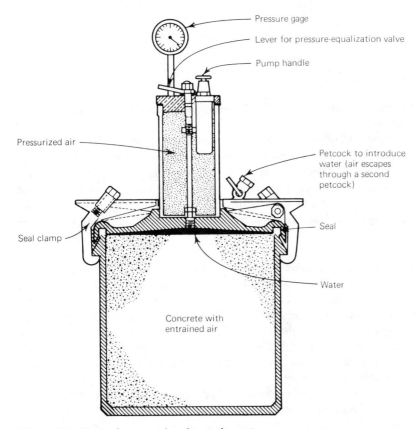

Figure 18.5 Device for measuring air-entrainment.

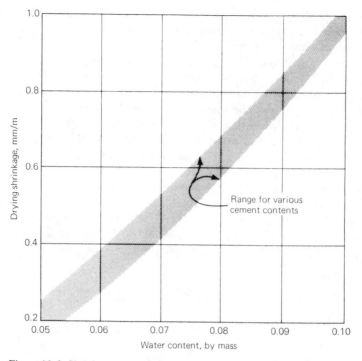

Figure 18.6 Shrinkage vs. water content.

the mix under water (ASTM C 173). Both these methods are subject to appreciable errors unless controlled very carefully.

A third method (ASTM C 231) has been developed in which the reduction in volume of the air voids in a sample of the fresh concrete (and hence a reduction in volume of the concrete) is effected by application of water pressure to the sample, the quantity of air being determined with Boyle's law. In general, this pressure method gives percentages of air about one point higher than the others and is the method most commonly used. Figure 18.5 shows a typical device for measuring air entrainment by this method.

Volume changes Various concretes experience significant change in volume during and after hardening. In addition to stresses, these changes may be caused by settlement of the fresh mass; chemical combinations of water, cement, and aggregates; and changes in moisture content or temperature. Although the volume changes may be small, they are significant because the concrete is generally constrained by surrounding structural elements, including itself, and steel reinforcing and prestressing devices. Volume changes are particularly serious when tensile stresses are developed, which tends to occur when the volume decreases.

Fresh concrete *settles* in the forms when bleeding and evaporation or drainage of the water takes place. This is an important consideration only under special conditions, such as in the case of pretensioned prestressed concrete.

The effect of *chemical reactions* depends on whether the combination occupies a greater or smaller volume than the original constituents. Hydration, the reaction of water with cement, normally results in a volume reduction unless special expansive cements are used. The reactions of high-alkali cements and certain aggregates result in expansion, which may be severe enough to disrupt the concrete, sometimes years after construction.

When hardened concrete dries, it shrinks, and when it gains moisture again, it expands. *Moisture shrinkage* is mainly a function of the water-cement ratio and increases with it. For a given water-cement ratio, the greater the proportion of cement paste, the greater the shrinkage. In any case, shrinkage increases with the water content of the mix. As suggested by Fig. 18.6, the relation is nearly linear. Various other factors intrinsic to the mix as well as some extrinsic ones also affect shrinkage.

Like most materials, concrete expands with rising *temperatures* and contracts with falling ones. The relation is approximately linear and is expressed by the coefficient of thermal expansion. The coefficient of thermal expansion is about 11 MK^{-1}, but it may range from 7 to 12 MK^{-1}. The value depends mostly on the coefficient of the aggregates; the aggregates tend to reduce the coefficient of neat cement, which is about 13.5 MK^{-1}. The normal value, fortunately for reinforced concrete construction, is quite similar to the coefficient for steel.

PROBLEM 18 Concrete Mix Preparation and Tests

Object: To prepare a concrete mix according to specification and to determine its properties.
Preparatory reading: ASTM C 143, C 231.
Materials: Type I cement (air entrained), well-graded fine and coarse aggregate, clean water.
Special apparatus: Mixer, slump-test cone, air-entrainment test device, specimen molds.
Procedure:
1. Weigh equal amounts of sand and gravel for a total of about 15 kg and place them into the mixer. Add 2 kg of cement and, with the mixer turning, add water, using a water-cement ratio of 0.45, 0.55, or 0.65, as assigned. Keep precise records of the amounts. It is convenient to measure the water in a beaker (1 L = 1 kg).
2. The aim is to produce a mix consistency with a slump of about 50, 100, or 150 mm, as assigned. Add water and cement in the correct proportion or aggregates until you think the mix proper.
3. Perform a slump test. If the slump is greater than the specified amount or smaller by more than 25 mm, mix the concrete again and make adjustments.
4. When the slump is satisfactory, perform a test for air entrainment as follows: Fill the container of the apparatus (tamp in the mix in three layers and each time tap the container to expel all voids). Clean the flanges, attach the top, and fill with water. Rotate and tap the device several times to rid the water of air bubbles. Follow the instructions furnished by the manufacturer of the particular apparatus used. Air must be pumped into the air chamber to a certain pressure and then released into the concrete container. The air content is obtained from the now equalized pressure.
5. Measure and weigh a specimen mold (shape and size as assigned), fill it with your mix, rodding three times and striking off the top, then weigh it again. Compute the density of the mix and observe its qualities, such as cohesiveness, workability, and tendency to bleed. Fill other molds, if instructed to do so. Tag your specimens with the pertinent identification and data.
6. Arrange for proper curing of your specimen(s) and clean the equipment. Do not wash the mixer if another team is going to use it.
7. At the specified age, test your specimens (see Problem 9).

Report:

1. Prepare a one-page data sheet for distribution to the members of the class. It should contain the mix proportions (using the mass of cement as unity), measured slump, proportion of entrained air, density of mix, and general observations.
2. Summarize all data and describe, in tabular or graphical form, the effects of the water-cement ratio and water content or slump.
3. Illustrate the effects of air entrainment.
4. Calculate the amounts of ingredients needed to make one cubic meter of concrete of each mix. What would be the yield of concrete in cubic meters per 50-kg sack of cement? Also calculate the cement content in each case.
5. From your own team's data, estimate the average relative density of the aggregates.

DISCUSSION

1. State the water-cement-ratio law or principle. Does it apply to all mixes? What considerations besides strength affect the selection of a water-cement ratio?
2. What factors, in addition to the water-cement ratio, affect the strength of concrete?
3. If a concrete has a density of 1.2 Mg/m^3 and a strength of 15 MPa, what is the expected modulus of elasticity?
4. Why is it that a rich mix may be weaker than a leaner mix, both mixes being of the same lots of cement and aggregate?
5. What is the effect of the age of the concrete on the water-cement-ratio vs. strength curve?
6. If you were in charge of construction and found that certain portions of the structure require a wetter consistency for proper placement, how would you modify the mix to maintain a uniform quality of hardened concrete?
7. Discuss how the cost of concrete of a given required strength is affected by the consistency.
8. In selecting the best of several mixes that have the same water-cement ratio and the same consistency, why is it logical to select the one that gives the greatest yield?
9. For concretes of a given water-cement ratio and given consistency, what would be the effect on yield of overwashing a fine aggregate so that is has few fine particles? Explain.
10. What would be the effect on water gain under similar circumstances? Explain.

READING

Causes, Mechanism, and Control of Cracking in Concrete, Symposium, American Concrete Institute, Detroit, 1968.

Feld, J.: *Construction Failure*, Interscience-Wiley, New York, 1968.

Illingworth, J. R.: *Management and Distribution of Concrete*, McGraw-Hill, Maidenhead, U.K., 1972.

Larson, T. D.: *Portland Cement and Asphalt Concretes*, McGraw-Hill, New York, 1963.

Lea, F. M.: *The Chemistry of Cement and Concrete*, 3d ed., E. Arnold, London, 1970.

Manual of Concrete Practice, Part 1: Materials and General Properties of Concrete, American Concrete Institute, Detroit, 1980.

Mindess, S., and J. F. Young: *Concrete*, Prentice-Hall, Englewood Cliffs, N.J., 1981.

Neville, A. M.: *Properties of Concrete*, Halsted Press, New York, 1973.

Troxell, G. E., and H. E. Davis: *Composition and Properties of Concrete*, 2d ed., McGraw-Hill, New York, 1968.

19.1 NATURE AND USES OF BITUMENS

In current technical usage, the term *bitumen* refers to a class of amorphous, colloidal, sticky substances composed principally of complex, high-molecular-mass hydrocarbons soluble in carbon disulfide.[†] Bitumens are black or dark-brown, and they may occur in a solid, semisolid, or viscous liquid state. They can be found in materials forming natural deposits, such as bituminous coal, some petroleums, and native asphalts, or they can exist in manufactured products such as tars and petroleum asphalts. They are a primary and distinguishing constituent of tars and asphalts.

In the solid and near-solid state at atmospheric temperatures, tars and asphalts exhibit strong adhesive and water-repellent properties, so that they can be used effectively as cementitious binders or as water-repellent membranes. They may exist in the viscous liquid state due to the presence of volatile oils (the evaporation of which reduces the material to a solid state). Tars and asphalts can be liquefied by fluxing the solid materials with suitable oils, by emulsifying them with water, or by heating the solids (i.e., they are "thermoplastic").

Bituminous materials in the form of asphalts were known to ancient civilizations and have been used through the ages for paving, waterproofing, painting, and preserving; the ancient Egyptians used an asphaltic material in the mummification process. The early sources of asphalts were probably pools ("tar" pits) and lakes of the material that had oozed to the earth's surface. Asphalts occurring as impregnations in the pores

---

[†]This description generally follows ASTM standard definitions. Some authorities, e.g., Abrahams [113], feel that the solubility in carbon disulfide is too restrictive a criterion for this class of materials.

of rock ("rock asphalt") or as seams in rock strata (asphaltites such as gilsonite) apparently have been used for paving purposes only within the past two centuries.

The bituminous materials now termed *tars* do not occur in natural deposits but are formed as condensates during the destructive distillation of coal, wood, some petroleums, oil shale, and other organic materials. Such raw condensates are usually liquid and are generally further refined to produce the viscous tars in commercial use. The more or less solid residues that result from distilling the volatile constituents from tars are known as *pitches*; pitches liquefy when heated and have cementing properties (ASTM D 8). It is unlikely that the ancients made or used tars, although tarlike substances are found under some conditions in the burning of wood, peat, and other fuels. It was not until gas and coke were made from coal that tars came into fairly common use.

Contemporary *asphaltic* materials are mostly obtained as residues in the refining of some types of petroleum, although natural asphalts are still used in limited amounts for some purposes. The manufacture of asphalt from petroleum permits the production of various types and grades of products suited to particular applications. As world petroleum supplies dwindle, it is likely that efforts will be made to derive asphaltic materials from other hydrocarbon sources.

Although both tars and asphalts have bitumens as their primary ingredient, the variety and relative proportions of the bituminous substances they contain differ. In chemical terms, the molecules of the bitumens in tars tend toward being aromatic and unsaturated (i.e., having a ring type of molecular structure and a higher degree of potential chemical activity), whereas the bitumen molecules in asphalts tend toward being aliphatic and saturated (i.e., having a straight chain molecular structure and being relatively nonreactive). In tars, the nonbitumen portion (i.e., the part insoluble in carbon disulfide) tends to be principally free carbon, whereas the insoluble portion of asphaltic materials tends to be largely mineral matter [113, 114].

Tars and asphalts also exhibit some differences in behavior. For example, with respect to temperature susceptibility (see Art. 19.3), tars tend to soften more readily in summer temperatures and become more brittle when cold than do asphalts. Thus, while tars seem to have a higher resistance to the disintegrative influence of water, asphalts seem to have the edge in weather resistance when applied in thin films [114].

Still, the cementitious, adhesive, and water-repellent properties of tars and asphalts are similar in many respects, and in some areas of the world, for reasons of cost, custom, or behavior under local conditions, one may be used in preference to the other. However, for purposes that are of particular interest here, such as paving, asphalts are by far the most widely used. Hence, the remainder of this chapter is directed to the uses and control of asphaltic materials. It may be noted in passing that the concepts and principles as well as many of the tests involved in the use and control of asphalts are the same as, or very similar to, those that pertain to the use of tars for similar purposes.

Asphaltic materials, in various forms and consistencies, find widespread use in the construction and maintenance of pavements, water-containing facilities, and roofs, as well as in the manufacture of a variety of specialty products. About three-fourths of the asphalt produced from petroleum in the United States in the 1970s was used for paving and allied work, about one-fifth in connection with roofing, and the remainder for other purposes. The discussion in the rest of this chapter mainly emphasizes the use of asphaltic materials in connection with road work.

In the construction of pavements and other surfacing (for example, canal linings), asphaltic materials of suitable types and characteristics may serve as

1. A binder for mineral aggregates in mixtures used in paving-type structures
2. A waterproofing and binding penetrant in compacted-in-place wearing courses of open-graded crushed stone (penetration macadam pavement)
3. Sheet asphalt surfacing (a mortar made of sand and a suitable grade of asphalt)
4. Seal coatings (an application of a liquid asphalt to an existing pavement surface, on top of which is usually placed a thin layer of fine mineral aggregate)
5. Surface treatments of various kinds, to seal against moisture penetration, to provide bond between existing and subsequently placed layers of paving, to prevent raveling of loose material, to allay dust
6. A crack filler or undersealant for portland cement concrete pavements
7. Soil stabilizers

19.2 MANUFACTURE AND TYPES OF ASPHALTS

Petroleum, or crude oil, is a variable mixture of hydrocarbons, the molecules of which vary in size (number of atoms) and structural arrangement, and the component fractions of which vary in the nature of their hydrocarbon constituents and in volatility. For convenience, petroleums are often characterized as having an asphalt base, a mixed base, or a paraffin base, although each one of these classes also comprises a complex mixture of hydrocarbons as well as varying amounts of the characterizing "base" component.

In the refining of petroleum, the crude is commonly first put through a "straight-run" distillation process in which the lighter fractions, such as gases, gasolines, naphthas, kerosenes, diesel oils, lubricating oils, and so on are successively distilled off, leaving a heavy residue that is sometimes called "topped crude." This *residuum*, from an asphalt-based crude, contains a high proportion of asphaltic bitumens along with heavy oils (such as heavy fuel oils) and other compounds of very low volatility. Mixed-base crudes are now also commonly used to obtain asphalts by means of additional processing (e.g., solvent extraction) that separates the waxy paraffins and other non-asphaltic, commercially useful components from the asphalt-bearing residue.

Figure 19.1 is a schematic illustration of fractions of an asphalt-containing petroleum, classified according to relative volatility. These are the fractions that would tend to be separated in a straight-run fractionating process. The diagram is not drawn in proportion to the relative amounts produced.

There are no sharp dividing lines between these fractions; they merge into each other. Distillates designated by the names gasoline, kerosene, and so on, have a range in boiling points rather than a single boiling point. The asphaltic residues generally include compatible oils of very low volatility, in varying relative amounts depending

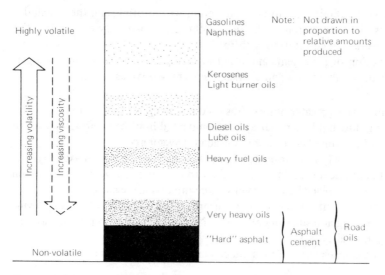

Figure 19.1 Diagrammatic representation of fractions of an asphalt-base liquid petroleum.

on the nature of the crude and the completeness of the distillation process. Most asphaltic products derived from these residues are made to include some amount of these oils. The fraction that would remain if all volatile oils were removed, sometimes called "hard" asphalt [114], would be too hard and brittle to be useful for paving purposes.

There are three main classes of asphaltic materials produced from the asphaltic residue of the petroleum refining process: asphalt cement, liquid asphalt, and air-blown asphalt.

Asphalt cement, as commercially produced, is a highly viscous, semisolid material available in several ranges of consistency (ASTM D 3381). It may be obtained directly from the topped crude residue if this residue is of an appropriate composition and purity; more commonly it is obtained by further processing the residue to remove waxy and undesired oily components. Asphalt exhibits thermoplastic properties; that is, it becomes less viscous when heated and returns to a near-solid state on cooling. An important use of asphalt cement is as a binder in hot-mix paving mixtures, described in Art. 19.6. After such mixes have been laid in place, compacted, and properly cured, some but not all of the residual oils will have evaporated, so that the material in the cementitious asphalt will be of a firmer consistency than that of the original asphalt cement.

Liquid asphalt is a term applied to the asphaltic materials that behave essentially as liquids at normal atmospheric temperatures. They are made, however, to have a considerable range in viscosity and certain other properties. The liquid state under these conditions may be obtained by (a) selecting an asphaltic residue from the fractionating process that already contains a suitable amount of a fluxing oil, (b) blending asphalt cement with a petroleum distillate to make "cutbacks," or (c) emulsifying asphalt cement (or a heavier cutback) and water.

Cutback asphalts are made by mixing asphalt cement of suitable composition with one of the lighter fractions (called the "solvent" or "diluent") from the distillation process. When the cutback asphalt is exposed to the atmosphere, the solvent evaporates, leaving the relatively hard asphalt cement. This is called "curing." Depending on the solvent used, as well as on the ambient temperature, the thickness of the layer, and the like, the curing process has variable durations. The blends or types of cutbacks are classified on the basis of the rate of curing.

Rapid-curing asphalts (RC) are made of asphalt cement cut back with a solvent of relatively high volatility, such as a gasoline or naphtha (ASTM D 2028).

Medium-curing asphalts (MC) are made of asphalt cement and a solvent of medium volatility, such as kerosene (ASTM D 2027).

Slow-curing asphalts (SC) are made either of asphalt cement cut back with a solvent of low volatility, such as diesel oil, or of a properly constituted asphalt-oil mixture obtained directly from the residue (ASTM D 2026). One of the SC grades of the latter type is still often called road oil.

The cutbacks are used primarily, but not exclusively, in road construction, as the binder in various types of mixtures, for surface treatments, and for patching.

An emulsified asphalt is one in which an asphalt cement has been emulsified in water with the aid of a small amount of an emulsifying agent. Normally, water is the continuous phase of the system, and the discontinuous or dispersed phase consists of very small asphalt globules. The globules may be anionic (negatively charged), cationic (positively charged), or nonionic, depending on the type of emulsifying agent used (ASTM D 977 for anionic and nonionic emulsified asphalts; ASTM D 2397 for cationic emulsified asphalts).

Three general types of asphalt emulsions are commercially available, designated as rapid setting, medium setting, and slow setting. Setting occurs when the water evaporates (or the charge on the surface of the globules is neutralized) and the asphalt globules coalesce to form continuous films of asphalt cement. The setting characteristics are influenced by the nature and amount of the emulsifying agent and the proportions of asphalt and water in the emulsion.

Inverted emulsified asphalts are those in which the water, in the form of small globules, is the dispersed phase and a rapid- or medium-curing cutback asphalt is the continuous phase. Such emulsions may also be anionic or cationic.

Emulsified asphalts find many of the same uses in road work as do cutback asphalts, such as for binders in mixtures, for surface treatments, and for patching. The use of emulsified asphalts has been increasing in recent years relative to cutbacks, owing to less restriction on field use because of unfavorable ambient temperatures, simpler processing techniques for some uses, fewer environmental and health hazards (as no volatile solvents are involved), and lower energy costs.

Air-blown (also known as blown or oxidized) asphalt is prepared by blowing air through an asphaltic residue while the residue is heated in a still. Such treatment gives the resulting materials a higher softening temperature than asphalt cements have—a property that is desirable for various industrial applications. Made in several grades, air-blown asphalts find uses in roofing, as pipe coatings, and as linings of canals and reservoirs; they are also used for undersealing portland cement concrete pavements.

19.3 ASPHALT PROPERTIES AND TESTS

From what has been said about the nature and types of bituminous materials, it can be concluded that asphaltic products, even within major designated types, are materials with varied and imprecisely defined compositions and an incompletely known chemistry. Under these circumstances, an important problem in using these materials is how to identify them in such a manner that materials with the desired quantities can be specified and procured. If a material's response to external influences varies with its composition, when the composition is complex and rather uncertain, a few simple, single-valued material properties cannot be chosen that would make possible unique definition and prediction of behavior. As a result, numerous test procedures, some empirically based and some quite arbitrary, have evolved.

In this article, the general features of current tests and procedures to measure properties pertinent to the selection and control of asphaltic materials are described. Measures of these various properties used as a basis for defining the grades of asphalt in common use are given in Art. 19.4. Properties and tests pertaining to mineral aggregates and to asphalt-aggregate mixtures are discussed in Arts. 19.5 and 19.6 respectively.

Inasmuch as many of the tests for asphaltic materials employ arbitrary procedures that must be followed meticulously in order to obtain comparable and meaningful results, the pertinent standards should always be consulted for detailed and current descriptions of the procedures.

For convenience, the several properties and tests are discussed here under the following headings:

- Physical characteristics
- Composition
- Consistency
- Stability (of emulsions)
- Ductility

Not all the tests described are used for every type of asphalt material; the applicability of the tests is indicated as they are described. Some tests associated with one particular property may be used in a procedure to determine whether or not a change in some other property has occurred; for example, penetration (consistency) tests might be made before and after heating a given sample in an oven, to determine whether the *composition* has changed.

Physical characteristics Two characteristics are of particular interest: density and flash point.

The *density* or the relative density (still sometimes called the "specific gravity") of asphalts is usually determined by the pycnometer method. This procedure is used for asphalt cements (ASTM D 70 or D 2170), as well as for cutback asphalts (ASTM D 3142). Density values are needed for making various mass-volume calculations in connection with the conduct of certain other asphalt tests and with the proportioning of asphaltic mixtures. Also, since the densities of asphaltic materials vary appreciably with temperature, the mass-volume relations at various temperatures are useful in connection with shipping and purchasing.

The relative density of asphalt cements at room temperature is in the range of 1.00 to 1.05. The relative density of cutback asphalts may range from about 0.92 to a little over 1.00, depending on the kind and amount of diluent.

In order to find the temperature to which the material may be safely heated in the presence of open flames, the *flash point* is determined. This is done by heating a specimen in a brass cup and playing a small flame over its surface, until sufficient vapors are released to produce a sudden flash. One test, the *Cleveland open cup* (*COC*) test (ASTM D 92), is used for asphalt cements and slow-curing cutback asphalt. For faster-curing cutback asphalts a procedure that is quite similar but provides for indirect heating is used; it is known as the *Tag open cup* (*TOC*) test (ASTM D 1310). Another test similar to the Cleveland open cup test, but which involves continuous stirring of the sample, is the Pensky-Martens test (ASTM D 93). To give a margin of safety, in job practice an asphalt is heated only to a temperature somewhat below its flash point.

Composition Several tests are aimed at determining the amount or conditions of the cementing (bitumen) content, the amount of the liquefying medium (if any), and/or the presence of inert or unwanted matter in a freshly manufactured material. Other tests are aimed at determining changes in composition caused by heating or by other factors. Usually, procedures involving changes in composition also call for determination of other properties of the altered material. Tests of interest here are solubility tests, distillation tests, tests for the presence of water, the spot test, and the particle charge test.

The amount of bitumens, the basic cementing constituents, present in a sample of asphaltic material is important, as is the amount of useless foreign matter. By ASTM definition, bitumens are those hydrocarbons soluble in carbon disulfide, but the use of this solvent involves certain hazards. The *solubility* test (ASTM D 2042) now used instead employs trichloroethylene, which appears to give a fair measure of the amount of cementitious material present. In this test the insoluble residue is collected by filtering. The test may be applied directly to asphalt cements, or to the asphalt cement portion remaining after distilling off the diluents from cutbacks or evaporating the water from emulsified asphalts.

For liquid asphalts, standardized *distillation* procedures are used to separate the asphaltic and the liquefying constituents and thus determine their relative amounts. As used for cutback asphalts (ASTM D 402), the relative amount of solvent that distills out at various temperatures gives a measure of the volatility of the evaporable fraction, and it is considered a rough indication of the curing characteristics of the material in the field. In the distillation tests, the residue that remains in the flask at a temperature of 360°C is considered to be asphalt cement.[†] The distillation procedure

---

[†]For SC cutbacks, the residue having a penetration value of 100 (10 mm) is taken as the equivalent of asphalt cement. (See following discussion of consistency.)

for emulsified asphalts (ASTM D 244) is similar to that for the cutbacks, but the apparatus is somewhat different. This test can also provide information on the amount of oily substances in the emulsion. Solubility, consistency, and ductility tests on the asphaltic cement residues of the cutbacks or emulsions may be called for.

The presence of more than some critical amount of water in asphalt cement or cutback asphalts may cause foaming above 100°C (the boiling point of water)—a hazardous condition. In a *water-in-asphalt* test (ASTM D 95), a solution of a sample of asphalt and a petroleum naphtha solvent is heated in a specially designed still and the water collected in an attached trap.

Whether or not the colloids in an asphalt cement have been detrimentally affected by overheating or cracking during the manufacturing process is a question of some concern. To obtain some indication of this condition, special (selective) solubility tests have been used. Some agencies have also employed a *spot* test, but there is disagreement as to the reliability of this procedure. Essentially the spot test consists of placing a drop of a solution of the asphalt and a standard naphtha solvent on a filter paper and observing the shape of the spot formed; a nonuniform spot is taken to indicate that some change may have occurred [114].

Sometimes it is necessary to determine whether or not an asphalt emulsion is of the cationic type. A *particle charge* test (ASTM D 244) is used for this purpose. Positive and negative electrodes are immersed in a sample of the emulsion and a direct current is applied to them. If, at the end of a specified period, the cathode has accumulated an appreciable layer of asphalt, the sample is a cationic emulsion.

Consistency The concept of consistency has to do with the degree of resistance to continuing and/or permanent deformation. Terms such as the "thickness" of a viscous liquid or the "firmness" of a plastic material denote qualitative aspects of this rather complex property of many substances. Viscosity in its scientific sense is a basic expression of consistency. But because of practical difficulties in measuring viscosity of asphalts in a near-solid state, arbitrary measures of behavior, such as resistance to penetration or extrusion, were devised and used over the years for special purposes.

Asphalts decrease in viscosity with increasing temperature. This feature of asphalt behavior is called temperature susceptibility and can be expressed as the rate of change of viscosity (or some arbitrary measure thereof) with respect to temperature. If all asphaltic materials had the same temperature susceptibility, a measurement of viscosity at some convenient temperature, together with the known rate of change with temperature, would allow the calculation of viscosity at any temperature level, and thus provide a guide to behavior under a variety of processing and service conditions in the field. However, asphaltic materials differ in their temperature susceptibility, so that measurements of viscosity at more than one temperature are needed to estimate field performance.[†] One of the shortcomings of, say, a penetration test developed for use

†Temperatures commonly specified for various types of asphalt tests include the following:

Celsius scale:	25°C	50°C	60°C	135°C	260°C	360°C
Fahrenheit equivalent:	77°F	122°F	140°F	275°F	500°F	680°F

with a material of rather firm consistency at normal temperatures, is that it becomes impractical to use when the material becomes rather fluid at a higher temperature.

Temperature susceptibility may be desirable or undesirable. For example, a given pavement may experience temperature variations between $-40°C$ and $+60°C$, and it would be clearly unsuitable if great changes in the consistency of the asphalt binder were to take place under these service conditions. Yet when the hot asphalt mix is placed during construction, it is obviously advantageous if the necessary fluidity can be achieved without the need for maintaining extremely high temperatures of the mix. Similar considerations that require compromise apply to uses of asphalts as sealants and the like.

Exposure to elevated temperatures or the mere passage of time at normal temperatures may also change the consistency of asphaltic materials by altering its composition. Changes in composition may be caused by evaporation of some volatile constituents, by chemical changes brought about by external influences (e.g., oxidation), or by internal reorganization of the hydrocarbon molecules. Information about the consistency of the altered material may serve as a clue to the type or extent of the change in composition, or it may provide a more realistic basis for judging the performance of the material under field conditions.

In the early stages of development of modern asphalt technology, three empirical consistency tests were devised to serve particular purposes: the penetration test, the softening point test, and the float test. More recently, viscosity tests have become common.

In the *penetration* test (ASTM D 5), a needle of prescribed dimensions weighted with a specified mass is brought into contact with a semisolid asphalt sample and allowed to settle for a prescribed duration. Penetration is measured in units of 0.1 mm; the lower the penetration value, the harder the asphalt. For most penetration measurements, a mass of 100 g is allowed to bear on the asphalt sample for 5 s at a temperature of $25°C$. For special purposes, a test is sometimes made at a lower temperature with a mass of 200 g and a penetration time of 60 s. In past practice, the penetration test was used extensively for defining the several grades of asphalt cement and in testing the asphalt cement residues from other test procedures. While this test has the advantage of simplicity, it has important shortcomings: (1) it is unsuited to testing the material in a very soft or liquid condition and (2) it fails to indicate the temperature susceptibility in the temperature ranges in which asphalt cement is hot mixed, rolled into place, or exposed to the sun in hot climates. It has been superseded by viscosity measures as a basis for defining the grades of all classes of asphalts used for paving purposes.

Asphalts harder than those used for paving are often subjected to a test to determine the *softening point* (ASTM D 2398 and D 36). Samples of asphalt are loaded with a steel ball and supported by a confining brass ring. This assembly is suspended in water, which is gradually heated. When the softened asphalt and the ball strike a plate suspended a certain distance (25 mm) below the ring, the temperature of the bath is measured and designated the "ring and ball" softening point of the asphalt.

Some asphaltic materials, such as the residue from slow-curing liquid asphalts after the distillation test, tend to be too soft for making the penetration test but too firm for a standard viscosity test to be performed easily at low temperatures. To obtain a measure of consistency, a *float* test (ASTM D 139) was formerly used, in which the time, in seconds, was determined for water to break through a plug of asphalt in the bottom of a float in a water bath kept at 50°C.

Viscosity tests in some form are now widely used for measuring the consistency of asphaltic materials. They can be performed at temperatures that correspond to mixing or to field conditions and can provide quantitative information about temperature susceptibility. Their purposes include serving both as a basis for defining asphalt grades (see Art. 19.4) and as an assessment of the condition of asphaltic materials after the materials have undergone field or simulated service conditions.

Viscosity, the resistance of a liquid or viscous substance to flow, may be expressed as dynamic (absolute) viscosity or as kinematic (relative) viscosity. Fluids of stiffer consistency require either higher pressure or more time to achieve a given flow, and *dynamic viscosity* is dimensionally expressed as the product of pressure and time. Under certain conditions of gravitational flow it is convenient to divide this value by the density of the fluid; the resulting ratio is known as the *kinematic viscosity*, which is dimensionally equal to length squared per unit of time. For a discussion of the units see App. A.

Depending on the specific type of asphaltic material and the test temperature, various kinds of apparatus are used. Each involves measuring the time of flow at a stipulated temperature and pressure.

A *dynamic* viscosity test (ASTM D 2171) for asphalt cements and cutback asphalts is conducted at a temperature of 60°C, which is about the maximum to which asphalt pavements are normally subjected. The test uses a *capillary tube viscometer*. Two standard types, the Asphalt Institute vacuum viscometer and the Cannon-Manning vacuum viscometer, are shown in Fig. 19.2. A calibration factor (in dPa) is obtained by using standard oils and is usually supplied by the manufacturer. To control the

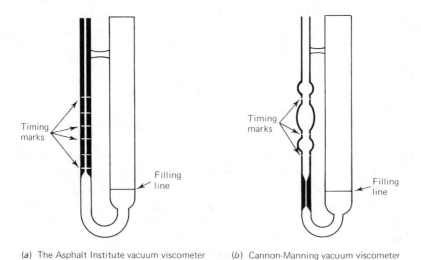

(*a*) The Asphalt Institute vacuum viscometer   (*b*) Cannon-Manning vacuum viscometer

Figure 19.2 Vacuum viscometers. (*From The Asphalt Institute, Ref. 118.*)

temperature, the viscometers are immersed in a water bath. The warm asphalt is poured into the large tube to the filling line, and the temperature is allowed to reach equilibrium. A specified partial vacuum is then applied to the small tube, and the time (s) of flow from one mark to the next is measured with a stopwatch. Multiplying this time interval by the calibration factor yields the dynamic viscosity (dPa · s). Current specification limits are still often expressed in terms of the poise (P), which equals 1 dPa · s.

A *kinematic* viscosity test (ASTM D 2170) is performed at 135°C, a normal temperature for mixing and laying hot mixtures for pavements. The asphalts are fluid enough at this temperature, so that a partial vacuum is not required and can be replaced by a hydraulic head on the influx side. Such a device is the Zeitfuchs cross-arm viscometer shown in Fig. 19.3. The test temperature exceeds the boiling point of water, and so oil must be used in the constant-temperature bath. Asphalt is poured into the large opening and allowed to remain in the "cross-arm" tube until the correct temperature is reached. Slight pressure is then applied to the influx tube (or a vacuum to the efflux tube) so as to start flow over the cross arm. As gravity pulls the asphalt into the smaller diameter tube, the time of flow from the first to the second mark is measured and multiplied by the calibration factor of the device. The result is the kinematic viscosity in mm^2/s or the obsolescent equivalent, centistokes (cSt). This value may be multiplied by the asphalt's density at the test temperature (Mg/m^3) to obtain the dynamic viscosity (mPa · s).[†] It should be evident that in this test the density affects

[†]1 mPa · s = 0.01 dPa · s = 0.01 P = cP.

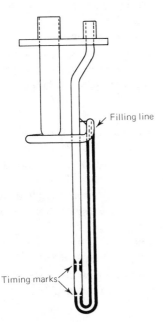

Filling line

Timing marks

Figure 19.3 Zeitfuchs cross-arm viscometer. (*From The Asphalt Institute, Ref. 118.*)

the results inasmuch as the pressure is supplied by gravity, while in the vacuum test the gravitational effect on the pressure is negligible.

Yet another viscosity test, the *Saybolt Furol test* (ASTM E 102 and D 244), is used primarily for emulsified asphalts of both the anionic and the cationic type. The asphalt is heated to 25°C or 50°C in a tube with a specified orifice in the bottom. A stopper in the orifice is removed when the test temperature is reached, and the time is measured for 60 mL of the asphalt to flow out. The time interval in seconds is used as an empirical measure (termed SSF, for Saybolt Furol seconds) of viscosity. Tables for converting Saybolt values to kinematic viscosity are published in ASTM D 2161.

In order to assess the effect on consistency of hardening conditions such as may exist in hot mix plants, a procedure called the *thin film oven* test is sometimes followed. In this procedure (ASTM D 1754), a 50-cm^3 sample of asphalt cement is placed on a flat dish 140 mm in diameter, then put on a rotating shelf in an oven and kept at 163°C for five hours. Viscosity, penetration, ductility or solubility tests are performed before and after this hardening process. A slightly different procedure with the same purpose is the *rolling thin film oven* test (ASTM D 2872). A so-called loss on heating test (ASTM D 6) is similar to the thin film oven test, differing mainly in the specimen dimensions, and is used for blown asphalt for a similar purpose.

Stability Since the emulsified asphalts are suspensions of asphalt globules in water, the emulsions may have a tendency to separate or break down, owing to a variety of causes. Instability may take place in storage or in mixing. Diverse tests, not all of them wholly satisfactory, have been devised in attempts to obtain some indications as to tendencies toward instability or to other undesirable changes during storage.

Details of procedures for common existing tests are described in ASTM D 244. Only brief mention of some of them can be made here. A *settlement* test determines the tendency of asphalt globules to settle, and a *sieve* test detects the presence of undesirably large globules. The *demulsibility* and *cement mixing* tests indicate the rate at which the colloidal asphalt particles will coalesce, that is, break, when the emulsified asphalts are spread in thin films over aggregate. Other tests include the *aggregate coating-water resistance* test, the *particle charge* test, the *storage stability* test, and the *freezing* test.

Ductility The ability of asphaltic mixtures in a layered pavement system to undergo flexure without cracking is an important quality. Although a ductility requirement is included in the specifications for asphalt cements and for the distillation residues of liquid asphalts, it is not considered a good indicator of the degree of flexibility of a pavement mixture under service conditions. It is, however, regarded as a guide as to whether or not a ductility characteristic is present. Tests involving tensile strength (or fatigue resistance) of cured mixtures may prove to be a better guide to the desired quality because in mixtures the asphalt is present in thin films.

The *ductility* test (ASTM D 113) is made by subjecting a standardized briquet molded under stated conditions to a kind of horizontal tension test, at a specified strain rate and temperature (usually 25°C). The total elongation to the breaking point, in centimeters, is reported.

19.4 CHARACTERISTICS OF COMMONLY USED PAVING ASPHALTS

The properties of asphaltic materials vary over a wide range. The selection of an asphalt for a particular use involves consideration of the properties that will promote desired behavior under a variety of conditions, such as storage, mixing, spreading and compacting in place, vehicle loading, weathering, and environmental impact.

Various types and grades of asphalts to serve various purposes have been designated by concurrence of users and producers. These types and grades are defined in terms of limiting values of desired and detrimental characteristics. Tests of the kinds described in the previous article provide the basis for determining such characteristics, which are of two sorts: (1) those that measure an essential property and hence by limits define the desired material, and (2) those that provide a basis for excluding undesirable substances or behavior.

Specifications prepared by standardizing agencies such as ASTM, AASHTO, and the Asphalt Institute have evolved and changed over the years as knowledge of the material has improved, as new technology of production and testing has developed, and as external conditions have altered.

Tables 19.1, 19.2, and 19.3 indicate the magnitudes of several selected properties of commonly used paving asphalts as designated by several current specifications. They also illustrate how a combination of characteristics can be used to classify and define complex products for various purposes.

The formal specifications issued by agencies such as those mentioned contain various requirements in addition to limits on the major properties illustrated here. They may define grades of asphalt other than those shown, or they may provide complete alternative requirements. It is certain that as the state of the art and the conditions of use change, such specifications will be modified. Users should consult the latest documents for both tests and materials, issued by a pertinent agency, for precise and up-to-date requirements.

Table 19.1 shows ranges in or limits to values of selected properties, as stated in current specifications, for two series of *asphalt cements* graded on the basis of viscosity. In the AC series the grades are designated on the basis of the dynamic viscosity at 60°C of the asphalt cement as originally produced. In the AR series the grades are designated on the basis of the dynamic viscosity at 60°C of so-called aged residue derived from the rolling thin film oven procedure. This procedure was devised to produce an age-hardened condition of an asphalt cement similar to the condition that results from hot-mixing and placing in the field; the objective is to be able to specify asphalt cements with characteristics that have some relation to field performance [122].

The numerical grade designations in both series represent median values of the viscosity ranges that define the grades. In each series the dynamic viscosity of the most viscous (hardest) grade is 16 times that of the least viscous (softest) grade.

When asphalt cements in the AC series are subjected to the RTFO procedure, they undergo considerable hardening, as indicated by the maximum values for dynamic viscosity at 60°C permitted for the aged residue. Incidentally, these values coincide with the maximum values for corresponding grades of the AR series.

Table 19.1 Selected characteristics of commonly specified grades of asphalt cement

AC series (graded by viscosity of original asphalt cement)

(1)	Property (2, 3)	Unit	Grade designation				
			AC-2.5	AC-5	AC-10	AC-20	AC-40
OA	Dynamic viscosity, 60°C	dPa · s (P)	200–300	400–600	800–1200	1600–2400	3200–4800
OA	Kinematic viscosity, 135°C, minimum	mm²/s (cSt)	80	110	150	210	300
OA	Penetration, 25°C, minimum	0.1 mm	200	120	70	40	20
OA	Flash point, COC, minimum	°C	163	177	219	232	232
AR	Dynamic viscosity, 60°C, maximum	dPa · s (P)	1250	2500	5000	10 000	20 000
AR	Ductility, 25°C, minimum	cm	100	100	50	20	10

AR series (graded by viscosity of aged residue from rolling thin film oven procedure)

(1)	Property (2, 3)	Unit	Grade designation				
			AR-1000	AR-2000	AR-4000	AR-8000	AR-16000
AR	Dynamic viscosity, 60°C	dPa · s (P)	750–1250	1500–2500	3000–5000	6000–10 000	12 000–20 000
AR	Kinematic viscosity, 135°C, minimum	mm²/s (cSt)	140	200	275	400	550
AR	Penetration, 25°C, minimum	0.1 mm	65	40	25	20	20
AR	Ductility, 25°C, minimum	cm	100	100	75	75	75
OA	Flash point, COC, minimum	°C	205	219	227	232	238

Note: The data shown in this table are selected to illustrate the magnitude of some of the properties of currently used types and grades of asphalts specified by agencies such as ASTM, AASHTO, and the Asphalt Institute. The latest specification documents of these and other agencies should be consulted for complete and up-to-date requirements that may be applicable in a particular jurisdiction.

(1) OA = test made on original asphalt; AR = test made on aged residue from RTFO procedure; see Art. 19.3.
(2) For description of properties and tests, see Art. 19.3.
(3) Temperature equivalents: 25°C = 77°F; 60°C = 140°F; 135°C = 275°F.

Specifications for asphalt cements are promulgated by ASTM (D 3381), AASHTO (M-226), and the Asphalt Institute. Specifications require that the minimum solubility of the original asphalt in trichloroethylene be 99.0 percent.

Table 19.2 shows values for selected properties, as stated in current specifications, for a series of grades of slow-curing (SC), medium-curing (MC), and rapid-curing (RC) *cutback asphalts*. For these liquid asphalts the grades are designated on the basis of kinematic viscosity at 60°C. The numerical grade designations represent the minimum values of the viscosity ranges that define the grades.

Within each series, the viscosity is controlled by the amount of diluent in relation to the base asphalt cement. The more viscous grades contain more asphalt cement and less diluent. In the SC and MC types this is also reflected by higher flash points for the more viscous grades.

The three types of cutbacks (RC, MC, and SC) differ in the hardness of the base asphalts and in the volatility of the diluents. The hardness of the base asphalt for the RC grades tends to run in the center portion of the viscosity spectrum for asphalt cements. The diluent for the RC grades (of a naphtha/gasoline type) is chosen so that the cutback can meet the distillation requirements. The relative amount of diluent distilled off at any given temperature is a function of the volatility of the diluent as well as of the proportion originally present.

The base asphalt for the MC grades runs somewhat softer than for the RC grades; this is reflected by the values for dynamic viscosity of the distillation residues of the MC and RC grades. The volatility of the diluent (a kerosene type) used in the MC grades is, of course, lower than that for the RC grades, and is reflected in the relative amounts distilled off at given temperatures (illustrated at 260°C in Table 19.2).

The diluents in the SC grades are oils of low volatility that were either present in the residue from the first-cut refining process or have been added to an asphalt cement. The base asphalt tends to retain some oil of extremely low volatility, even after the standard distillation test is completed, and the residue from the distillation tends to have a variable degree of hardness.

Specifications for cutback asphalts are promulgated by ASTM (D 2026, D 2027, D 2028) and AASHTO (M-82). All specifications require that specified residues from the distillation test have a minimum solubility in trichloroethylene of 99.0 percent and a minimum ductility of 100 cm.

In *asphalt emulsions* used for road work, the asphalt is the dispersed (discontinuous) phase. Emulsifying agents are chosen so as to produce emulsions in which the dispersed asphalt globules carry either negative or positive charges.[†] The anionic (−) emulsions are characterized as being alkaline, while the cationic (+) emulsions are acidic in nature. Table 19.3 shows values of a few important properties of selected grades of slow-setting (SS and CSS), medium-setting (MS and CMS), and rapid-setting (RS and CRS) emulsified asphalts; the prefix C in the designations indicates a cationic type.

The readiness with which the asphalt globules coalesce and form films on the surfaces or mineral particles (called the rate of setting) is largely controlled by the

†Nonionic emulsions can be manufactured for other purposes.

Table 19.2 Selected characteristics of commonly specified grades of cutback asphalt

Slow-curing cutbacks

(1)	Property (2, 3)	Unit	Grade designation			
			SC-70	SC-250	SC-800	SC-3000
OA	Kinematic viscosity, 60°C	mm²/s (cSt)	70–140	250–500	800–1600	3000–6000
OA	Flash point, COC, minimum	°C	66	79	93	107
OA	Total distillate to 360°C	% by volume	10–30	4–20	2–12	5 max.
OA	Asphalt residue having 100 penetration, minimum	% by mass	50	60	70	80

Medium-curing cutbacks (4)

(1)	Property (2, 3)	Unit	Grade designation (4)			
			MC-70	MC-250	MC-800	MC-3000
OA	Kinematic viscosity, 60°C	mm²/s (cSt)	70–140	250–500	800–1600	3000–6000
OA	Flash point, TOC, minimum	°C	38	66	66	66
OA	Distillate to 260°C	% by vol. of dist. to 360°C	20–60	15–55	35 max.	15 max.
OA	Residue for distillate to 360°C, minimum	% by vol.	55	67	75	80
DR	Dynamic viscosity of residue from distillate, 60°C	dPa · s (P)	300–1200	300–1200	300–1200	300–1200

Rapid-curing cutbacks

| (1) | Unit | Grade designation | | | |
Property (2, 3)		RC-70	RC-250	RC-800	RC-3000
OA Kinematic viscosity, 60°C	mm²/s (cSt)	70–140	250–500	800–1600	3000–6000
OA Flash point, TOC, minimum	°C	–	27	27	27
OA Distillate to 260°C	% by vol. of dist. to 360°C	70 min.	60 min.	45 min.	25 min.
OA Residue from distillate to 360°C, minimum	% by vol.	55	65	75	80
DR Dynamic viscosity of residue from dist., 60°C	dPa · s (P)	600–2400	600–2400	600–2400	600–2400

Note: The data shown in this table are selected to illustrate the magnitude of some of the properties of currently used types and grades of asphalts specified by agencies such as ASTM, AASHTO, and the Asphalt Institute. The latest specification documents of these and other agencies should be consulted for complete and up-to-date requirements that may be applicable in a particular jurisdiction.

(1) OA = test made on original cutback asphalt; DR = test made on residue from distillation test.
(2) For description of properties and tests, see Art. 19.3.
(3) Temperature equivalents: 60°C = 140°F; 260°C = 500°F; 360°C = 680°F.
(4) MC grades in addition to those shown here are included in specifications by some agencies.

Table 19.3 Selected characteristics of commonly specified grades of emulsified asphalt

Slow-setting (SS and CSS) emulsion

(1)	Property (2, 3)	Unit	Grade designation			
			SS-1	SS-1h	CSS-1	CSS-1h
OE	Viscosity, Saybolt Furol, 25°C	s	20–100	20–100	20–100	20–100
OE	Cement mixing test, max. amt. of coag.	% by mass	2.0	2.0	2.0	2.0
OE	Residue from dist. test, minimum	% by mass	57	57	57	57
DR	Penetration on dist. residue, 25°C	0.1 mm	100–200	40–90	100–250	40–90

Medium-setting (MS and CMS) emulsions (4)

(1)	Property (2, 3)	Unit	Grade designation (4)			
			MS-2	MS-2h	CMS-2	CMS-2h
OE	Viscosity, Saybolt Furol, 25°C	s	100 min.	100 min.	–	–
OE	Viscosity, Saybolt Furol, 50°C	s	–	–	50–450	50–450
OE	Residue from dist. test, minimum	% by mass	65	65	65	65
DR	Penetration of dist. residue, 25°C	0.1 mm	100–200	40–90	100–250	40–90

Rapid-setting (RS and CRS) emulsions

(1)	Property (2, 3)	Unit	Grade designation			
			RS-1	RS-2	CRS-1	CRS-2
OE	Viscosity, Saybolt Furol, 25°C	s	20–100	–	–	–
OE	Viscosity, Saybolt Furol, 50°C	s	–	75–400	20–100	100–400
OE	Residue from dist. test, minimum	% by mass	55	63	60	65
DR	Penetration of dist. residue, 25°C	0.1 mm	100–200	100–200	100–250	100–250

Note: The data shown in this table are selected to illustrate the magnitude of some of the properties of currently used types and grades of asphalts specified by agencies such as ASTM, AASHTO, and the Asphalt Institute. The latest specification documents of these and other agencies should be consulted for complete and up-to-date requirements that may be applicable in a particular jurisdiction.

(1) OE = test made on original emulsion; DR = test made on residue distilled from emulsion.

(2) For description of tests for these properties see Art. 19.3.

(3) Temperature equivalents: 25°C = 77°F; 50°C = 122°F.

(4) Grades designated as MS-1, HMFS-1, -2, -2h, and -2s are included in some specifications, but values for these grades are not shown here.

relative amount and nature of emulsifying agent and any stabilizing agent that may also be used. The ratio of water to base asphalt and the nature of the latter also have some influence. For the SS and CSS grades, often characterized as having greater mixing stability, it is thought that the slow evaporation of water is the major feature of the setting process, while in RC emulsions the relatively rapid breaking takes place as the ionic charges are neutralized and the globules coalesce in the presence of water.

The CRS and CMS emulsions may be made from cutbacks, so that the specifications for these types commonly include limitations on the amount of oil present.

Within a series characterized by rate of setting, the grade carrying the higher numerical designation has the higher viscosity. (In Table 19.3 only the RS series shows this.) The viscosity is controlled by the proportion of water to base asphalt, which is reflected in the amount of residue from the distillation test in the RS series.

Grades in the same rate-of-setting series may also differ in the hardness of the base asphalt. The letter h following the numeral in the grade designation indicates that the emulsion contains a harder base asphalt; this is reflected in the penetration values for the residues from the distillation test. Similarly, a suffix s (not illustrated in Table 19.3) indicates a softer than normal base asphalt.

In the medium-setting (MS) series, current specifications also include grades designated as HFMS-1, HFMS-2, HFMS-2h, and HFMS-2s; these are not shown in Table 19.3. The HF prefix indicates that the float-test value of the base asphalt is higher than in the case of the other grades. The high float grades were formulated in order to permit thicker asphalt films on aggregate particles, with less likelihood of drainage before setting.

Specifications for emulsified asphalts are promulgated by ASTM (D 977 for anionic and nonionic types, D 2397 for cationic types) and AASHTO (M-140 and M-208). All the specifications require that the residue from distillation for all types and grades have a minimum solubility in trichloroethylene of 99.5 percent and a minimum ductility of 40 cm.

19.5 MINERAL AGGREGATES

The mineral aggregates used in asphalt mixtures may be classified as to source and/or preparation. *Pit-run* or *bank-run aggregates* include sands and gravels found in natural deposits. *Processed aggregates* are made by crushing natural rock or gravel; they are more angular in shape and usually have rougher particle surfaces than the more-rounded natural sands and gravels. *Synthetic* and *artificial aggregates* that are sometimes used in road construction and in mixtures for some purposes include slags, mine tailings, cinders, and fused clays.

Aggregates for asphalt mixtures are usually screened into various size ranges, described later. For quality mixtures, aggregates are often washed to remove deleterious coatings from the particles.

In specifying aggregates for particular uses, or for making calculations for proportioning mixes, the following characteristics are pertinent:

• Density, moisture content and absorption, void content

- Size and gradation of particles
- Particle shape, surface texture, and surface area
- Toughness and abrasion resistance
- Soundness
- Cleanness
- Mineral composition and surface chemistry

Density, voids, porosity, and absorption Although density is basically defined as mass per unit volume, in discussing properties of particulate materials one must distinguish between measures of density relating to an aggregated mass of particles and measures relating to solid particles considered as discrete pieces of matter.

The interconnected spaces *among* the particles (interparticle spaces) that make up an aggregation are here termed the *voids*. The *bulk density* [†] of the aggregation is the mass of particles (in some stated condition) in a unit volume made up of volumes of the solid particles plus the volume of the voids. A quantity of aggregate, for the purpose of batching or estimating, may be stated either in terms of (bulk) volume or in terms of the mass of the particles. Conversion from one basis to the other may, of course, be made by knowing the bulk density of the aggregated mass.

The spaces wholly or partially enclosed *in* a particle (intraparticle spaces) are here termed the *pores*. The impermeable pores are those within the particle that are not connected to the exterior and thus cannot be filled with water (or another liquid). The permeable pores are those that are connected to the exterior so that they can be filled with water. Interrelations of these volumes are represented schematically in Fig. 19.4. The permeable pore space is taken to be the volume of water determined from the difference in the saturated surface-dry condition and the oven-dry condition of the particle (Fig. 19.4*b*). The bulk volume of a particle, V_β, is the volume that would be enclosed by a membrane stretched tightly over a saturated surface-dry particle so as to enclose all pore spaces. The apparent volume, V_α, is the volume that would be determined by displacement for a completely dry particle and hence includes only the impermeable pores. Thus, using the letter symbols in Fig. 19.4:

$$\text{Bulk density} = \frac{M_s}{V_s + V_{ip} + V_{pp}} = \frac{M_s}{V_\beta}$$

$$\text{Apparent density}^\ddagger = \frac{M_s}{V_s + V_{ip}} = \frac{M_s}{V_\alpha}$$

The bulk density of most commonly used natural mineral aggregate particles falls within the range of 2.55 to 2.75 Mg/m^3. A representative hard limestone might have a bulk density of 2.70 Mg/m^3 and an apparent density of 2.74 Mg/m^3.

If the aggregate is nonabsorptive, $V_{pp} = 0$, and the bulk density equals the apparent density.

[†]The term unit weight has been commonly (if inaccurately) applied to the density of the aggregation of particles.

[‡]In Chap. 18, in connection with portland cement concrete, this was termed the "true density."

V_s = vol. of mineral matter; no pores
M_s = mass of mineral matter
V_{ip} = vol. of impermeable pores
V_{pp} = vol. of permeable pores
V_α = apparent vol. of particle = $V_s + V_{ip}$
V_β = bulk vol. of particle = $V_\alpha + V_{pp}$
V_{bp} = vol. of asphalt that could be absorbed in the pores
V_e = effective vol. of particle = $V_\beta - V_{bp}$

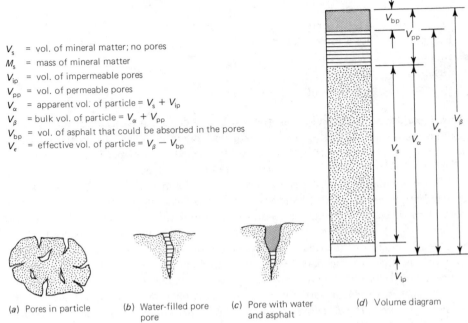

(a) Pores in particle (b) Water-filled pore pore (c) Pore with water and asphalt (d) Volume diagram

Figure 19.4 Schematic representation of volume relations in a particle of mineral aggregate.

In the design and control of asphaltic mixes it is useful to know the volume of air space that is contained along with the asphalt binder and mineral aggregate in a compacted mixture. When porous aggregates[†] are used, calculation of the air content on the basis of either bulk or apparent density can result in some error, because asphalt may partially penetrate the externally connected pores, as indicated schematically in Fig. 19.4c. One approach to dealing with this situation is to use an intermediate density, termed the "effective density," the determination of which is explained in Art. 9.6. The value of effective density is intermediate between that of the bulk and apparent densities.

The term absorption, as used in connection with aggregates, refers to the amount of water contained in the permeable pores when they are completely filled. It is expressed as a percentage of the oven-dry mass of the aggregate particles. Average gravels and crushed limestone may have absorption values on the order of one-half to 1 percent, some sandstones may run to 5 to 7 percent, very light porous materials, up to 25 percent or even more.

Standardized methods of testing relating to density have been issued as follows:

	ASTM	AASHTO
• Unit weight (bulk density) of aggregate	C 29	T-19
• Specific gravity (relative density) and absorption of coarse aggregate	C 127	T-85

[†]Sometimes defined as aggregates that can absorb more than 2 percent water.

	ASTM	AASHTO
• Specific gravity (relative density) and absorption of fine aggregate	C 128	T-84
• Specific gravity (relative density) of mineral filler	D 854	T-100
• Surface moisture in fine aggregate	C 70	
• Moisture in mineral aggregate	C 566	

No method of determining effective density has been standardized. Some methods that have been used are described in Refs. 118, 125, 126, and 127.

Size and gradation of particles The size distribution or gradation of particles in a mineral aggregate is determined by passing a representative sample of aggregate through a standard series of sieves in which each successively smaller size sieve usually has clear (square) openings one-half those of the next larger size.[†] While the sieve analysis gives the amount (mass) of material retained on each sieve, the gradation is usually expressed in terms of the percentage of the total sample smaller than the clear opening of each successive sieve (the "cumulative percent passing").

The maximum size of aggregate is stated in terms of the clear opening of the smallest sieve in the series that will just pass 100 percent of the sample; the nominal maximum size is defined by the size of the sieve that will pass between 90 and 100 percent of the sample. The size of a fraction is defined as a range in size of particles all of which would pass a given sieve and all of which would be retained on a designated smaller sieve.

Standardized methods of making sieve analysis have been issued as follows:

	ASTM	AASHTO
• Sieve analysis of fine and coarse aggregate	C 136	T-27
• Sieve analysis of mineral filler	D 546	T-37
• Material finer than No. 200 (74-μm) sieve by washing	C 117	T-11

In asphalt paving practice, all particles that would be retained on a 2.38-mm (No. 8) sieve are classed as coarse aggregate, and those that would pass that sieve are classed as fine aggregate. However, to avoid the segregation of particles in aggregates having a large range in particle size, coarse aggregates are usually processed so as to have limited size ranges. Specifications giving grading limitations within each of several designated size groups for coarse mineral aggregates have been issued (ASTM D 448, AASHTO M-43). Specifications for fine aggregates generally give permissible ranges in sizes of material smaller than the 2.38-mm sieve (ASTM D 1073). In some asphaltic mixes, a certain amount of material finer than the 74-μm sieve is desirable; if available aggregates lack fines of this nature, a *mineral filler* of specified (ASTM D 242) grading limits may be incorporated in the total mix. Limits may be placed on the amount of claylike (plastic) material. Some aggregate specifications carry general requirements pertaining to properties such as soundness, abrasion, crushed pieces, and

[†]Sieves with size ratios of $1 : \sqrt{2}$ are often used for the coarser aggregates. Information on sieve sizes is given in App. B.

so on, in addition to the gradation requirements (ASTM D 242, filler; D 692, coarse aggregate; D 693, crushed stone; D 1073, fine aggregate).

Asphaltic mixtures used for pavement construction may call for different types of overall aggregate grading. Two commonly used types of grading are characterized as dense graded and open graded. In a *dense-graded* aggregate, all sizes of particles, from the maximum specified to those passing the 74-μm sieve, are present in relative proportions such that the mixture will tend to have a fairly high density and low void content. In an *open-graded* aggregate, there may be a preponderance of some of the coarse fractions, discontinuous grading and/or little or no mineral filler, so that in the resulting compacted mixture and void space may be relatively large.

In Table 19.4 are shown some examples of grading requirements for dense- and open-graded aggregates used in wearing and base courses of pavements; these gradations are for use with mixtures made with appropriate grades of asphalt cement or emulsified asphalt. As discussed in Art. 19.6, specifications for asphalt *mixtures* for particular purposes or uses usually include grading requirements.

Since aggregates are often supplied in two or more size ranges, they must be combined in such proportions that the overall grading requirements are met. Probably the simplest approach to selecting the proportions is by trial and error, using a gradation chart to aid in visualizing the problem. Usually the gradations of the separate aggregates and of the specification gradation limits are plotted on a graph with percent passing as the y axis and sieve size from small to large as the x axis. Trial percentages of each separate aggregate are combined to achieve a composite grading that will fall within the specified limits [114, 118].

Table 19.4 Examples of grading limits for selected maximum sizes of aggregates

Sieve size		Dense graded		Open graded	
mm	in/No.	% passing		% passing	
38.1	$1\frac{1}{2}''$	100	–	100	
25.4	$1''$	90–100	–	95–100	100
19.0	$\frac{3}{4}''$	–	100	–	90–100
12.7	$\frac{1}{2}''$	60–80	90–100	25–60	–
9.51	$\frac{3}{8}''$	–	–	–	20–55
4.76	No. 4	25–60	45–70	–	–
2.38	No. 8	15–45	25–55	0–10	0–10
1.19	No. 16	–	–	0–5	0–5
0.595	No. 30	–	–	–	–
0.297	No. 50	3–18	5–20	–	–
0.149	No. 100	–	–	–	–
0.074	No. 200	1–7	2–9	0–2	0–2

Source: Ref. 121.

Particle shape, surface texture, and surface area Aggregate particles suitable for use in asphaltic mixtures should tend toward somewhat equidimensional proportions, in contrast to flaky, thin, or elongated particles, which should be avoided. However, in compacted mixtures, angular-shaped particles exhibit greater particle interlock and internal friction, and hence result in greater mechanical stability than do rounded particles; on the other hand, fresh mixtures containing aggregates made up of rounded particles have better workability and require less compactive effort than do fresh mixtures made with angular aggregates. With respect to mechanical stability, angularity may be just as important in the fine aggregates as in the coarse aggregates.

Somewhat similar influences derive from *surface texture*. Rough surfaces, such as are found on most crushed stones, tend to induce higher internal friction in the mixture and often better adherence of asphalt than do the smooth surfaces found on many river gravels and sands; on the other hand, smooth surfaces tend to result in a lower asphalt requirement for the same workability.

In selecting aggregates for mixtures designed for use in high-performance surface courses, some compromises are obviously necessary. When gravels are used in such mixtures, specifications may call for some substantial percentage of the aggregate to have crushed faces [119]. In some parts of the pavement structure, such as base courses, less high-quality aggregates may be used.

Measures of *surface area* of aggregates are used in some procedures as a basis for determining the amount of asphalt to be used in trial mixes and are also used for estimating the thickness of asphalt films on aggregate particles. Calculations of the total surface area in a graded aggregate are made basically by summing up the estimated surface areas for each of several fractions of the aggregate. Since particles of aggregate are rarely spherical, empirical factors are applied to calculate the surface areas attributable to each of the several fractions.

For determining asphalt coating requirements, absorption of asphalt by the aggregate and the nature of the surface itself may have to be taken into account in addition to the surface area. In the *centrifuge kerosene equivalent* (*CKE*) test, described in the following paragraph, the evaluation includes the combined effect of absorption and of the nature of the surface. A modification of the CKE test separates these two effects [139].

In the CKE procedure [118, 119] for determining the amount of asphalt required for trial mixes, a composite surface area value is first calculated for the entire aggregate by applying "surface area factors" to the proportion of the aggregate passing each sieve in a specified series. A factor K_f is then determined to bring in the effect of texture and absorption for the material passing the 4.76-mm (No. 4) sieve; it is based on the amount of kerosene retained on a sample that has been centrifuged after immersion in kerosene. A factor K_c for the material coarser than 4.76 mm is based on the amount of a lubricating-grade oil retained on a drained sample after immersion in the oil.

The values of the surface area and of the two surface capacity factors (K_f and K_c) for a given aggregate are used as inputs to a rather intricate prescribed calculation to obtain a value of the amount of a given grade of liquid asphalt to be used in a trial mix.

Toughness and abrasion resistance In a pavement subjected to repeated traffic loads, the aggregate is a principal load transmitter, and thus must be strong and tough enough to resist crushing and breakdown or abrasion at points or surfaces in contact. Degradation of the particles can also occur as the mixture is worked during the mixing and compaction processes. Whatever the combination of properties is that provides resistance to degradation under such conditions, a test called the *Los Angeles abrasion* test (ASTM C 131, AASHTO T-96) is employed to obtain an indication of the desired characteristics. The test is performed by tumbling a specified type of coarse-aggregate sample, along with a charge of steel balls, in a steel drum. The abraded material that passes a 1.68-mm (No. 12) sieve is weighted and is reported as the percentage of wear (by mass of the original sample). The maximum wear for natural mineral aggregates used in high-type mixtures for surface courses is limited by current specifications to 40 percent. Specifications for the materials that can be used in most base courses permit a maximum limit of 50 percent. This test does not appear to be fully satisfactory for use with slags, cinders, and some synthetic aggregates; for these, minimum density requirements are often applied.

Soundness Some materials, such as shale or certain sandstones, deteriorate or disintegrate under the action of wetting and drying, freezing and thawing (weathering). A test to determine the presence of particles that lack resistance to weathering (called soundness) may be desirable for aggregates that do not have a record of adequate performance in service, especially if they are to be used in highway-quality pavements and/or under severe conditions of exposure.

A commonly used *accelerated soundness* test (ASTM C 88; AASHTO T-104) consists of subjecting sized fractions of an aggregate to a specified number of cycles of alternate wetting in a saturated solution of sodium or magnesium sulfate and oven drying. The loss in mass for each size fraction is determined by sieving, and the weighted average percent loss for the entire sample is computed.

Cleanness In general, cleanness refers to the absence of substances that are damaging to the desired quality of an aggregate or to the performance of an asphalt mixture in pavement service. Undesired substances may occur as inclusions of deleterious foreign material or as contamination in the form of coatings on the aggregate particles.

In addition to mineral particles subject to degradation, as discussed in preceding sections on toughness and soundness, undesirable foreign materials include claylike fines; reactive chemicals; reactive particles (especially in slags) that expand appreciably due to hydration, oxidation, or carbonation; and vegetation or other undesirable organic matter.

A test commonly used to indicate the relative proportion of plastic fines in a fine aggregate is the *sand equivalent* test (ASTM D 2419; AASHTO T-176). In this test, a sample of aggregate passing the 4.76-mm sieve is agitated in a water-filled transparent cylinder together with a flocculating agent, so that after settling, the sand is separated from the flocculated clay. The sand equivalent is expressed as the ratio of the height of the sand to the total height of sand plus floc. Specifications for aggregates for paving mixtures often place a sand equivalent of 25 percent minimum on sands and of 35 percent minimum for processed dense-graded aggregate.

In another approach to controlling detrimental plastic fines, some specifications call for *plasticity* tests (ASTM D 423 and D 424) of the type used for clay soils, on the fraction of aggregate passing the 74-μm sieve.

The presence of other types of inclusions is determined by visual examination or by applicable chemical tests.

Coatings of fine dust may interfere with the adhesion of asphalt to aggregate surfaces. Simple washing tests or asphalt coating tests may be made to appraise this condition. Sometimes the surfaces of aggregates carry chemical coatings precipitated from exposure to brines, soil stabilizing agents, etc. These may interfere especially with the setting of emulsified asphalts. Coating tests made with the asphalts proposed for use in the given situation can provide indication of such contaminants.

In practice, some undesirable material will be found in most aggregates. Limits, such as those that have been indicated, are specified to hold their presence to tolerable levels. Such permissible limits can usually be higher for base courses than for wearing courses.

Mineral composition and surface chemistry The properties of the solid particles in aggregates are intimately associated with the mineralogical nature of the particles. Basic physical and mechanical properties such as density and strength are functions of the composition and structure of the minerals themselves. Other properties, such as shape, surface texture and gradation, reflect in part the response of the constituent minerals to external influences, natural and artificial. Properties such as these, having significance for the use of aggregates in asphaltic mixtures, were discussed in previous sections.

The nature of the minerals composing the particles also has an important influence on the physicochemical reactions that can occur on the particle surfaces. The adhesion of asphalt to the surface of the particles greatly affects the long-term stability or durability of asphalt mixtures in service; for example, the coating of asphalt on the particles of some aggregates can become detached after considerable exposure to water, an action called "stripping." Theories advanced thus far for explaining the interactions between asphalt and mineral surfaces are not fully adequate. It is an observed fact, however, that mineral aggregates having preponderantly siliceous composition strongly exhibit this preferential affinity for water (unless countermeasures are taken).

Many commercially available aggregates are of variable mineralogical composition. Further, even with aggregates of fairly uniform mineralogical composition, the surface chemistry may be altered by oxidation, hydration, foreign coatings, and so on. Thus, whether stripping will take place cannot be reliably predicted from mineral identity alone.

One type of test for detecting the compatability of asphalt and aggregate involves coating a sample of loose particles with asphalt, curing, and then, after a period of immersion in water, visually estimating whether 95 percent of the aggregate surfaces retain the asphalt film (ASTM D 1664).

Another type of test involves determination of the compressive strength of specimens of compacted asphalt mixtures after they have been cured and immersed in water (ASTM D 1075, AASHTO T-165).

Tests such as these may be replaced by methods based on the physical chemistry of the surface interactions as better understanding of these phenomena develops.

The stripping tendency of aggregates is presently being controlled by either (1) making the asphalt mixture, as compacted in the pavement, less permeable to water through mix design or (2) using a suitable surface-active agent [114]. With an emulsified asphalt, an emulsifier of appropriate ionic charge and composition is incorporated in the emulsion; with asphalt cements and cutback asphalts, oil-soluble surface-active agents have been used.

19.6 ASPHALT MIXTURES

A highway pavement is a layered structure that serves to transmit vehicle loads to the subgrade in such a way that the basement soil is not displaced or ruptured. Among the various qualities that a pavement structure should possess, appropriate load-bearing and stress-transmitting characteristics are paramount. In addition, the mixtures[†] that compose the pavement layers must in themselves be resistant to the localized effects of load (displacement, rupture, rutting, abrasion, cracking, bleeding, raveling) and to the disintegrative influences of weathering and chemical reactions (stripping, hardening of asphalt films, temperature or shrinkage cracking, breakdown of aggregate particles, and the like).

The design of a mixture is a problem in devising a composite material having properties that are inputs to the larger problem of pavement design. Mixture design and pavement design are thus interrelated. Within practical and economic limitations, the nature and magnitude of mixture characteristics are specified for pavements intended for particular uses and ambient conditions. At the same time, pavement design is limited by the state of knowledge concerning, and the availability of, mixtures that can be produced.

The characteristics of mixtures that affect the performance of pavements in service may be characterized as follows: internal structure, stability (stiffness and strength), durability, permeability, and surface characteristics. Characteristics of freshly compounded mixtures, such as workability, are important during the construction process.

The properties categorized under *internal structure* include density, porosity, absorption, and moisture content.

The term *stability* has been used loosely and variously, but as applied here to asphalt mixtures, it is a complex mechanical property that has to do essentially with the integrity and the nature of the change in form of a body under applied forces. The change in form may be elastic (recoverable) or plastic (permanent). Stability thus involves both *stiffness* (a function of dimensional change under load) and *strength* (the

---

[†]The term *mixture* is used here to designate a composite substance having particular and unique properties. The term *mix*, as used here, emphasizes the makeup or proportions of the ingredients of a mixture, especially as they are freshly compounded.

limiting value of load that a material can sustain without rupture or excessive deformation). Stiffness is commonly expressed in terms of the ratio of stress to strain, but this relation for asphaltic materials may be complicated especially by the effects of the time rate of loading and of temperature. Sometimes, however, stiffness is characterized by the amount of deformation under a specified load. The term *consolidation* is often applied to the decrease in volume of a material under sustained loads owing to expulsion of fluids (gas or liquid) from interparticle spaces. Particular aspects of strength are compressive strength, tensile strength, flexural strength, strength under repeated loading (fatigue strength), strength under sustained loading (creep strength), and abrasion resistance.

The concept of *durability* here concerns the preservation of the integrity of a mixture under the influence of weathering, that is, the resistance to the deleterious effects of heat, water, and other atmospheric influences. For example, the hardening of the asphalt films on aggregate particles, owing to oxidation and/or to the excessive loss of solvent, can markedly affect the durability of a mixture.

The ease with which a fluid can flow into or through a porous solid, the *permeability*, may be advantageous or detrimental. Under some conditions, a degree of permeability permits water to drain away from a pavement structure. On the other hand, a relatively impermeable mixture reduces the access of water or water vapor, and thus serves to inhibit the stripping of asphalt from the aggregate, undesirable effects on the aggregate or subgrade, etc.

Characteristics of the surface of a mixture used in a wearing course affect skid resistance, hydroplaning, noise produced by moving traffic, and under some conditions, visibility as influenced by the absorption of light.

An ever-present consideration in the design of the pavement structure and its component mixtures is economy. This must be determined for each project, of course, by comparing choices among materials, processes of construction, and resulting mixtures.

Mix design The objective of the mix-design process is to obtain an appropriate and economic blend of ingredients that will result in a product having characteristics needed to serve certain purposes. The development of some properties may be at the expense of others. For example, a mixture made of a hard asphalt may have relatively high strength at early ages of service but may exhibit impaired durability in the longer run. Thus, design involves compromises that must be resolved by considering overall uses and costs.

In selecting mix proportions, general physical requisites of a fresh mix are:

- There should be enough asphalt to coat the aggregate particles thoroughly so that it serves effectively as a strong binder and fills the interparticle space enough to give desired impermeability. However, the amount of asphalt should not be so great as to allow the mixture to behave in a viscous mode at higher atmospheric temperatures.
- The gradation of the aggregate should be such as to provide an effective load-transmitting skeletal structure, but the interparticle space should be adequate to allow for sufficient asphalt plus air voids needed for the particular kind of mixture.

- The amount of air voids in the compacted mixture in place should be adequate to allow for some additional consolidation to take place under longer term traffic loads, so as to avoid flushing of asphalt to the surface (a condition known as "bleeding").
- The amount and gradation of the aggregate and the amount and viscosity of the asphalt, in combination, should make for a workability of the fresh mix that will avoid segregation but not require excessive effort in laying and compacting.

Based on field experience and observation of performance of pavements, a number of agencies have formulated specifications for mixtures of various types for various purposes. These specifications state general requirements for materials, proportions, and/or properties of mixtures. Even so, the range in available constituent materials and their behavior is such that a mix must be designed for a special job if optimal performance and economic use of materials is to be achieved.

While empirically derived recipes are used for some purposes, for high-quality, high-performance pavements a trial-mix approach is generally used. This approach involves laboratory curing and testing of compacted specimens as well as observations and tests of ingredients and fresh mixes.

Steps in the mix design process may be summarized as follows:

1. Selection of type of mixture to be designed
2. Selection of aggregate
3. Selection of asphalt
4. Selection of trial mix proportions
5. Manufacture and testing of trial mixes
6. Selection of optimal mix

Selection of type of mixture The nature of the traffic and that of the underlying soils obviously influence the kind of pavements to be provided; the kind of pavement needed to support high volumes of traffic with heavy vehicles will differ from one for a low-volume secondary road. A pavement to be built on dense, gravelly soils will differ from one to be built on loosely composed loams. Severity of climate can influence the design of pavement and hence the mixtures employed. The layers or courses (wearing course, base course, subbase) employ different types of mixtures. And local availability of materials, locally available "know-how," or other reasons may make for a choice of dense-graded, open-graded, macadam, or another kind of general composition. Considerations such as these narrow the range of choice.

For the purpose of illustrating the general approach and principles involved, the remainder of this discussion will concern mainly dense-graded mixes for use in wearing courses and will focus mainly on mixes using asphalt cement (AC) or emulsified asphalt (EA).

Selection of aggregate Specifications for asphalt mixtures are generally written to prescribe acceptable characteristics of mixtures of particular types or for particular purposes. For example, ASTM D 3515 is a specification for "Hot-Mixed, Hot-Laid Bituminous Paving Mixtures." Such specifications commonly contain provisions concerning aggregate grading, as well as limits on abrasion resistance, crushed faces, unsound particles, plastic fines (by the sand equivalent test), and other deleterious inclusions (described in Art. 19.5).

In this kind of specification, mixtures may be designated according to maximum size of aggregate, and within such designations the overall grading is specified in terms of permissible ranges of the percent passing a selected series of sieves, down through the 74-μm (No. 200) sieve. The maximum size to be chosen is some fraction of the thickness of the layer to be laid.

One of the kinds of mixture commonly used in high-quality wearing courses is a dense-graded mixture. The grading ranges for such mixtures, with several maximum sizes, are included in ASTM D 3515. These gradings can be employed for mixes using either asphalt cement or emulsified asphalt [119, 121, 129]. An example of a dense grading is given in Table 19.4 in Art. 19.5.

Commercially available aggregate sizes are blended to produce an overall grading that may vary within the specification limits to meet needs for workability of the mix or to provide appropriate interparticle void space (called "voids in mineral aggregate," VMA).

Selection of asphalt Considerations in choosing the type and grade of asphalt include type of construction, type of aggregate to be used, method of mixing and placing, atmospheric temperature and moisture at time of placing, and rate at which setting and curing can or will take place, as well as conditions to which the mixture will be subjected in long-term service.

While the compatibility of the asphalt and the mineral aggregate may be judged in a preliminary way from the mineral nature of the aggregate (see comments in Art. 19.5), trial mix procedures with emulsified asphalt commonly include tests to provide evidence of how well the asphalt aggregate bond is made and will be maintained.

Asphalts commonly considered suited to use in dense-graded mixtures for wearing courses include the middle grades of the AR-type of asphalt cements and the slow-setting (SS or CSS) asphalt emulsions. Suggestions and guidelines for other uses and types of mixtures maybe found in Refs. 119, 121, 137, and 138 and in ASTM D 3628.

Selection of trial mix proportions In preparing trial mixes for test and evaluation, it is desirable first to make an estimate of what an optimal asphalt content of the mix is likely to be. Several trial mixes are then prepared having a range in asphalt contents that bracket the estimated value.

In some trial mix procedures, preliminary estimates of initial asphalt contents are chosen with the aid of tabulations compiled from previous practice. For dense-graded mixtures, in a number of current procedures, the estimate is based on some function of the surface area of the aggregates to be used (see comments on CKE procedure in Art. 9.5) [118, 139].

The asphalt contents for stable mixtures made with comparable aggregate gradings vary with the maximum size of aggregate—the smaller the maximum size, the more asphalt required. For dense-graded mixes of 25-mm-maximum-size aggregate made with asphalt cement, acceptable asphalt contents might range from 4 to 8 percent by mass of the total mixture for normal aggregate densities (ASTM D 3515) [137]. A similar type of mix made with emulsified asphalt would require about the same amount of "residual" asphalt (the asphaltic substance remaining after the water in the emulsion has evaporated). For expressing batch quantities for an emulsified asphalt mix, the amount to be used would, of course, be expressed as the mass (or volume) of the emulsion. In making up trial mixes, the proportion of asphalt or emulsion is stated in terms of its ratio to the mass of aggregate, while for other purposes it may be given as the ratio to the mass of the total mixture; the basis should be clearly stated.

When emulsified asphalts are used, some aggregates could absorb so much water from the emulsion that the mix would stiffen prematurely and would be difficult to compact to suitable density. To deal with this, additional water may be added, preferably to the aggregate until its appearance just begins to darken. Added water might be on the order of 3 to 5 percent [129] ; excess water can induce stripping.

Manufacture and testing of trial mixes Several approaches to the design of mixtures have been developed that use laboratory tests as a basis; some have been superseded. All of them essentially evaluate the fresh mix and many of them make some kind of test for stability of the mixture; most of them also attempt to shed some light on the effect of curing, the effect of water, etc. The ultimate gauge of the merit of any such procedure is how well it can predict actual field behavior.

In the batching stage and at key points in the subsequent process, measurements are made on materials, fresh mixes and compacted test specimens to provide information concerning mass-volume relations and other physical properties.

In making laboratory mixes, thorough mixing is of great importance. Observations on workability are made during the mixing stage.

For emulsified asphalt mixtures, a number of tests and observations are made at the time the fresh mixes are in preparation to determine whether the emulsion used with the given aggregates will perform satisfactorily. These include tests to gauge whether uniform dispersion of the emulsion can take place, determination of completeness of aggregate coating, and, for open-graded mixes, determination of "run-off" (the fresh mix is allowed to drain over a screen); a "wash-off" test of open-graded mixes may also be made after 24 hours [119].

Strength or stability measures from any method of test are very sensitive to the nature and degree of compaction. For meaningful and comparable results, a specified compaction procedure must be closely adhered to. Compaction methods in use include: static pressure, drop-hammer impact (ASTM D 1559), kneading compaction (ASTM D 1561), and gyratory compaction. Various curing schedules are followed, and in some procedures an accelerated soaking procedure may be called for to determine whether stripping of the aggregate or aggregate deterioration has taken place.

Based on the performance of asphaltic mixtures in pavement service, certain limiting ranges in air void content, ratios of volume of asphalt to volume of voids in mineral aggregate, and similar relations, are considered desirable for satisfactory performance; values of such quantities for a set of trial mixes are used in making judgments as to choice of job mix.

Information required for an analysis of *mass-volume relations* of any given compacted mixture basically includes: composition of the mixture in terms of proportions by mass of asphalt and aggregate; density of asphalt; weighted average bulk density of aggregate particles, bulk density of the compacted mixture; and the "theoretical maximum density" of the mix.

The volumetric relations in a compacted mixture are represented schematically in Fig. 19.5, where the symbols used here are also defined. If the aggregate can absorb some of the asphalt, the space occupied by the aggregate, V_β and the asphalt not absorbed in the pores of the aggregate, V_{be} (subscript b refers to bitumen), is V_{mm}. The "theoretical maximum density," $D_{mm} = (M_a + M_b)/V_{mm}$, is found by measurement (ASTM D 2041, AASHTO T-209). From D_β, D_b, and D_{mm} the following quantities can be computed in turn: the effective density of the aggregate, absorbed asphalt, "effective" asphalt volume, and air voids [118]. The space in the compacted mixture not occupied by the bulk volume of the aggregate, called voids-in-mineral-aggregate (VMA), is obtained from $V_m - V_\beta$; it is expressed as a percentage of the volume of the compacted mix. Under the Marshall mix design procedure (described subsequently) a

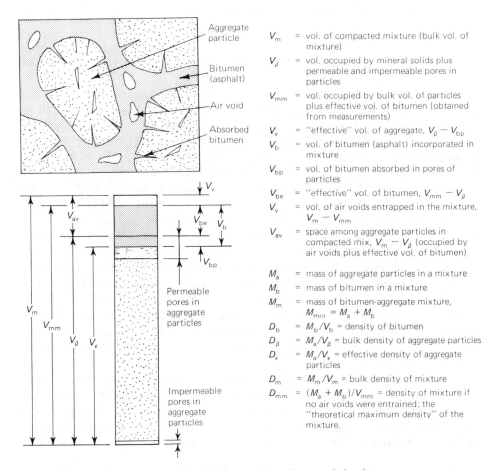

Figure 19.5 Schematic representation of volume relations in an asphalt mixture.

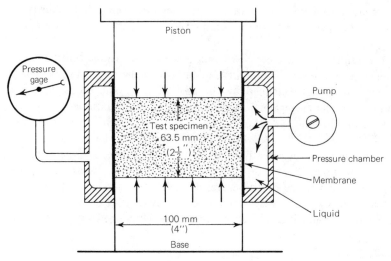

Figure 19.6 Diagram of Hveem stabilometer. (*From The Asphalt Institute, Ref. 118.*)

minimum **VMA** is specified for various maximum sizes of aggregate in dense-graded mixtures.

Tests for *stability* on laboratory-cured specimens are distinguishing features of most of the trial mix procedures. Stiffness tests are commonly supplemented by auxiliary measurements or tests that help avoid the selection of mixtures that may have high stiffness but inadequate cohesion. All of the tests described here employ compacted cylindrical specimens that are 100 mm in diameter by 64 mm in height.[†] Thus the maximum size of aggregate in the test mixture is limited.

Two approaches to evaluating stability that have attained wide usage and have reached the stage of standardization are the Hveem procedure (ASTM D 1560 and D 1561) and the Marshall procedure (ASTM D 1559) [118]. These are procedures in the sense that they involve a sequence of tests, the results of which are combined, with judgment, to make a decision about the mix proportions. Particular agencies have tended to modify some of the methods, tests, and the criteria of these two procedures, to meet their own needs or to take advantage of newly developing techniques [119, 129, 130]. Inasmuch as the criteria for judging the selection of a job mix presumably reflect a particular agency's experience in successfully tying test values to field performance, it is advisable that the many detailed steps in a given procedure be carefully followed. Only some of the principal and more or less common features of current practice are outlined here.

In the *Hveem* procedure (ASTM D 1560) two major tests are used: a "stabilometer" test which is a kind of triaxial compression test, and a "cohesiometer test," which obtains a measure of tensile strength by breaking a specimen in flexure. For the stabilometer test, a compacted specimen, heated to 60°C, is placed into the type of apparatus shown in Fig. 19.6. Vertical pressure is exerted by the piston, while resulting

[†]Or 4 in by $2\frac{1}{2}$ in.

lateral pressures are measured by the gage connected to the surrounding liquid-filled chamber. The specimen is not tested to failure. The "stabilometer S value" is an arbitrary relative value that is a function of a given vertical pressure and the corresponding lateral pressure; the equation for this relative stability is so formed that S would be zero if the specimen were liquid, and 100 if the specimen were completely rigid (no lateral pressure). This S value is used for evaluating dense-graded mixtures for surface courses. For base-course and subgrade materials, a similar stabilometer R value is computed, using a different loading schedule and a different equation (ASTM D 2844).

The cohesiometer test is performed by clamping the same specimen, also at 60°C, into the kind of flexure device illustrated in Fig. 19.7. The specimen is actually tested as the fixed end of a cantilever beam, with the load being applied by dropping shot, at a rate of 30 g/s (1.8 kg/min) into a bucket supported by the free end of the cantilever. From the mass of shot causing failure and the specimen's dimensions the "cohesiometer C value" is calculated.

The Hveem S values for dense-graded good-quality mixtures may range from about 20 to 45. C values may run between about 40 and 200.

In the *Marshall* procedure (ASTM D 1559), a compacted specimen, heated to 60°C, is tested in the apparatus illustrated in Fig. 19.8. With the circular faces placed vertically, the load is applied to portions of the circumference of the cylinder through split-ring heads as shown in the figure. Load is applied at a constant deformation rate of 50 mm/min until failure occurs. The "Marshall stability" of the mixture is the maximum load sustained. The amount of vertical deformation from zero to maximum load is called the "flow value," in units of 0.25 mm (or 0.01 in).

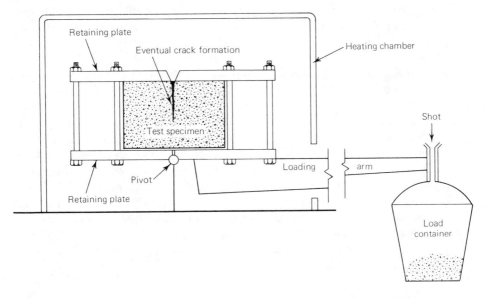

Figure 19.7 Diagram of Hveem cohesiometer. (*From The Asphalt Institute, Ref. 118.*)

Figure 19.8 Marshall stability and flow apparatus. (*Courtesy of Forney's Inc.*)

Marshall stability values obtained from tests on dense-graded good-quality mixtures may run from about 7 to 9 kN[†] and the flow values may range from 5 to 30 units.

A somewhat different approach to obtaining measures of strength and stiffness has been developed using the same kind of specimens as used in the Hveem and Marshall tests, but with loading applied through narrow strips located along opposite elements of the cylindrical surface, as indicated by the diagram in Fig. 19.9. The contact strip is curved to conform to the curvature of the cylinder.

For uniform compressive line-loading applied in the plane of a diameter (y-y in Fig. 19.9), theoretical analysis of an idealized thin elastic disk deduces an almost uniform tensile stress distribution across that diametral plane [131, 134]. A 64-mm-thick disk of nonhomogeneous viscoelastic material under narrow strip loading does not conform to the assumptions of the theory of elasticity. Nevertheless, numerous experiments on asphalt mixtures have led to the use of the *indirect tensile* test which yields a useful measure of tensile strength [133, 135]. When loaded in the manner described, a specimen (if not too viscous) fails by splitting across the vertical diameter.

If the horizontal strain distribution along the horizontal diameter (x axis), normal to the plane of the loads, is deduced by such theoretical analysis, and is integrated, an expression may be obtained for the total diametral deformation Δ in terms of the load P, modulus of elasticity E, and Poisson's ratio ν; transposing, E may be expressed as a function of P, Δ, and ν [134].

[†]1 kN = 225 lbf.

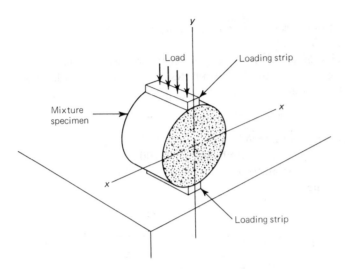

Figure 19.9 Specimen arrangement for indirect tensile and diametral resilient modulus tests.

Under slow (static) loading, the measured deformations of asphalt mixtures include components of viscous strain. However, if the loading (or unloading) times are short enough, the viscous effect is small. Under repeated, dynamic, pulse-type[†] loading, by means of sufficiently sensitive transducers, diametral deformation can be recorded which yields a fair measure of "resilient" (instantaneous, or elastic recoverable) strain [134, 135]. Using the theoretically derived expression for E and an assumed value of ν (often taken to be equal to 0.35 for these conditions), a value of stiffness called the *diametral resilient modulus*, M_R, is computed from such pulse-type loading tests.

Measures of fundamental properties such as these can provide direct inputs to procedures for analyzing the structural behavior of layered pavement systems, and thus lead to new methods of pavement systems, and thus lead to new methods of pavement design. The use of cyclical loading in tests of this kind can also simplify the acquisition of information concerning fatigue cracking and accumulation of permanent (plastic) deformation.

Selection of optimal mix The mix to be used on the job is selected by a consideration of the various characteristics of the trial mixes. An optimal mix is one that is at or near maximum stability, and that has near-maximum density and a percentage of air voids considered suitable for the type of mix and type of service. Other limitations may be imposed by particular job specifications, for example: swell due to soaking in water; minimum VMA; some measure of flow or cohesion; percent coating, runoff and washoff (emulsified asphalts); loss of strength or stability after soaking. Too much or too little asphalt adversely affects performance, often very seriously.

[†] A pulse-type loading schedule that has been used involves a 3-s cycle with a 0.1-s sinusoidal load duration [119, 134].

An example of data from a series of trial mix tests made under the Marshall procedure is given in Fig. 19.10.

Representative examples of some of the requirements for dense-graded paving mixtures are summarized here to indicate the magnitudes of the properties of interest. Values such as these would be desired from tests having specific prescribed procedures and test conditions.

Hveem stabilometer, S value, minimum	37
Hveem cohesiometer, C value, minimum	100
Marshall stability (heavy traffic), minimum	3.5 kN
Marshall flow value	8–16 units

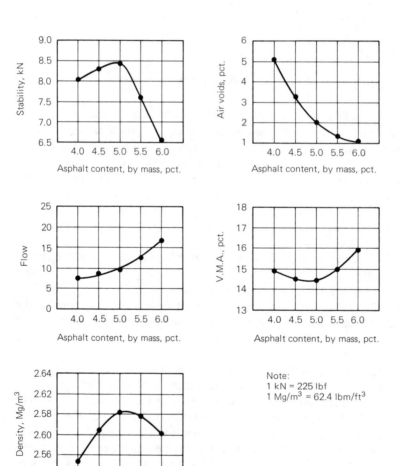

Note:
1 kN = 225 lbf
1 Mg/m³ = 62.4 lbm/ft³

Figure 19.10 Typical plots of trial mix properties of Marshall procedure. (*From the Asphalt Institute, Ref. 118.*)

Air voids in mixture	3-5%
Voids in mineral aggregate, minimum	13%

Complete specifications used by pertinent agencies should be consulted concerning other and detailed requirements for these and other types of mixtures for a variety of uses.

PROBLEM 19: Marshall Test of an Asphaltic Mixture

Object: To obtain the bulk density, the stability, and the flow of an asphalt concrete specimen.
Preparatory reading: ASTM D 1559.
Specimen: Standard Marshall specimen with given asphalt content.
Special apparatus: Marshall test apparatus.
Procedure:
1. Measure the dimensions of the specimen. Weigh the specimen in air, then coat it with paraffin and weigh the coated specimen in air as well as in water. Compute the bulk density. (Allow for paraffin coating.)
2. Remove the paraffin from the specimen and place the latter into a water bath at 60°C for 30 to 40 min.
3. Clean the testing head, oil the guide rods lightly, and familiarize yourself with the machine so that you can insert the specimen and start the test without wasting time.
4. Remove the specimen from the bath, dry it quickly but carefully, and place it into the lower head. Position the upper head and the flow meter. This step should take but a few seconds so that the specimen does not cool appreciably.
5. Immediately apply the load, at a deformation rate of 50 mm/min, and continue until the load starts dropping off. Read the load dial at this point, remove the flow meter, and turn off the machine.
6. Record the load dial and flow-meter readings, convert the former to compressive load by means of the calibration chart for the proving ring; if the thickness t of the specimen varies from the standard 63.5 mm (2.5 in) by more than a millimeter, correct the load by a factor equal to t $(t/63.5)^2$, increasing the value for undersized specimens and vice versa.

Report:
1. Exchange information with the other laboratory groups that tested specimens with different asphalt contents.
2. Plot stability, flow, and density as ordinates on graphs with asphalt content as the abscissa.
3. Comment on the appearance of the curves. What other values should be plotted?

DISCUSSION

1. Are asphalts used only for paving? What are some other uses that you have observed?
2. Draw a diagram showing the classification of asphalts.
3. Why is viscosity so important in asphalt design? Name some common viscosity tests.
4. Distinguish between dynamic and kinematic viscosities.
5. How do the grade designations of asphalt cements, cutbacks, and emulsions differ?
6. Explain why aggregates are essential to asphalt mixes.
7. What aggregate properties are beneficial to the finished pavement, and which facilitate placement of the mix?
8. What is the role of testing in the design of asphalt mixes?
9. Compare the properties involved in the Marshall method with those considered in the Hveem method.
10. What justification is there for finding and using the resilient modulus M_R, which is really an elastic modulus, when asphalt mixtures are viscoelastic materials?

READING

Abraham, H.: *Asphalts and Allied Substances*, 6th ed., Van Nostrand, Princeton, N.J., 1960.

Barth, E. J.: *Asphalt: Science and Technology*, Gordon & Breach, New York, 1962.

A Brief Introduction to Asphalt and Some of Its Uses, Manual MS-5, 7th ed., The Asphalt Institute, College Park Md., 1977. (Also see other manuals in series.)

Chilingarian, G. V., and T. F. Yen (eds.): *Bitumens, Asphalts and Tar Sands*, Elsevier, Amsterdam, 1978.

Deacon, J. A.: "Materials Characterization–Experimental Behavior," HRB Special Report 126, Natl. Research Council, Highway Research Board, Washington, pp. 150–179.

Emulsion Mix Design, Stabilization and Compaction, Transportation Research Record, Natl. Research Council, Transportation Research Board, Washington, 1980.

Hatherly, L. W., and P. C. Leaver: *Asphaltic Road Materials*, Edward Arnold, London, 1967.

Krebs, R. D., and R. D. Walker: *Highway Materials*, McGraw-Hill, New York, 1971.

Larson, T. D.: *Portland Cement and Asphalt Concretes*, McGraw-Hill, New York, 1963.

Traxler, R. N.: *Asphalt: Its Composition, Properties and Uses*, Reinhold, New York, 1961.

Wallace, H. A., and J. R. Martin: *Asphalt Pavement Engineering*, McGraw-Hill, New York, 1967.

Woods, K. B. (ed.): *Highway Engineering Handbook*, McGraw-Hill, New York, 1960 (Sec. 18, "Bituminous Materials and Mixture").

Zakar, P.: *Asphalt*, Chemical Publishing Co., New York, 1971.

TWENTY
WOOD

20.1 TYPES AND PRODUCTION

Wood has served human needs longer than any other engineering material, but it is not at all obsolescent. Although serious study of wood dates back to the invention of the microscope in the seventeenth century, comprehensive research was left to the twentieth century and is continuing. Better understanding combined with advances in the development of tools and adhesives has caused considerable progress in wood utilization. Plywood and particle board production had to await the development of synthetic resins (see Chap. 21).

Wood comprises the tissue of stems and roots of various plants, bu. ɪne wood used as an engineering material is obtained from the trunks of forest trees. We distinguish between (1) *conifers* or *cone-bearing* trees, which produce what are called softwoods, although some pines are very hard, and (2) *broadleaf* or *deciduous* trees, which produce what are called hardwoods, although they can be quite soft. ASTM D 1165 lists a nomenclature for softwoods and hardwoods grown in the United States. In temperate climates almost all constructional wood is softwood.

In contrast to the other materials described in this book, wood is a renewable resource. By the "sustained yield" method, each year a volume of wood equal to the annual growth of all trees is cut.

After the trees to be harvested are marked, they are felled, cut into logs, and limbed. They are then skidded by tractor, helicoptor, or elephant to concentration yards. Chips may be transported by pipeline to pulp mills, and the trunks, which are of main interest to engineers, are transported to mills by rail, truck, or waterway.

At the mill, some trunks are shaved and treated to become "roundwood" products, such as piles, poles, or posts. Others are softened in hot water, then rotated on a lathe while a knife peels off the wood in thin layers called "veneer." Still others are processed as "sawnwood," and become lumber, structural timber, and railroad ties. The term "timber" sometimes refers to standing trees, but it is used here to denote lumber with large dimensions (over 100 mm). Terms relating to timber are defined by ASTM D 9.

When the logs arrive at the sawmill, they are sprayed with or stored in water until they are fed to a "headsaw" to be cut into rough slabs. These are then resawn, ripped, and crosscut to make lumber of varying sizes, typically with thicknesses of 20 to 100 mm, widths of over 800 mm, and lengths of 2 to 6 m. The amount of waste ranges up to 50 percent, but this "waste" is used in many ways.

The lumber is dried, or "seasoned," by stacking it in such a way as to permit air to circulate among the boards, either in the open (air drying) or in heated chambers (kiln drying). The temperature in the kilns ranges from 40 to 75°C, and for green lumber to dry to 6 percent moisture content (based on the mass of the dry wood) takes several weeks. Air drying might reduce the moisture content to 20 percent in several months and be followed by kiln drying for several days. A moisture content appropriate to the intended environment is desired. For special uses, some lumber is chemically treated.

The lumber is finally planed and then graded according to a manufacturer's association's grading scheme, the purpose being to provide some quality assurance to the user. Lumber destined for structural applications is "stress graded," meaning that the grades are established on the basis of strength or stiffness properties. This may be done by visual or machine methods. In visual stress grading, which was practiced in Scandinavia as early as the eighteenth century, trained inspectors armed with written guidelines, such as ASTM D 245, segregate the lumber on the basis of grain configuration, presence of knots, and other apparent characteristics. In machine grading, which came into use in the 1960s, the lumber is subjected to nondestructive flexure testing and graded on the basis of stiffness. The wood is then ready for constructional use.

Veneer, lumber, and by-products such as shavings may also be bonded with synthetic resins, i.e., plastics, to form other constructional materials, for example plywood, particle board, flake board, laminated wood beams and frames (called "glulam"), and various laminates and sandwich panels that may include wood as either the core or the cladding. Many of these products have some characteristics superior to those of wood itself.

20.2 STRUCTURE

Wood is an organic substance composed principally of cellulose (40 to 55 percent), hemicellulose (15 to 25 percent) and lignin (15 to 30 percent).

The unit of wood structure is called the *cell,* a general term for diverse wood elements such as tracheids or fibers, rays, and resin ducts or vessels. The characteristic size, shape, wall thickness, structure, and arrangement of cells in different species account for variations in density and strength. The longitudinal cells are composed principally of cellulose cemented together by lignin. They range from 1 to 3 mm in length, with diameters about 1/100 of their lengths; they have pointed ends and are usually placed with their long dimensions lengthwise with the tree. Radially oriented cells are called *rays*. Cells are the principal constituent of wood.

Most of the cells are arranged longitudinally in the direction of the trunk or branch, so that wood can be visualized as resembling a bundle of straws. This arrangement makes it obvious that the material is *anisotropic*, meaning that its properties in one direction differ from those in another. The radially arranged rays, which constitute 3 to over 10 percent of the mass of softwood and 5 to over 30 percent of hardwood, account for further anisotropism. Figure 20.1 shows the arrangement of the cells. In softwood the longitudinal cells, called *tracheids*, are arranged in neat radial rows, while in hardwood, where they are called *fibers*, they form a more irregular pattern. In softwood, *resin ducts* are formed by the space between several cells. The cells are connected by various *pits* that allow fluids to flow through thin membranes. In hardwood, on the other hand, *vessels* are formed by series of single cells joined together end to end, with *perforation plates* at the junctures.

Trees grow in diameter and height by adding new layers of wood cells. Growth in both conifers and broad leaf trees occurs in the same way: by the formation of new cells in the *cambium layer*, which is just beneath the bark. During each growing season new wood is formed in a continuous layer directly on the previous growth, covering the entire tree. No growth occurs in the wood already formed.

This somewhat brief discussion of the structure of the structure of wood, that is, the arrangement of the cells, is sufficient to explain why wood behaves in an approximately orthogonally anisotropic, or *orthotropic*, manner. At a particular point in the wood one can expect different characteristics along three mutually perpendicular axes, namely in the longitudinal (L), radial (R), and tangential (T) directions. For reasons to be discussed later the R and T directions are often lumped together and called "perpendicular to the grain" and the L direction is called "parallel to the grain."

Annual growth rings or growth layers, which are usually concentric about the innermost ring or "pith," can be observed on any cross section of a tree. They are caused by variations in rate of growth as a result of seasonal differences in moisture and temperature (see Fig. 20.1). An annual growth ring usually consists of two bands, one darker than the other, the difference being quite distinct in most woods. *Earlywood* (formerly called "springwood") is the inner light-colored layer in the ring. It contains cells that have relatively large openings and thin walls. *Latewood* (formerly called "summerwood") is the outer dark-colored layer in the ring. In most native woods with well-defined differences in seasonal growths, latewood is heavier, harder, and stronger but shrinks more than earlywood. The difference in appearance of the growth ring between the beginning and end of a growing season is usually well defined in *ring-porous* hardwoods, such as white oak, in which large cells develop during the rapid growth of the earlywood. *Diffuse-porous* woods, such as yellow poplar (shown in Fig. 20.1c) and woods from trees that grow continuously throughout the year, as in the tropics, have no well-defined rings. In softwoods, such as pine, the apparent difference in color of the layers is principally due to differences in density.

Rate of growth is measured by the number of annual growth rings per meter counted along a line perpendicular to the rings across a right section of the tree. It is quite variable, even in the same species, as shown by old slowly growing original-

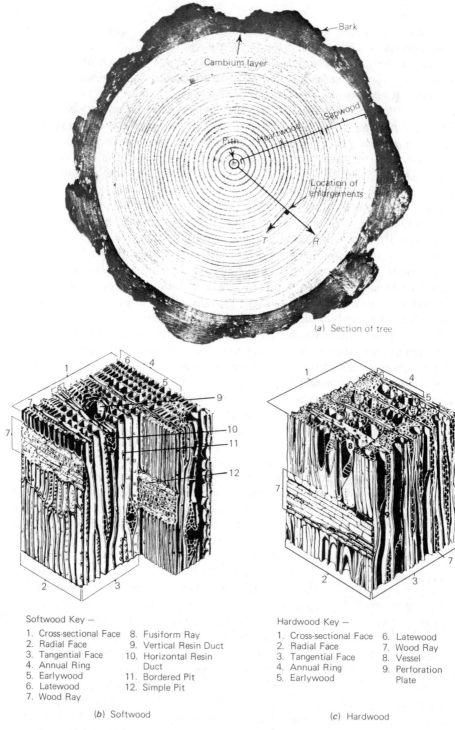

(a) Section of tree

Softwood Key —

1. Cross-sectional Face	8. Fusiform Ray
2. Radial Face	9. Vertical Resin Duct
3. Tangential Face	10. Horizontal Resin Duct
4. Annual Ring	
5. Earlywood	11. Bordered Pit
6. Latewood	12. Simple Pit
7. Wood Ray	

(b) Softwood

Hardwood Key —

1. Cross-sectional Face	6. Latewood
2. Radial Face	7. Wood Ray
3. Tangential Face	8. Vessel
4. Annual Ring	9. Perforation Plate
5. Earlywood	

(c) Hardwood

Figure 20.1 Structure of wood. (*Courtesy of U.S. Forest Service.*)

growth redwoods, called virgin trees, which may add half a millimeter of wood a year, compared with second-growth redwood, which may add five or more millimeters a year. Note that "rate of growth" is really a misnomer in that it becomes smaller the faster the tree grows. It would make more sense to express this quantity in terms of the width of the growth rings, i.e., as increase in radius per year, but that is not the custom.

Heartwood is the physiologically inactive inner portion of a tree. Usually it is darker in color than *sapwood*, which in an old tree is the relatively narrow band of wood (about 40 to 75 mm wide in Douglas fir) between the heartwood and the bark. Sapwood contains the active cells that participate in the life processes of the tree. In living trees heartwood is more susceptible to decay than sapwood, but in cut lumber it is less susceptible. There is no consistent difference in strength or density between dry sapwood and heartwood.

Grain is a term usually employed to indicate the direction of fibers with respect to the main axis or the surface of a piece of wood; the grain may be cross, diagonal, edge, and flat or slash. It also refers to the arrangement of the fibers themselves, which may be straight, curly, or spiral grain.

Defects are any irregularities in wood that decrease its strength, durability, or utility, such as knots, checks (seasoning cracks), shakes (separation of annual rings in the living tree), cross grain, decay, wane (bark or lack of wood on a corner or edge of a piece), warp, and pitch pockets.

Besides these specific defects, to which might be added certain inherent weaknesses such as relativley low strength perpendicular to the grain, wood has other overall shortcomings such as hygroscopicity, combustibility, and susceptibility to boring organisms, for example teredos and termites, and to attack by fungi and other destructive parasitic growths.

20.3 PHYSICAL PROPERTIES

Woods exhibit considerable variations in physical properties. One characteristic, namely hygroscopicity, or the tendency to absorb water, influences most of the other properties.

Hygroscopicity Wood attracts water to both the cell walls and the cell cavities. Inbibed water is in the cell walls; free water is in the cell cavities. The cell walls must be saturated before free water can exist. The moisture content at the critical stage when the cell walls are saturated but no free water exists is called the *fiber-saturation point* (FSP), which for most woods varies from about 25 to 30 percent. *Moisture content* is expressed as the percentage of water, by mass, in wood dried at a temperature of $103 \pm 2°C$. The ranges in moisture content of four characteristic conditions are: green, 30 to 250 percent; air dry, 12 to 15 percent; kiln dry, 6 to 7 percent; and oven dry, 0 percent. The moisture content of wood tends to come to equilibrium with the relative humidity and temperature of the surrounding atmosphere at an "equilibrium moisture content" point.

Moisture is usually determined by oven-drying or by an electric-resistance method. *Oven-drying* is reliable, but it is slow and is not applicable if the wood cannot be cut for a sample. Oven-drying is widely used in laboratory testing practice. The moisture sample should be a clear, solid piece (no knots or splinters) about 25 mm long in the direction of the grain and taken from the test specimen near the break. The sample should be left in a well-ventilated oven until dried to constant mass. Portable *electric-resistance* instruments have small terminals that can be embedded in the wood without seriously injuring it; the terminals are connected to a battery circuit and a resistance meter. Factors for converting observed resistances to moisture content have been determined for each species of wood. This method is most satisfactory for moisture contents from about 7 to 24 percent and for lumber less than 50 mm thick. It is rapid and is often used as a basis for sorting lumber.

Shrinkage Shrinkage of wood is due to drying of the cell walls. No shrinkage occurs until the moisture content reaches the fiber-saturation point. The shrinkages of air-dry and kiln-dry wood are about one-half and three-fourths the oven-dry shrinkage respectively. Shrinkage is a reversible phenomenon, since imbibing of water by the cell walls causes wood to swell. Tangential shrinkage (parallel to the growth rings) is approximately double the radial shrinkage (normal to the growth rings); longitudinal shrinkage is practically zero. The predominance of radial seasoning checks is due to the high ratio of tangential to radial shrinkage. Shrinkage is a direct function of the relative density; the approximate percentages of radial, tangential, and volumetric shrinkages are respectively 9, 17, and 28 times the relative density. For Douglas fir the percentages are 4, 7.5, and 12 respectively. Warping of lumber is due to unequal shrinkage of various portions. This may result from unequal loss of moisture from different parts of the piece or to variations in structure throughout the piece.

The dimensional instability of wood, which is due to shrinkage and swelling, is a deterrent to its use in many applications. Shrinkage and swelling can be controlled to some extent by coating and impregnation, but the preferred procedure is to season the wood before use to the approximate moisture content it will have during service.

Density The density of wood is, of course, greatly influenced by the moisture content. In order to obtain densities for comparison, a specific moisture content must be specified at which the volume is to be measured. Density tests are specified by ASTM D 2395.

The wood substance, which is the cell-wall material, has a relative density of about 1.5 regardless of species. Of greater practical interest are the macroscopic relative densities. These very much depend on the species and range from 0.3 to 0.9 for woods that grow in temperate climates. Some tropical woods fall outside that range, such as lignum vitae (1.23) and balsa (0.13). Density in turn affects the maximum possible moisture content; for balsa this is 800 percent!

Other properites The coefficient of thermal expansion for woods is quite small and becomes negligible when compared with the dimensional changes brought about by moisture variations. Temperatures below 0°C may cause some surface checking.

Wood also has low heat conductivity. In addition to making the material useful as an insulator, this property has a beneficial influence on its resistance to fire. Although wood ignites readily at temperatures of about 400°C, the interior of heavy timbers is protected for quite a long time. The heat conductivity is at least twice as large longitudinally as transversely. It increases with density and moisture content.

The one important electrical property is electric conductivity. Electric conductivity varies with moisture content, increasing spectacularly from the oven-dry condition to the fiber-saturation point, then much more slowly to the maximum moisture content. This makes possible the use of the electric moisture meter mentioned earlier.

20.4 MECHANICAL PROPERTIES

As wood is an orthotropic material, its mechanical properties vary greatly with direction. Moisture content is also of considerable importance. Finally, wood is unique among engineering materials in that its properties are determined by nature, with little opportunity available to correct deficiencies.

The greatest influence on the growth of trees is exerted by soil, climate (moisture and temperature), and growing space. Hence, clear wood of any species will exhibit variations in characteristic mechanical properties, such as strength and stiffness, because of changes in uncontrollable but easily determined factors: density and the closely related percentage of latewood and rate of growth. Some defects are almost unavoidable in large pieces, so that full-size members are not homogeneous. Important test factors, under more control, are the size and shape of the specimen, test procedures such as the rate and duration of loading, temperature, and the amount of moisture in the wood. When comparing test results with other reported values, these factors should be taken into account.

Tests on small, clear specimens are specified by ASTM D 143 and on large members by ASTM D 198; ASTM D 2555 addresses methods of establishing clear-wood strengths.

Effects of density The density-strength relation within any species is very definite; the correlation is even high for all normal wood, irrespective of species. Approximate values are given in Table 20.1.

Effects of growth The percentage of latewood, when the difference between it and the adjacent earlywood is well marked as in Douglas fir, is a good index of strength: the more latewood, the greater the strength. This follows from the close relation between the percentage of latewood and density. Rate of growth is not a reliable indicator of strength.

Large wood beams and columns always have some defects, such as checks and knots, whereas the small standard specimens are clear, straight-grained wood; therefore, the effects of size and defects must be considered together. The possible variations in defects that may occur in large timbers make it impossible to assign specific reduction factors for any particular defects, but the strength of large timbers is considerably lower than that for small, clear test pieces.

Table 20.1 Density-strength relation[†,‡]
(Regardless of species)

Property [§]	Moisture condition	
	Green	Air dry (12%)
Static bending		
Fiber stress at proportional limit, MPa	$70\ G^{1.25}$	$115\ G^{1.25}$
Modulus of rupture, MPa	$121\ G^{1.25}$	$177\ G^{1.25}$
Modulus of elasticity, GPa	$16.3\ G$	$19.3\ G$
Total work, kJ/m^3	$710\ G^2$	$500\ G^2$
Impact bending (50-lb/22.7-kg tup)		
Height of drop at failure, m	$2.90\ G^{1.75}$	$2.41\ G^{1.75}$
Compression parallel to grain		
Proportional limit, MPa	$36\ G$	$60\ G$
Ultimate strength, MPa	$46\ G$	$84\ G$
Modulus of elasticity, GPa	$20.1\ G$	$23.3\ G$
Compression perpendicular to grain		
Proportional limit, MPa	$21\ G^{2.25}$	$32\ G^{2.25}$
Shearing strength		
Parallel to grain, MPa	$19\ G^{1.33}$	$28\ G^{1.33}$

[†]Based on *Wood Handbook*, Forest Products Laboratory, U.S. Department of Agriculture, 1974.

[‡]G represents the relative density of oven-dry wood, based on the macroscopic volume at the moisture condition indicated. It varies between about 0.3 and 0.9 and is numerically equal to the density in Mg/m^3.

[§]1 MPa = 145 psi.

When trees grow abnormally or lean, *reaction wood* is formed. In softwoods it is found on the underside of an inclined tree and is called "compression wood," whereas in hardwoods the abnormal tissue forms on the upper side and is called "tension wood." Reaction wood is up to 40 percent denser than normal wood, and its longitudinal shrinkage may be up to 10 times that of normal wood. Its physical properties are somewhat unpredictable, which makes it desirable to eliminate reaction wood from lumber.

Effects of moisture Moisture variations above the fiber-saturation point do not affect the strength properties, but decreases in moisture below that condition produce increases in strength and stiffness of small, clear specimens. This relation is shown in Fig. 20.2 for the compressive strength parallel to the grain and the modulus of rupture for clear specimens of Sitka spruce. Stiffness and other mechanical properties show somewhat similar relations with moisture content, although the percentage effects are not the same. These relations do not hold for large timbers, since any increase in strength due to seasoning is offset by defects that develop during the drying process.

Effects of orientation The properties of clear wood depend greatly on the orientation—longitudinal (*L*), radial (*R*), or tangential (*T*)—with respect to the growth rings, as shown in Fig. 20.3. Twelve constants (nine are independent) are needed to describe

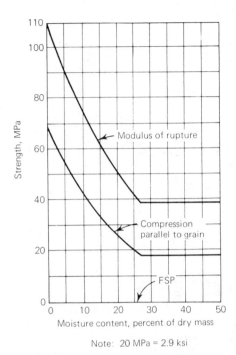

Figure 20.2 Effect of moisture on strength of clear Sitka spruce. (*From Markwardt and Wilson, Ref. 60.*)

the elastic behavior of wood: three moduli of elasticity, E, three moduli of rigidity, G, and six Poisson's ratios, μ.

The three *moduli of elasticity* denoted by E_L, E_R, and E_T are respectively the elastic moduli along the longitudinal, radial, and tangential axes of wood. These moduli are usually obtained from compression tests, but they are also obtained from flexure tests.

The three *moduli of rigidity* denoted by G_{LR}, G_{LT}, and G_{RT} are the elastic constants based on the shear strains in the LR, LT, and RT planes respectively.

The six *Poisson's ratios* are denoted by μ_{LR}, μ_{RL}, μ_{LT}, μ_{TL}, μ_{RT}, and μ_{TR}. The first letter of the subscript refers to the direction of applied stress and the second letter refers to the direction of lateral deformation. For example, μ_{LR} is the Poisson's ratio for deformation along the radial axis caused by stress along the longitudinal axis.

Table 20.2 shows some representative ratios of the radial and tangential moduli of

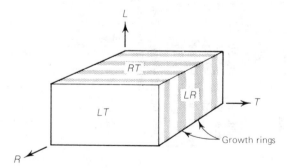

Figure 20.3 Principal axes of wood.

Table 20.2 Elastic constants of various species†

Species	Approx. density, Mg/m³‡	Approximate moisture content, %	Modulus of elasticity ratios		Ratio of modulus of rigidity to modulus of elasticity			Poisson's ratios					
			E_T/E_L	E_R/E_L	G_{LR}/E_L	G_{LT}/E_L	G_{RT}/E_L	μ_{LR}	μ_{LT}	μ_{RT}	μ_{RL}	μ_{TR}	μ_{TL}
Balsa	0.13	9	0.015	0.046	0.054	0.037	0.005	0.229	0.488	0.665	0.217	0.011	0.007
Birch, yellow	0.64	13	0.050	0.078	0.074	0.068	0.017	0.426	0.451	0.697	0.447	0.033	0.023
Douglas fir	0.50	12	0.050	0.068	0.064	0.078	0.007	0.292	0.449	0.390	0.287	0.020	0.022
Spruce, Sitka	0.38	12	0.043	0.078	0.064	0.061	0.003	0.372	0.467	0.435	0.240	0.029	0.020
Sweetgum	0.53	11	0.050	0.115	0.089	0.061	0.021	0.325	0.403	0.682	0.297	0.037	0.020
Walnut, black	0.59	11	0.056	0.106	0.085	0.062	0.021	0.495	0.632	0.718	0.379	0.052	0.035
Yellow poplar	0.38	11	0.043	0.092	0.075	0.069	0.011	0.318	0.392	0.703	0.329	0.029	0.017

†*Wood Handbook*, Forest Products Laboratory, U.S. Department of Agriculture, 1974.
‡Based on oven-dry mass and volume at the moisture content shown. 1 Mg/m³ = 62.4 lbm/ft³.

elasticity (based on relatively few tests), as well as of the moduli of rigidity, to the longitudinal modulus of elasticity. The Poisson's ratios are also listed. These values vary not only among species but also within species and with density and moisture content.

Resistance to *tension* is offered primarily by the cell fibers, so that the tensile strength in the longitudinal direction is by far the greatest. Because of the rays, the tensile strength in the radial direction is slightly greater than in the tangential direction. In longitudinal *compression*, the cell walls first buckle, after which the cells themselves fail by column action. Therefore longitudinal compressive strength tends to be only about half of the longitudinal tensile strength, and the stress-strain curve in compression is less linear that that in tension. In the radial and tangential directions, failure occurs by crushing of the cells. The stress level at which this takes place can be defined as the compressive strength, which is about 8 percent of the longitudinal compressive strength. The compressive strength may be slightly higher in the radial than in the tangential direction or vice versa, depending on the species.

Shear failure may occur on one of the faces—*LR*, *LT*, or *RT*—shown in Fig. 20.3, in each case in either direction. Both shear strengths on the *RT* face (across the grain) are very high, because the primary bonds in the cell walls must be broken; in a test, in fact, the cells would probably be crushed before shear failure occurs. The shear actions on the *LR* and *LT* faces are similar to each other and depend greatly on the direction of the shear. When the shear stress acts in the *L* direction (parallel to the grain), we have the kind of action illustrated by horizontal shear in a beam. When the shear stress acts in the *R* or *T* directions (perpendicular to the grain), we have the kind of shear illustrated by torsion; it is called "rolling shear," because the tracheids (longitudinal cells) can be envisioned to be rolling over each other. The rolling shear strength is much smaller than the longitudinal shear strength.

From this discussion it is clear that differences in properties between the tangential and radial directions tend to be slight, but that these properties tend to vary a great deal from those in the longitudinal direction. For this reason the former are often lumped together and called perpendicular to the grain while the latter is called parallel to the grain, as noted earlier.

Occasionally, the directions of important stresses may not coincide with the natural axes of fiber orientation in the wood. This may occur by choice in design, by the way the wood was removed from the log, or because of grain irregularities that formed during growth.

Elastic properties in directions other than along the natural axes can be obtained from elastic theory. Strength properties in directions ranging from parallel to perpendicular to the fibers can be approximated using a Hankinson-type formula [105]:

$$N = \frac{PQ}{P \sin^n \theta + Q \cos^n \theta}$$

where N represents the strength property at an angle θ from the fiber direction, Q is the strength across the grain, P is the strength parallel to the grain, and n is an empirically determined constant. The formula has been used for estimating the modulus of elasticity as well as strength properties. Values of n and associated ratios of Q/P have been tabulated below from Ref. 105.

Property	n	Q/P
Tensile strength	1.5–2.0	0.04–0.07
Compressive strength	2.0–2.5	0.03–0.40
Bending strength	1.5–2.0	0.04–0.10
Modulus of elasticity	2.0	0.04–0.12
Toughness	1.5–2.0	0.06–0.10

Time effects The rate of loading and the *duration* over which the load is sustained have a marked effect on the strength of wood. Under long-time loading, beams have been broken at stresses only slightly greater than 50 percent of the ultimate strength value given by tests in which the load was applied within a few minutes. Conversely, the values for strength properties after very short durations of loading may be considerably greater than those obtained by standard tests. It has been suggested that the strength increases or decreases by approximately 7 or 8 percent whenever the load duration is decreased or increased respectively by a factor of 10 [105].

This phenomenon is, of course, a manifestation of creep, which in wood is intensified not only by higher temperatures but also by higher humidity.

Summary A summary of the principal mechanical properties for a few representative native woods at certain relative densities is given in Tables 20.3 and 20.4. Standard specimens and methods of testing were used. Since the values of modulus of elasticity shown were derived from beam tests and include some effect of shear deflection, they should be increased by about 10 percent for application to problems involving axial loads.

Wood is strongest in tension parallel to the grain, but values for this property are not given in Tables 20.3 and 20.4 since it is difficult to design end connections to develop the tension. On the other hand, wood is weakest in tension across the grain. Values of these two tensile strengths for air-dry Douglas fir are about 100 and 2 MPa respectively.

The shearing strengths parallel to the grain for standard specimens that are given in Tables 20.3 and 20.4 are for shear planes tangent to the growth rings.

The compressive strengths parallel to the grain compare favorably with those of portland cement concretes. Tensile strengths in the same direction are roughly twice as large, so that they approach those of some metals. Consequently, woods tend to have a very favorable strength-to-density ratio—an economically important fact for some applications.

20.5 DURABILITY

Wood is not a very durable material; being a natural organic substance, it forms part of the food chain and is susceptible to attack by predators. No species is immune to attack, although denser woods tend to have greater resistance. Wood is, of course, also combustible. There are methods, albeit imperfect ones, of protecting wood against these hazards.

Table 20.3 Mechanical properties of some softwoods grown in the United States[†],[‡]

Common names of species	Density, Mg/m³	Static bending — Modulus Rupture, MPa	Static bending — Modulus Elasticity,[§] GPa	Static bending — Work to maximum load, kJ/m³	Impact bending—height of drop causing complete failure, m	Compression parallel to grain—maximum crushing strength, MPa	Compression perpendicular to grain—fiber stress at proportional limit, MPa	Shear parallel to grain—maximum shearing strength, MPa	Tension pendicular to grain—maximum tensile strength, MPa
Cedar, western red cedar	0.31	35.9	6.5	34	0.43	19.1	1.7	5.3	1.6
	0.32	51.7	7.7	40	0.43	31.4	3.2	6.8	1.5
Douglas fir, coast	0.45	53.0	10.8	52	0.66	26.1	2.6	6.2	2.1
	0.48	85.0	13.4	68	0.79	49.9	5.5	7.8	2.3
Fir, white	0.37	41.0	8.0	39	0.56	20.0	1.9	5.2	2.1
	0.39	68.0	10.3	50	0.51	40.1	3.7	7.6	2.1
Hemlock, western	0.42	46.0	9.0	48	0.56	23.2	1.9	5.9	2.0
	0.45	78.0	11.3	57	0.58	49.0	3.8	8.6	2.3
Pine, longleaf	0.54	59.0	11.0	61	0.89	29.8	3.3	7.2	2.3
	0.59	100.0	13.7	81	0.86	58.4	6.6	10.4	3.2
Pine, ponderosa	0.38	35.0	6.9	36	0.53	16.9	1.9	4.8	2.1
	0.40	65.0	8.9	49	0.48	36.7	4.0	7.8	2.9
Redwood, old growth	0.38	52.0	8.1	51	0.53	29.0	2.9	5.5	1.8
	0.40	69.0	9.2	48	0.48	42.4	4.8	6.5	1.7
Redwood, young growth	0.34	41.0	6.6	39	0.41	21.4	1.9	6.1	2.1
	0.35	54.0	7.6	36	0.38	36.0	3.6	7.6	1.7
Spruce, Engelmann	0.33	32.0	7.1	35	0.41	15.0	1.4	4.4	1.7
	0.35	64.0	8.9	44	0.46	30.9	2.8	8.3	2.4
Spruce, Sitka	0.37	39.0	8.5	43	0.61	18.4	1.9	5.2	1.7
	0.40	70.0	10.8	65	0.64	38.7	4.0	7.9	2.6

[†]*Wood Handbook*, Forest Products Laboratory, U.S. Department of Agriculture, 1974.

[‡]Results of tests on small, clear, straight-grained specimens. Values in the first line for each species are from tests of green material; those in the second line are adjusted to 12 percent moisture content. Density is based on mass when oven dry and volume when green or at 12 percent moisture content.

[§]Modulus of elasticity measured from a simply supported, center-loaded beam, on a span-depth ratio of 14:1. The modulus can be corrected for the effect of shear deflection by increasing it 10 percent.

Table 20.4 Mechanical properties of some hardwoods grown in the United States[†],[‡]

Common names of species	Density, Mg/m³	Static bending — Modulus Rupture, MPa	Static bending — Modulus Elasticity,[§] GPa	Static bending — Work to maximum load, kJ/m³	Impact bending—height of drop causing complete failure, m	Compression parallel to grain—maximum crushing strength, MPa	Compression perpendicular to grain—fiber stress at proportional limit, MPa	Shear parallel to grain—maximum shearing strength, MPa	Tension pendicular to grain—maximum tensile strength, MPa
Ash, Oregon	0.50	52.0	7.8	84	0.99	24.2	3.7	8.2	4.1
	0.55	88.0	9.4	99	0.84	41.6	8.6	12.3	5.0
Basswood, American	0.32	34.0	7.2	37	0.41	15.3	1.2	4.1	1.9
	0.37	60.0	10.1	50	0.41	32.6	2.6	6.8	2.4
Beech, American	0.56	59.0	9.5	82	1.09	24.5	3.7	8.9	5.0
	0.64	103.0	11.9	104	1.04	50.3	7.0	13.9	7.0
Birch, yellow	0.55	57.0	10.3	111	1.22	23.3	3.0	7.7	3.0
	0.62	114.0	13.9	143	1.40	56.3	6.7	13.0	6.3
Hickory, true, mockernut	0.64	77.0	10.8	180	2.24	30.9	5.6	8.8	
	0.72	132.0	15.3	156	1.96	61.6	11.9	12.0	
Locust, black	0.66	95.0	12.8	106	1.12	46.9	8.0	12.1	5.3
	0.69	134.0	14.1	127	1.45	70.2	12.6	17.1	4.4
Maple, red	0.49	53.0	9.6	79	0.81	22.6	2.8	7.9	
	0.54	92.0	11.3	86	0.81	45.1	6.9	12.8	
Oak, red, southern red	0.52	48.0	7.9	55	0.74	20.9	3.8	6.4	3.3
	0.59	75.0	10.3	65	0.66	42.0	6.0	9.6	3.5
Oak, white, swamp chestnut	0.60	59.0	9.3	88	1.14	24.4	3.9	8.7	4.6
	0.67	96.0	12.2	83	1.04	50.1	7.7	13.7	4.8
Walnut, black	0.51	66.0	9.8	101	0.94	29.6	3.4	8.4	3.9
	0.55	101.0	11.6	74	0.86	52.3	7.0	9.4	4.8

[†] Wood Handbook, Forest Products Laboratory, U.S. Department of Agriculture, 1974.
[‡] Results of tests on small, clear, straight-grained specimens. Values in the first line for each species are from tests of green material; those in the second line are adjusted to 12 percent moisture content. Density is based on mass when oven dry and volume when green or at 12 percent moisture content.
[§] Modulus of elasticity measured from a simply supported, center-loaded beam, on a span-depth ratio of 14 : 1. The modulus can be corrected for the effect of shear deflection by increasing it 10 percent.

Fungi are microorganisms that attack both the cellulose and the lignin. If the cell walls are damaged, decay occurs, a most important type of deterioration. Fungi need air, water, and favorable temperatures. Thus fungi will not live in dry wood (with a moisture content below 20 percent) or in wood submerged in water. Prehistoric wood structures have been found intact in both types of environment.

If wood must be subjected to moist air, it must be coated with a moistureproof film, such as paint or varnish, or impregnated with a preservative, such as coal-tar creosote, pentachlorophenol, or zinc chloride. The service life of railroad ties can be increased from 5 to 30 years by treatment with creosote. Appropriate test procedures are described by ASTM D 1860.

Termites are insects that feed on cellulose. Dry-wood termites, found in warm climates, live in wood structures and do not require moisture. More common are subterranean termites, which can stand colder climates but need moisture and abhor light. They normally live below ground in pieces of dead wood. They will undertake excursions to wood structures above ground, traveling in dark crevices or through tunnels they build. Such access should be prevented. If termites are detected, treatment with insecticides is possible, but it must be very thorough.

Submerged structures, such as waterfront piling, are subject to *marine borers*, which are also insects. One genus, *Teredos*, attacks the wood near the bottom. Teredos bore into the pile radially and then turn in a longitudinal direction, leaving tunnels. Their presence cannot be detected until failure occurs. Another genus, *Limnoria*, attacks a pile at the waterline, perpendicular to the grain. As the weakened wood wears away, the limnorias penetrate deeper. The effect is visible as an annular indentation around the pile. A possible remedy is to sheathe the pile with an impenetrable material, but the integrity of the sheathing must be maintained. Prevention in this case may be more expensive than the cure: periodic replacement of the wood.

The *fire resistance* of wood may be improved by impregnation with foam-forming organic substances or aqueous solutions of chemicals such as borax. The treatment is best accomplished under pressure in large cylinders. ASTM E 160 describes a test method.

20.6 PLYWOOD

Plywood, in some form at least, apparently dates back to antiquity, but in the contemporary sense it is a product of the twentieth century. Plywood consists of several thin sheets of wood, called veneers in the trade, bonded with synthetic resin. The resin accounts for only 4 or 5 percent of the total mass, but it is an essential element. Thus modern plywood had to await the development of plastic adhesives.

Engineering uses of plywood include applications as concrete forms, walls, floors, roofs, and gusset plates. It is available in the form of rectangular panels of various standard dimensions up to about 3.6 m in length, 1.5 m in width, and 30 mm or more in thickness. A traditional size in the United States is 4 by 8 ft (1.22 by 2.44 m).

The panels are typically composed of an odd number of plies, with the grain direction alternating at right angles. The grain of the outside layers usually runs in the direction of the longer dimension of the panel. This arrangement tends to equalize the properties in the two directions, but does not do so fully in that the longitudinal properties of the wood tend to predominate in the longitudinal direction of the panel and the tangential properties in the lateral direction. Plywood terms are defined by ASTM D 1038.

Adhesives The adhesives have the combined purpose of preventing the separation of the veneer plies and of preventing the shear distortion of one ply with respect to another. Moisture causes wood to swell by different amounts parallel and perpendicular to the grain, so that shear stresses on the order of 50 MPa can be developed. Thermoplastics do not satisfactorily prevent the shear displacements, and so thermosetting plastics are preferred (see Chap. 21).

The characteristics of plywood, especially moisture resistance, depend to a large extent on the adhesive chosen. Phenol-formaldehyde (PF) is very strong, but it is expensive. It also causes staining, a problem that can be overcome by using melamine-formaldehyde (MF) instead, but this is more costly yet. For plywood not subjected to weather, less expensive urea-formaldehyde (UF) is generally used, which may be fortified with melamine-formaldehyde.

Wood plies Since the panels consist mainly of wood, the strength characteristics depend considerably on the species. To simplify design, various species with similar properties are combined as *groups*. Group 1 has the highest strength characteristics and group 5 the lowest (not of structural grade).

Here are a few examples from each group:

Group 1: American beech, sugar maple, southern pine, Douglas fir 1 (Pacific coast)
Group 2: Port Orford cedar, white fir, western hemlock, Virginia pine, Sitka spruce, Douglas fir 2 (Rocky Mountains)
Group 3: Alaska cedar, eastern hemlock, redwood, ponderosa pine
Group 4: Bigtooth aspen, western red cedar, eastern white pine
Group 5: Basswood, balsam fir

Also important is the integrity of the veneers: their freedom from knots, knotholes, plugs, and patches. On this basis the industry classifies veneers into *veneer grades*, Grade A being the best and D the worst (not allowed for exterior applications). The grades are as follows:

Grades A
  and N: No knots, restricted patches (N indicates natural finish, A painted finishes)
Grade B: Solid surface; small, round knots; patches; round plugs
Grade C
  plugged: Improved grade C
Grade C: Small knots, knotholes, patches
Grade D: Larger knots and knotholes

The front, back, and interior plies need not all be of the same grade.

The plies are usually about 3 mm thick, but range in thickness from 2 to 8 mm. Occasionally laminated cores are used, which may consist of sawnwood glued together or of particle board. In some instances two or more plies are laminated with their grain directions parallel; they are then jointly referred to as a "layer." Thus there may be more plies than layers in a panel (see Fig. 20.4).

Production The veneer is usually obtained by rotary shaving of a log. To form the finished plywood, the prepared plies are arranged symmetrically with respect to the core as regards species, thickness, and direction of grain. The grade, however, may be different for the front and back face veneers.

Adhesive is spread on the veneers, which are then subjected to cold-pressing (at room temperature) or hot-pressing (at 75 to 190°C). Thermoset adhesives may be set at the same time. Most panels are pressed flat, but forms curved in one or both directions can be used to make "molded plywood." Because the presses are very expensive, the time of set becomes an economically critical factor.

Some panels may, and very long ones necessarily must, be jointed. That is accomplished by scarf or finger jointing, as shown in Fig. 20.5. This kind of jointing does not reduce the strength characteristics of the plywood.

Specifications Because plywood is subject to so many variations and manufacturing defects, rigorous product standards have been established. In the United States, such a specification is U.S. Product Standard PS1, which is published by NBS and was

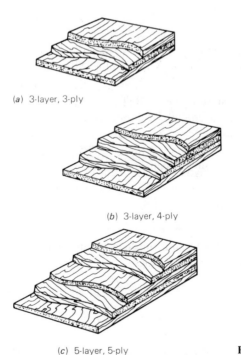

(a) 3-layer, 3-ply

(b) 3-layer, 4-ply

(c) 5-layer, 5-ply

Figure 20.4 Typical plywood panel constructions.

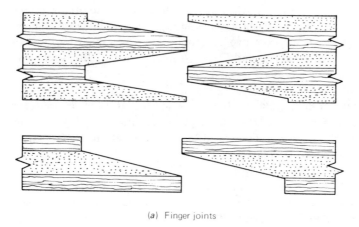

(a) Finger joints

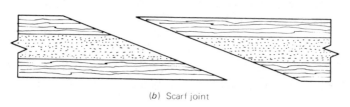

(b) Scarf joint

Figure 20.5 Plywood joints.

accepted by ANSI as American National Standard A199.1. Individual panels are marked with the species group, type, and veneer grade.

PS1 establishes detailed requirements for wood veneers, adhesives, and manufacture, and it specifies standard qualification tests for glue bond. The standard classifies plywood into two *types*, interior and exterior, according to durability (i.e., exposure capability). Within each of the exposure capability types there are several *panel grades*, based on the grade of the veneers and the panel construction. The panel grades include terms that are descriptive of suitable uses for the panels, such as

Marine
Decorative
Underlayment
Concrete form
Structural
Interior type bonded with exterior glue
Special exterior
Overlays

Of primary importance in an engineering context is the structural grade. Structural panels may be of interior or exterior type, and within each type they are designated

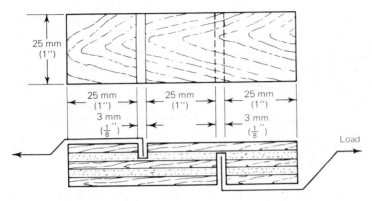

Figure 20.6 Specimen for ply shear test.

structural I or structural II. The former kind is of higher quality, the veneers being restricted to species of group I, whereas for the latter kind groups 1, 2, and 3 are allowed.

Typical plywood *tests* include shear tests on specimens such as the one shown in Fig. 20.6. The specimens are first subjected to soaking in a partial vacuum or to cyclic boiling, drying, and cooling, after which they are tested in a testing machine to shear failure in the one ply under investigation. The test is intended to check the adequacy of bonding, and the criterion of this is that a minimum percentage of the failure occur in the wood, the percentage being 85 for the exterior type.

Tension tests are specified by ASTM D 3500; compression tests by D 3501; flexure tests by D 3043; various shear tests by D 2718, D 2719, and D 3044; and toughness tests by D 3499.

Physical Properties Since plastic constitutes only 4 or 5 percent of plywood by mass and perhaps a third of that amount by volume, the general properties of the material, such as relative density, primarily depend on the species of wood. Typical densities are around 0.5 Mg/m^3. The coefficient of thermal expansion is quite small, as in the case of wood.

Because of the way the veneer is made, *shrinkage* normal to the grain equals the tangential shrinkage of the wood, which is the largest of the three values, whereas the shrinkage in the direction of the grain equals the longitudinal shrinkage, a negligibly small value. A major advantage of plywood is that these two values average out, so that the aggregate shrinkage is approximately the same in all directions. Warping and splitting are much reduced.

The *strength* and *stiffness* properties of plywood also have values intermediate between those for the basic wood in the directions perpendicular and parallel to the grain, depending on the direction of stress with respect to the various layers. If, for example, a dry panel has seven equal layers of Douglas fir with tensile strengths of 100 and 2 MPa, and if compatible deformations are assumed, then the tensile strength of the panel parallel to the face veneers is about $[(4)(100) + (3)(2)]/7 = 58$ MPa. For bending and shear the situation is more complicated.

Because of the multiplicity of factors affecting strength properties—type of adhesive, species of wood, moisture content, direction of grain, variation among plies, defects, and the like—comprehensive strength data would be prohibitively voluminous. The problem is further complicated by the fact that the material is neither homogeneous nor isotropic, although it is essentially orthotropic. If, for example, the grain of the face veneers run at right angles to the span of a bending panel, the extreme fibers (where the strains are largest) contribute practically nothing to the flexural strength.

For these reasons design tables are furnished by industry organizations such as the American Plywood Association (APA). The procedure works in three steps:

1. For certain panel grades and thicknesses, effective cross-sectional properties are established.
2. For panel types and veneer grades, allowable stresses are established.
3. The designer proportions the element using the appropriate effective properties and allowable stresses.

PROBLEM 20 Compression Tests of Wood

Object: To study the action of wood under compressive loading in the three principal directions and to determine some of its physical and mechanical properties for each specimen and load condition as follows:
1. Moisture content
2. Approximate density, both as received and when dry
3. Average width of annual growth rings
4. Percentage of latewood
5. Percentage of sapwood
6. Elastic strength
   a. Proportional limit
   b. Yield strength at 0.05 percent offset
7. Ultimate strength
8. Modulus of elasticity
   a. Initial tangent
   b. Secant at 0.05 percent offset
Preparatory reading: ASTM D 2555 and D 143.
Specimens: Clear wood, about 40 by 160 mm, with the long dimensions oriented respectively in the longitudinal, radial, and tangential directions. (It is also possible to stack three or four cubes cut from a standard specimen.)
Special apparatus: Compressometer, compression machine, balance, drying oven.
Procedure:
1. Note any defects in the specimens. Measure the cross section and length of each specimen to the nearest 0.1 mm and weigh to the nearest 0.1 g. Determine the average width of the annual growth rings, the percentage of latewood, and the percentage of sapwood.
2. Determine the gage length and multiplication ratio of the compressometer. Determine the strain corresponding to the least reading of the dial. Attach the compressometer to one specimen in the longitudinal direction, and remove the spacer bars. Center the specimen on the table of the testing machine. Adjust the compressometer dial to read zero and make certain that most of its range is available.
3. Apply the load continuously until failure, at a speed that will allow you to read the compressometer dial and the load at about 10 intervals. Draw a sketch, in perspective, indicating the grain of the wood and the manner of failure.

4. Repeat the procedure for the other two specimens.
5. Cut a moisture sample about 25 mm in length from each specimen near the failure. Remove all splinters and weigh to the nearest 0.1 g. Place in a drying oven controlled at $103 \pm 2°C$. After the moisture specimens have dried to constant mass, which may take two to four days, again weigh them.

Report:
1. For each test, construct a stress-strain curve and obtain the strength values asked.
2. Calculate the moisture contents as ratios to the oven-dry masses.
3. Calculate the densities on the basis of the original masses and dimensions. Subtract the moisture contents and compute the dry density on the basis of the original volume.
4. Discuss the differences between these tests and standard ASTM compression tests on wood.

DISCUSSION

1. What is the effect of moisture on the strength properties of wood?
2. What is the relation between density and strength?
3. Are the values obtained from tests on small, clear beams applicable to beams of large size? Explain.
4. What effect does the time element have in the loading of wood?
5. What is the effect of the percentage of latewood on strength?
6. What methods may be employed for determining the moisture content of wood?
7. Would you expect the neutral axis to remain at the mid-depth of a rectangular wood beam as the ultimate load is approached? If not, which way would it shift, and why?
8. Discuss the relation between the modulus of rupture of a small wood beam and the ultimate compressive strength of a short block loaded parallel to the grain.
9. How does the speed of load application affect the observed strength?
10. Account for the differences among strength properties in the longitudinal, radial, and tangential directions of clear wood.

READING

Desch, H. E.: *Timber, Its Structure and Properties*, 5th ed., Macmillian, London, 1974.
Hoyle, R. J.: *Wood Technology in the Design of Structures*, 4th ed., Mountain Press, Missoula, Mont. 1978.
Stamm, A. J.: *Wood and Cellulose Science*, Ronald, New York, 1964.
Timber Construction Manual (AITC), Interscience-Wiley, Somerset, N.J., 1974.
Tsoumis, G.: *Wood as a Raw Material*, Pergamon, New York, 1968.
Wangaard, F. F.: *The Mechanical Properties of Wood*, Wiley, New York, 1950.
Wood Handbook, Agricultural Handbook No. 72, Forest Products Laboratory, U.S. Department of Agriculture, Washington, 1974.

TWENTY ONE

PLASTIC

21.1 TYPES AND CHARACTERISTICS

Plastics constitute a major contribution of the twentieth century to the storehouse of constructional materials. Since Leo Hendrik Baekeland combined phenol and formaldehyde in 1910 to form a commercial plastic that was to be known as Bakelite, progress in the development of the synthetic materials known as plastics has been vigorous. As construction materials they find application not only as sealants and insulators but also as structural components. Like metals, they can be given properties suited to many applications. Some are reinforced with fibers.

Ironically, the term "plastics" is a bit elastic. Derived from the Greek *plastikos* ("to form"), the meaning sometimes includes natural materials, such as rubber, concrete, asphalt, and glass. This broadened meaning may in part be attributed to the similar behavior of these materials, but also perhaps to the fact that some of them, as well as many other materials, are *polymers*, since there is a misconception that all polymers are plastics.

Polymers (from the Greek for "many parts") have extremely large molecules, which are made up of smaller units called monomers. Most polymers are organic substances called *resins*, but some are inorganic. Some are synthetic materials, which are called plastics if they are pliable at some stage, and others are natural. Chemists seem unable, however, to devise a completely rigorous system of classifying polymers.

The materials to be discussed here are synthetic organic polymers capable of being molded and used in engineering applications. They are both plastics and synthetic resins, and in fact these two terms are often used interchangeably.

These plastic materials include two major types. *Thermoplastics* will soften when they are heated and harden when cooled, no matter how often the cycle is repeated. *Thermosetting plastics*, or "thermosets," harden when heated in an irreversible process. Other plastics harden by chemical reaction or by evaporation of a solvent. Some plastics cannot be definitively classified in this manner, as various materials can be made either in thermoplastic or thermosetting form.

Plastics are sometimes referred to by trade name, but in modern practice the generic terms are preferred. Since many of the latter are long and unwieldy, ASTM has suggested two- to five-letter abbreviations for most plastics (ASTM D 1600). For example, Bakelite is called phenol-formaldehyde or PF. Some common plastics and their abbreviations are listed in Table 21.1. The table also includes the names of some chemicals often added to the plastics. Terms relating to plastics are defined by ASTM D 883.

21.2 PRODUCTION METHODS

The basic raw materials for plastics are natural fibers, coal, and petroleum. By chemical reactions, washing, distilling, heating, crystallization, and the like, the polymers are obtained and usually emerge in the form of powders, granules, or pellets. Additives such as plasticizers and stabilizers can be incorporated, after which the plastics are heated and shaped by such processes as molding and extruding.

Raw materials Cellulose, a crystalline material in the form of long chains, is obtained from plant fibers, especially cotton and wood. When coal is carbonized, tars and light oils are obtained; the former in turn yield phenol, cresols, and naphthalene, and the latter benzene, toluene, and xylene. Petroleum yields ethylene and propylene, as well as most of the coal derivatives mentioned above. By various processes some of these compounds can be converted to others, for example benzene to phenol.

Polymerization Plastics are manufactured from these and other substances by building chains and networks of molecular units. The formation of these macromolecules is called polymerization, which may proceed in one of two ways. In *condensation polymerization* reactive molecules combine with one another in a stepwise fashion, while small molecules, such as H_2O, are eliminated; as the polymers build up, the process slows down. In *addition polymerization* small identical or somewhat diversified units are linked together, with no elimination of small molecules. An example of a condensation polymer is polyethylene terephthalate (PETP), or Teflon, and one of the few examples of addition polymers is polymethyl methacrylate (PMMA), or acrylic. An important object in polymerization is to maximize the molecular mass of the polymers, as this increases the strength of the material.

Additives Various substances can be added to the polymers to improve the properties of the plastic materials. The concept is quite analogous to alloying metals or adding aggregates and admixtures to concretes. Some additives play several roles.

Table 21.1 ASTM abbreviations of plastics

Plastics and resins		Plastics and resins	
Term	Abbreviation	Term	Abbreviation
Acrylonitrile-butadiene-styrene plastics	ABS	Poly(diallyl phthalate)	PDAP
Carboxymethyl cellulose	CMC	Polyethylene	PE
Casein	CS	Polyethylene terephthalate	PETP
Cellulose acetate	CA	Poly(methyl-α-chloroacrylate)	PMCA
Cellulose acetate butyrate	CAB	Poly(methyl methacrylate)	PMMA
Cellulose acetate propionate	CAP	Polymonochlorotrifluoroethylene	PCTFE
Cellulose nitrate	CN	Polyoxymethylene, polyacetal	POM
Cellulose propionate	CP	Polypropylene	PP
Chlorinated poly(vinyl chloride)	CPVC	Polystyrene	PS
Cresol-formaldehyde	CF	Polytetrafluoroethylene	PTFE
Diallyl phthalate	PDAP	Polyurethane	PU
Epoxy, epoxide	EP	Poly(vinyl acetate)	PVAC
Ethyl cellulose	EC	Poly(vinyl alcohol)	PVAL
Melamine-formaldehyde	MF	Poly(vinyl butyral)	PVB
Perfluoro(ethylene-propylene) copolymer	FEP	Poly(vinyl chloride)	PVC
Phenol-formaldehyde	PF	Poly(vinyl chloride-acetate)	PVCA
Poly(acrylic acid)	PAA	Poly(vinyl fluoride)	PVF
Polyacrylonitrile	PAN	Poly(vinyl formal)	PVFM
Polyamide (nylon)	PA	Silicon plastics	SI
Polybutadiene-acrylonitrile	PBAN	Styrene-acrylonitrile	SAN
Polybutadiene-styrene	PBS	Styrene-butadiene	SB
Polycarbonate	PC	Styrene-rubber plastics	SRP
		Urea-formaldehyde	UF

Plastic and resin additives

Term	Abbreviation
Dibutyl phthalate	DPB
Dicapryl phthalate	DCP
Diisodecyl adipate	DIDA
Diisodecyl phthalate	DIDP
Diisooctyl adipate	DIOA
Diisooctyl phthalate	DIOP
Dinonyl phthalate	DNP
Di-n-octyl-n-decyl phthalate	DNODP
Dioctyl adipate	DOA

Monomers

Term	Abbreviation
Diallyl chlorendate (diallyl ester of 1,4,5,6,7,7-hexachlorobicyclo-(2,2,1)-5-heptene-2,3-dicarboxylic acid)	DAC
Diallyl fumarate	DAF
Diallyl isophthalate	DAIP

Miscellaneous plastics terms

Term	Abbreviation
General purpose	GP
Single stage	SS

Plastic and resin additives

Term	Abbreviation
Dioctyl azelate	DOZ
Dioctyl phthalate	DOP
Dioctyl sebacate	DOS
Diphenyl cresyl phosphate	DPCF
Diphenyl 2-ethylhexyl phosphate	DPOF
Trichloroethyl phosphate	TCEF
Tricresyl phosphate	TCF
Trioctyl phosphate	TOF
Triphenyl phosphate	TPP

Monomers

Term	Abbreviation
Diallyl maleate	DAM
Diallyl orthophthalate	DAP
Methyl methacrylate	MMA
Monochlorotrifluoroethylene	CTFE
Tetrafluoroethylene	TFE
Triallyl cyanurate	TAC

Miscellaneous plastics terms

Term	Abbreviation
Solvent welded plastic pipe	SWP

Plasticizers, such as triphenyl phosphate (TPP), are added to polymers by heating and mixing. They increase the flexibility of thermoplastics by reducing the intermolecular contact and attraction of the polymer chains. The particular plasticizer used depends on the type of polymer and its intended application. All plasticizers should have minimum volatility so as to preserve the flexibility of the material as long as possible. Cellulose plastics always require plasticizers, and the flexibility of polyvinyl chloride (PVC) depends on it.

Heat, light (especially ultraviolet light), and oxidation tend to degrade polymers. These effects may be retarded by adding *stabilizers* and *antioxidants* to the materials. PVC, for example, suffers from emission of hydrochloric acid, which can be reduced by adding lead salts or soaps. Low-density polyethylene (PE) must be protected from oxidation, for which phenols can be used. Carbon black protects against oxidation caused by light. This, of course, is especially important for outdoor applications of plastics.

Fillers and *reinforcements* include powders, tiny fibers, and microspheres. The most common of these materials are wood flour and glass fiber, but asbestos, carbon black, cotton, carbon fiber, metal oxides, and other substances are also used. Like aggregates in concrete, the fillers provide cost-saving bulk and also reduce shrinkage. Reinforcements provide additional strength, much like steel strands in ferrocement.

Plastics may be colored with *dyes*, which are soluble organic substances, or with *pigments*, which are insoluble and may be organic or inorganic. Dyes are used for transparent plastics, and inorganic pigments for opaque ones.

Some additives, such as phosphates, bromine, and chlorine, serve as *flame retardants* and are used when nonflammability is important. Flexible PVC, for example, can be treated with phosphates, whereas other materials may require other retardants. Some polymers, such as rigid PVC and fluorocarbons, are nonflammable, and some, such as nylons and polycarbonates, are self-extinguishing. It is also possible, as in the case of polyesters, polymethane foam, and epoxy resins, to build bromine into the polymer molecules.

21.3 PROCESSING METHODS

A dozen or more basic techniques can be used to form the plastic powders or pellets into desired shapes. We will take a brief look at some methods important for engineering plastics.

Extruding For extrusions, the plastic granules are fed through a hopper into a long barrel, the temperature of which is controlled by heating-cooling tubes. Inside the barrel a conveyor screw drives the plastic toward the nozzle of the barrel, where a die imparts the desired shape to the material (see Fig. 21.1). The assembly resembles the familiar meat grinder. Thermoplastic pipes, T and H sections, sheets, wire insulation, and the like can be made. As the material leaves the nozzle it is cooled and removed by conveyor. If closed sections, such as pipe, are to be made, a mandrel is centered in the nozzle.

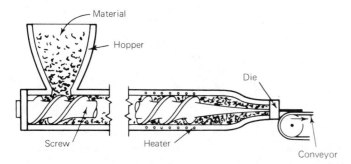

Figure 21.1 Extrusion. (*Courtesy of The Society of the Plastics Industry, Inc.*)

The conveyer screws may have various configurations so as to compress the plastic gradually and homogenize it. The length of the screw is perhaps 25 or 35 times the diameter. Nozzle diameters range up to several hundred millimeters.

Extrusion is sometimes followed by *blow molding* to make bottles and such. As a molten tube leaves the extruder it enters a hollow mold and is forced against the cold walls by air pressure.

In another variation the emerging tube is blown up to several times its original diameter. As the tube cools, rollers compress it to a double layer of film or sheet.

Compression molding Thermosetting plastics are introduced into a mold much like a waffle iron, often after preheating and/or preforming. They are then heated to about 700 to 750°C and subjected to pressures ranging from 10 to 50 MPa or so. Both parts of the mold, referred to as male and female, are heated. The quantity of material introduced must, of course, be carefully controlled to ensure complete filling and to avoid spillage. See Fig. 21.2.

Cold molding is also possible, but it results in inferior quality of the product. Thermoset powder is cured after being pressed into the desired shape.

Injection molding For injection molding, thermoplastic material is fed into a hopper, then measured into a heating chamber. The granules are melted and the viscous fluid is forced through a nozzle into a mold. The mold is cooled, and after the article solidifies it is ejected (see Fig. 21.3).

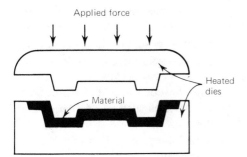

Figure 21.2 Compression molding. (*Courtesy of The Society of the Plastics Industry, Inc.*)

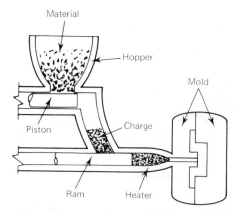

Figure 21.3 Injection molding. (*Courtesy of The Society of the Plastics Industry, Inc.*)

Casting Some plastics can be cast into a mold in liquid form. These include mainly thermosets such as polyesters, urethanes, silicones, and epoxies. The forms may be made by dipping a steel prototype into molten lead, a hardened crust of which becomes the mold after the mandrel is withdrawn. Once in the form, the plastic is cured for a few days at perhaps 75°C, after which the lead is knocked off and reused. The method is suitable for high-precision articles such as gears and lenses.

Foaming When chemicals that form gas bubbles are introduced into certain plastics, such as urethanes, while they are in the liquid state, expanded or foamed plastics are created. The bubbles are formed by a "blowing agent," which can be a volatile liquid such as pentane, used in expanding polystyrene, or an unstable chemical that releases gas when heated. Alternatively, a gas such as carbon dioxide can be dissolved in the hot plastic at high pressure; the bubbles form in the soft material when the pressure is released.

The foamed plastics may be "open-cell," in which the bubbles are interconnected, or "closed-cell," in which the individual bubbles are preserved. The first type makes a good sponge, the second, a safe life preserver.

Other forming methods Calendering, drawing, rotational molding, vacuum forming, and laminating are among other techniques used to process plastics. There are many variations of these methods, and new ones are developed every year.

Fabrication Plastics may be machined, sawed, drilled, theaded, and so forth, much like metals. They are, however, more flexible and soften under built-up heat, so that these mechanical operations are more difficult and less precise. For these reasons laser machining, during which a laser beam vaporizes the material, is helpful. Holes as small as 50 μm can be drilled, for example to create nozzles for aerosol cans.

Many plastics can be joined by heat. This may be done in a direct process by applying either a stream of a hot gas or a heated metal tool such as a soldering iron. PVC sheets can be welded together in either of these ways to form the joints in tank liners, for example. Heat can also be generated by high-frequency vibrations at about 20 kHz. This method is particularly effective for hard plastics.

Finishing methods for plastics include printing, leaf hot-stamping, vacuum metallizing, electroplating, painting, and other coating procedures.

Plastics may also be reinforced with cloth or filaments in the finished product, much as concrete is reinforced with steel bars. Quite often, for example, polyesters are reinforced with glass fibers to become fiberglass-reinforced plastics (FRP), which are used for roofs and boats.

21.4 ENGINEERING PLASTICS

Some plastics are particularly useful for constructional applications. Structural panels, curtain walls, pipe systems, tank liners, gears, skylights, concrete forms, insulating boards, and the like are familiar examples. Their common feature is that they are able to support loads to the extent required.

Thermoplastics Thermoplastics consist of long molecules resembling threads; "long," of course, is a relative term, the length being on the order of 20 to 30 nm[†]. At normal temperatures the molecules, which are intricately intertwined, resist displacements, but they become slippery at higher temperatures. In the heated state thermoplastics can be easily molded or extruded, but this also means that under service conditions they exhibit little heat resistance. The materials as a class are not well suited for structural uses, with certain exceptions.

ABS plastics are copolymers of acrylonitrile, butadiene, and styrene. They are strong and tough, resist heat, weather, and chemicals quite well, and are dimensionally stable. They are flammable, however, and can be dissolved by organic substances. Pipes, concrete forms, building and vehicle components, and appliance housings are often made of ABS.

Acetals are polymers and copolymers of formaldehyde. Their high strength, toughness, and stiffness, combined with resistance to fatigue, organic solvents, and thermal distortion, often make them suitable replacements for metals. They do not resist weathering or burning well, however. Applications include machine parts and pipe systems.

Acrylics such as polymethyl methacrylate (PMMA) can be colored or transparent, and they resist weathering. They are fairly strong but are subject to damage by abrasion. They make fine windows, wall panels, and lighting devices and also serve in the form of sealants and coatings.

Nylon, or polyamide (PA), has a very low coefficient of friction, is very strong and tough, and resists abrasion and chemical action. It is used not only for pantyhose but also for tracks, rollers, gears, and bearings.

Polyethylenes (PE) are either low-density, featuring branched molecules, or high-density, having linear molecules. The heavier type is stronger and more rigid. Both types resist chemicals well but are subject to weathering, mainly because of sensitivity to ultraviolet light, and they are flammable. Tubes, pipes, and tanks are made of PE.

Polypropylenes (PP) are similar to high-density PE but are stronger and stiffer. They are used for pipes and tanks as well as belting.

[†]Nanometer: 1 nm = 10^{-9} m = 0.000 000 039 in.

Polyvinyl chlorides (PVC) may be rigid or flexible. They resist water and solvents, but they tend to be sensitive to elevated temperatures, and the rigid type to ultraviolet light. PVC is used to make curtain wall panels, siding, window parts, pipes, ducts, gutters, and corrugated sheets. Of all the thermoplastics, PVC ranks at or near the top as building material.

Other thermoplastics include cellulosics (CA, CAB, and CAP), polycarbonates (PV), and polystyrenes (PS).

Thermosetting plastics The molecules of thermosets, like those of thermoplastics, resemble minute threads when the material is soft. Heating, however, causes the molecules to combine chemically at various points, making a strong cross-linked structure. Once cooled, thermosetting plastics can no longer be melted, so that they are quite resistant against heat, and they resist chemicals and solvents. Many are fine structural materials.

Epoxies (EP) can be cured at room temperatures. They are resistant to moisture, solvents, chemicals, and impact, and they have high strength and dimensional stability. They find application as adhesives, bonding agents, linings for pipes and tanks, and laminates for glass and carbon fiber.

Phenolics (PF), or phenol-formaldehyde resins, are resistant to heat and moisture, have good mechanical and electrical properties, are easily molded, and are dimensionally stable. They can be used as binders for various fillers, including wood flour, minerals, and glass fiber, and they serve as adhesives for plywood and other wood products.

Polyesters include many resins, but alkyds and especially unsaturated polyesters are the primary constructional types. These may be cured at normal temperatures, with the aid of a catalyst and an accelerator, after blending with styrene monomer. Often glasss fiber is used for reinforcement. Unsaturated polyesters are strong and resist environmental influences well. Applications include fiberglass boats and automobile bodies, pipes and tanks, roof panels, vent and ducts, adhesives, coatings, and laminates.

Silicones (SI) are excellent sealants, but they are very expensive. They are chemically inert and withstand weathering even at extreme temperatures.

Polyurethanes (PU) are mainly used as foams, which have a wide range of flexibility, some being highly rigid. Flexible urethane foams can substitute for foam rubber, at lower cost. In rigid foams the molecules are highly cross-linked so as to form closed cells. This causes not only the rigidity but also extremely low thermal conductivity and very low relative density (about 0.03). Since urethanes also adhere very tightly, they are perfectly suited to making insulated sandwich panels for curtain walls or flotation devices for boats. In a nonporous form urethanes are used as sealants, caulks, and coatings. The properties of urethanes in all forms are impaired by high temperatures.

Alkyds, *allylics*, *amines*, *melamines*, and *polysulfides* are among other thermosetting plastics used as engineering materials in various ways.

Adhesives Structural adhesives capable of withstanding high stresses are common in many industrial applications; sometimes plastic adhesives even replace welding and riveting of light-gage metallic components. Some are suitable for bonding wood or

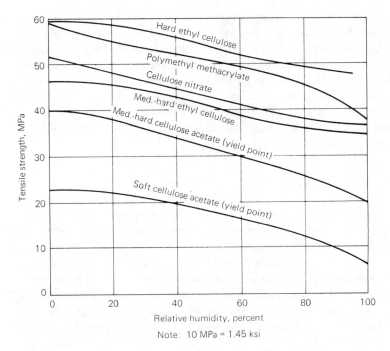

Note: 10 MPa = 1.45 ksi

Figure 21.4 Effect of relative humidity on tensile strength of thermoplastics. (*From ASTM Symposium on Plastics, Ref. 90.*)

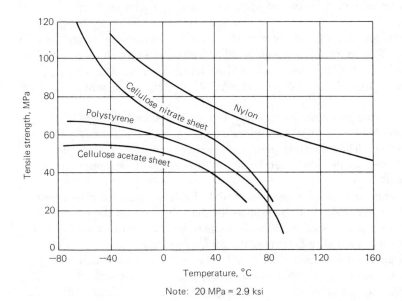

Note: 20 MPa = 2.9 ksi

Figure 21.5 Effect of temperature on tensile strength of thermoplastics. (*From ASTM Symposium on Plastics, Ref. 90.*)

concrete members and even for bonding rubber to metal, as well as bonding products made of plastics. The various organic synthetic resins used as adhesives have a wide range of mechanical properties depending on the composition of the base material and also on the additives that may be combined with it to improve certain characteristics. Their high strength results from the polymerization of their molecules. They are commonly sold under trade names, but are essentially based on epoxy, phenolic, polyester, allyl, acrylic, polyvinyl, polystyrene, or similar resins used singly or in combination. Their strength characteristics are determined in accordance with ASTM standards D 897, D 950, and D 1002.

21.5 PHYSICAL PROPERTIES

The range of the properties of plastics is considerable. Depending on their compositions and reinforcing materials, they can be extremely weak or very strong. Fillers and plasticizers greatly change the properties of plastics. They can modify brittle plastics, such as phenolic resin, which are difficult to mold and are rather costly, so that they are more moldable and less expensive to use. Laminates and reinforced plastics made of plastics combined with sheet or fibrous materials have properties superior to those of either material alone. The laminates make use of paper, fabric, wood, and other sheet materials, whereas the reinforced plastics usually incorporate glass fibers in the form of felted or woven fabrics.

One of the first engineering uses for large amounts of plastics was for electrical insulation. Other characteristics such as good resistance to acid corrosion and mechanical abrasion, together with a variety of optical properties, greatly widened their field of use.

Although plastics generally do not corrode under the influence of water, they do absorb moisture. This in turn has a significant effect on strength. As shown in Fig. 21.4, some thermoplastics may lose half of their tensile strength when the ambient humidity increases from 0 to 100 percent. An absorption test is standardized by ASTM D 570.

Thermoplastics are also sensitive to heat. As Fig. 21.5 shows, the tensile strength may diminish rapidly above normal atmospheric temperatures, and it tends to increase at low temperatures. Thermosetting plastics, however, may exhibit very good heat resistance. Some, such as PF, have been heated as high as $250°C$ without appreciable decrease in strength.

The coefficients of thermal expansion tend to be rather high for plastics, when compared to other materials. For thermoplastics the values are on the order of 100 MK^{-1}; for thermosets the values are typically more or less half of this amount. ASTM D 696 describes a test method to determine the coefficient for plastics.

Plastics are apparently weaker than metals, the materials with which they most often compete. Conversely, their relative densities, which are determined according to ASTM D 792, are considerably lower, ranging from approximately 1.0 to 2.0. Economically, the ratio of strength to relative density, sometimes called the *specific strength*, may become important. By this criterion reinforced epoxies lead all metals by a wide margin. They are followed in order by heat-treated titanium, reinforced

Table 21.2 **Properties of thermoplastics**[†]

Material[‡]	Density, Mg/m³	Tensile strength, MPa	Elonga- tion, %	Modulus of elasticity, GPa	Compressive strength, MPa	Modulus of rupture, MPa	Izod impact, per notch length, J/m	Rockwell hardness	Thermal expansion, MK⁻¹	Water-ab- sorption, 3 mm thick 24 h, %
ABS	1.03	35–45	15–60	1.7–2.2	25–50	–	160–320	R95–R105	100–110	0.2
CA	1.27–1.34	15–60	6–50	0.6–3.0	90–250	15–110	20–280	R50–R125	80–160	1.7–6.5
CN	1.35–1.40	50–55	40–45	1.3–15.0	150–240	60–75	270–375	R95–R115	80–120	1.0–2.0
PA	1.14	80	90	3.0	85	–	50	M79, R118	100	1.5
PMMA	1.18	50–70	2–10	–	80–115	90–115	21–27	M85–M105	90	0.3–0.4
PS	1.06	35–60	1–4	3.0–4.0	80–110	55–110	13–27	M65–M90	60–80	0.03–0.05
PVC, rigid	1.32–1.44	40–60	5	2.4–2.7	60	–	13–110	R110–R120	50–60	0.03–0.04
PVCAc, rigid	1.34–1.45	50–60	–	2.0–3.0	70–80	85–100	21–37	–	70–185	0.07–0.08

[†] Based on *Modern Plastic Encyclopedia*, Modern Plastics, Hightstown, N.J., 1980–81, and on other sources.
[‡] Molding compounds, except as noted.

Table 21.3 Properties of thermosetting plastics[†]

Material[‡]	Density, Mg/m³	Tensile strength, MPa	Elongation, %	Modulus of elasticity, GPa	Compressive strength, MPa	Modulus of rupture, MPa	Izod impact strength, per notch length, J/m	Rockwell hardness	Thermal expansion MK⁻¹	Water absorption, 3 mm thick 24 h, %
EP, reinforced with glass cloth	1.80	350	–	175	410	485	1000	–	90	
MF, alpha-cellulose filler	1.47–1.52	50–90	0.6–0.9	9	170–300	70–110	13–19	M110–M125	40	0.1–0.6
PF, no filler	1.25–1.30	50–55	1.0–1.5	5–7	70–200	80–100	11–19	M124–M128	25–60	0.1–0.2
PF, wood flour filler	1.32–1.55	45–60	0.4–0.8	6–8	160–250	60–85	13–32	M100–M120	30–45	0.3–1.0
PF, macerated fabric filler	1.36–1.43	25–65	0.4–0.6	6–9	100–160	60–100	40–425	M95–M120	10–40	0.04–1.75
PF, cast, no filler	1.30–1.32	40–65	1.5–2.0	3	85–115	75–115	13–21	M93–M120	60–80	0.3–0.4
Polyester, glass-fiber filler	1.90	35–65	–	11–14	140–175	95–115	650–1300	–	10–30	0.15–0.20
UF, alpha-cellulose filler	1.47–1.52	55–90	0.5–1.0	10	175–240	70–100	13–119	M115–M120	27	0.4–0.8

[†]Based on *Modern Plastic Encyclopedia*, Modern Plastics, Hightstown, N.J., 1980–81, and on other sources.
[‡]Molding compounds, except as noted.

phenolics and polyesters, ultra-high-strength steels, glass-reinforced nylon, hard aluminum alloys, cast epoxies, and Nylon 66. Methods of determining mechanical properties of plastics are described by such ASTM specifications as D 638 for tensile properties, D 695 for compression, D 790 for flexure, D 256 for impact, and D 785 for Rockwell hardness.

Tables 21.2 and 21.3 list various physical properties of thermoplastics and thermosettings plastics respectively. The method and direction of molding should be known when interpreting tests results because of the marked effect of molding pressure on the mechanical properties.

PROBLEM 21 SPECIFIC STRENGTH OF PLASTIC

Object: Determine the ratio of tensile strength to relative density for a plastic.
Preparatory reading: ASTM D 792, D 638, and D 651.
Specimens: Plastic (as assigned).
Special apparatus: Beaker, balance, testing machine.
Procedure:
1. Using your ingenuity, prepare a device that will allow you to weigh a specimen immersed in a beaker of de-aired, cool, distilled water. A possible arrangement would be to place the balance on a drafting stool and the beaker under it; the specimen would be supported by a thin wire, and this in turn by a lightweight frame that carries the load around the seat of the stool to the balance.
2. Clean the specimen and obtain its weight[†] both in air and in water. Subtracting the latter from the former yields the weight of the displaced water. Both measurements should be made to 0.01 g. The relative density equals the specimen weight in air divided by the weight of the displaced water.
3. Subject the dry specimen to a tension test to rupture, after determining its dimensions.
Report:
1. Calculate the tensile strength and its ratio to the relative density, i.e., the specific strength.
2. Study the effect of the density of water at the given temperature. Explain any errors due to the temperature.
3. Assess the effect of neglecting the fact that in weighing the submerged specimen a short piece of the supporting wire was also submerged.
4. Comment on possible effects of moisture.
5. Estimate the total elongation of the specimen by studying the broken pieces. Comment on the reliability of this result.

DISCUSSION

1. In general, are thermosets or thermoplastics better structural materials to make, say, a glider? Why?
2. What is the molecular difference between thermosets and thermoplastics?
3. What plastics might be used to make a drainpipe and how is such a pipe made? How can pipe elbows and T's be made?
4. What materials could be used to make "fiberglass"?
5. How do the physical properties of thermosetting plastics compare with those of hardwoods?

[†]The term "weight" here means "apparent mass," i.e., the reading in grams on the balance.

6. Name some devices that could be made of plastics or metals, and discuss the merits and short-comings of each material.
7. In addition to Bakelite, do you know any tradenames for various plastics? Try to determine their generic names.
8. Distinguish between "plastics" and "polymers."
9. Aside from plastics, what words do you know that start with "poly-"?
10. Assess the probable future use of plastics in the context of available natural resources.

READING

Brydson, J. A.: *Plastic Materials*, 2d ed., Van Nostrand, New York, 1970.
Cagle, C. V.: *Adhesive Bonding, Techniques and Applications*, McGraw-Hill, New York, 1968.
Ives, G. C., J. A. Mead, and M. M. Riley: *Handbook of Plastics Test Methods*, The Plastics Institute, London, 1971.
Lever, A. E., and J. A. Rhys: *The Properties and Testing of Plastics Materials*, 3d ed., Temple, London, 1968.
Schmitz, J. V. (ed.): *Testing of Polymers*, Vols. I and II, Interscience-Wiley, New York, 1965.
Shultz, J.: *Polymer Materials Science*, Prentice-Hall, Englewood Cliffs, N. J., 1974.
Skeist, I.: *Plastics in Buildings*, Reinhold, New York, 1966.

APPENDIXES

THE INTERNATIONAL SYSTEM OF UNITS (SI)

A.1 GENERAL FEATURES

Created by international agreement in 1960, the International System of Units (known as "SI" in all languages) has become the one worldwide measurement system. Its basis is the centuries-old metric system, which means that many of its units were already widely used, and also that various compromises had to be made. On balance it is far superior to all earlier measurement systems, particularly to the heterogeneous collection of units used in the United States. The principal advantages of SI are as follows:

1. For a particular physical quantity, only one single unit is used.
2. Standard symbols are used in all countries and with all languages.
3. The system is coherent, which means that the factor 1.0 replaces a multiplicity of conversion factors.
4. SI is based on the decimal system, i.e., the number system in common use throughout the world.

Units SI includes three general classes of units, namely base units, supplementary units, and derived units. Some of the latter have special names, and some do not. Table A.1 shows the base units and supplementary units, and Table A.2 the derived units with special names. The conversion table in Appendix B includes many derived units without special names as well, and others may be formed as needed.

Prefixes In order to form units larger or smaller than the units discussed above, prefixes corresponding to powers of 10 may be attached to them. These prefixes are shown in Table A.3. It should be noted that the prefixes h, da, d, and c are generally avoided in technical work.

Table A.1 SI base and supplementary units

Quantity	Name	Symbol	Class
Length	meter	m	
Mass	kilogram	kg	
Time	second	s	Base
Electric current	ampere	A	units
Temperature	kelvin	K	
Amount of substance	mole	mol	
Luminous intensity	canadela	cd	
Plane angle	radian	rad	Supplementary
Solid angle	steradian	sr	units

Table A2 Derived SI units with special names

Quantity	Name	Symbol	Definition
Frequency	hertz	Hz	s^{-1}
Force	newton	N	$kg \cdot m/s^2$
Pressure, stress	pascal	Pa	N/m^2
Energy, work, etc.	joule	J	$N \cdot m$
Power	watt	W	J/s
Electrical charge, etc.	coulomb	C	$A \cdot s$
Electric potential	volt	V	W/A
Electric capacitance	farad	F	C/V
Electric resistance	ohm	Ω	V/A
Electric conductance	siemens	S	A/V
Magnetic flux, etc.	weber	Wb	$V \cdot s$
Magnetic flux density	tesla	T	Wb/m^2
Inductance	henry	H	Wb/A
Luminous flux	lumen	lm	$cd \cdot sr$
Illuminance	lux	lx	lm/m^2
Activity (radioactive)	becquerel	Bq	s^{-1}
Absorbed dose	gray	Gy	J/kg
Dose equivalent	sievert	Sv	J/kg
Temperature interval	degree Celsius	°C	K
Temperature	degree Celsius	°C	$K - 273.15$

References For detailed and authoritative information, the latest editions of the following publications may be consulted:

ISO 1000: "SI Units and Recommendations for the Use of Their Multiples and of Certain Other Units"
NBS Special Publication 330: "The International System of Units (SI)"
ANSI Z 210.1: "American National Standard for Metric Practice"
ASTM E 380: "Standard for Metric Practice"
ASTM E 621: "Practice for Use of Metric (SI) Units in Building Design and Construction"
ISO 2955: "Information Processing—Representation of SI and Other Units for Use in Systems with Limited Character Sets"

Table A.3 SI prefixes

Prefix	Symbol	Factor
exa	E	10^{18}
peta	P	10^{15}
tera	T	10^{12}
giga	G	10^9
mega	M	10^6
kilo	k	10^3
hecto[†]	h	10^2
deka[†]	da	10^1
deci[†]	d	10^{-1}
centi[†]	c	10^{-2}
milli	m	10^{-3}
micro	μ	10^{-6}
nano	n	10^{-9}
pico	p	10^{-12}
femto	f	10^{-15}
atto	a	10^{-18}

[†]Avoid in technical use.

A.2 SPECIAL NOTES

The use of SI units for certain quantities deserves particular attention. One important aspect of SI is that it allows, and indeed demands, a clear distinction between the quantities mass and force; in this respect it differs from previously used "gravitational" systems. Other noteworthy aspects have to do with the fact that, for reasons of convenience, a few units that do not really fit into the system are nevertheless allowed to be used as part of SI or in connection with it.

Mass The mass of a body is independent of the body's location. Whether the body is at the equator or at a pole, submerged in water, resting on the surface of the moon, or orbiting around the earth does not affect its mass. In common parlance, the word "weight" is often misused to mean "mass." The unit of mass is the kilogram (kg), an SI base unit, equal to the mass of the international prototype. The prefix "kilo-" was retained because the kilogram is so widely known and used. In forming larger or smaller units, the prefixes are attached to the word "gram." A megagram (1 Mg = 1000 kg) may also be called a "ton" (t) in commercial contexts, but preferably not in technical work.

Density The mass per unit volume of a material is known as that material's density. The unit for density is kilograms per cubic meter (kg/m^3), but as these values are usually rather large, a more convenient unit is Mg/m^3. It happens that water in its densest state (at 4°C) has a density very nearly equal to 1 Mg/m^3 (actually 0.999 979 Mg/m^3). The ratio of the density of a given material to this reference density of water is called the "relative density" of the material. The term "specific gravity" is still widely used for this quantity, but it is better avoided because the concept is not related to gravity.

Force and weight If a force F imparts an acceleration a to a body of mass m, then according to Newton's second law $F = m \cdot a$. The *force* that imparts an acceleration of 1 m/s^2 to a mass of 1 kg is defined as a newton ($1 \text{ N} = 1 \text{ kg} \cdot \text{m/s}^2$). This conforms with the coherent nature of SI. The gravitational attraction of the earth gives a free-falling body an acceleration g, a value that varies by about 0.5 percent over the earth's surface but has been given the standard value of $9.806\,650 \text{ m/s}^2$. Thus a force of $9.806\,650 \text{ N}$ is required to support a mass of 1 kg; this is known as the *weight* of the body. The relation is somewhat confused by the fact that commonly the mass of a body is determined by "weighing," which may be accomplished by using a balance that compares the gravitational attraction of one mass with that of a known one, or by using a spring scale that really measures weight but is calibrated to indicate units of mass.

In both the metric and the inch-pound system, some force units had been defined as those giving a unit mass an acceleration of g. These units, known as the kilogram-force (or kilopond) and pound-force respectively, of course are incompatible with SI.

Moment Torque and bending moment are obtained as vector cross products of a position vector and a force. Their magnitudes are often expressed in terms of newton meters ($\text{N} \cdot \text{m}$). When moments turn through rotations expressed in radians (rad), work is done and expressed in joules, which also equal newton meters. Unless the radian (a supplementary unit) is considered dimensionless, this results in some confusion in dimensions. Therefore it is better practice to express moments in units of $\text{N} \cdot \text{m/rad}$, even if this appears awkward at times. The question still awaits official resolution.

Plane angle The unit of plane angle is the radian, defined as the angle between two radii of a circle that cut off on the circumference an arc equal in length to the radius. Because it could not be decided whether this is a base unit or a derived unit, it has been termed a supplementary unit. The degree, a non-SI unit, may also be used.

Energy and power The unit of energy and work, whether mechanical, electrical, thermal, or chemical, is the joule ($1 \text{ J} = 1 \text{ N} \cdot \text{m}$). Likewise, the unit for all types of power is the watt ($1 \text{ W} = 1 \text{ J/s}$).

Pressure or stress The unit of pressure or stress is the pascal ($1 \text{ Pa} = 1 \text{ N/m}^2$). As this is a rather small unit, prefixes are usually attached. Thus pressures are generally given in kPa, stresses in MPa, and moduli of elasticity in GPa. Normal atmospheric pressure is 101.325 kPa.

Time The unit of time is the second (s). Prefixes may be attached in the usual manner (ms, ks, Ms, etc.). Because of the use of clocks and calendars, it is sometimes convenient and permissible to use traditional non-SI units of time, such as the minute (min), hour (h), day (d), or year (a). Usually, times should be converted to seconds before they are used in computations.

Hardness Various indentation hardness values, such as Rockwell, Brinell, and Vickers numbers, are based on a force expressed in kilograms-force, an obsolete unit. Yet these values are so widely known and used that it has been internationally decided to retain them. It must be remembered, however, that such a force value really represents the weight of a mass of the stated number of kilograms. When testing machines calibrated in kilonewtons are used, it is permissible to multiply the mass value in kilograms by the rounded value 9.81 m/s^2 in order to obtain the force in newtons.

Toughness The *modulus of toughness* is expressed in units of energy per unit volume, or J/m^3. This is dimensionally equal to pascals. *Fracture toughness* is given in units of stress times the square root of a crack length, that is, usually in MPa $\cdot$ m$^{1/2}$ or MN $\cdot$ m$^{-3/2}$.

Temperature The SI unit of thermodynamic temperature is the kelvin (K), a base unit. The degree Celsius ($^\circ$C), a derived unit, is also used. One degree Celsius (formerly also called centigrade) of *temperature interval* equals exactly one kelvin. The zero point on the Celsius scale corresponds to 273.15 K, by definition. The *coefficient of linear thermal expansion* is expressed as strain per temperature interval; since the strains per kelvin are quite small, a convenient unit is MK^{-1}, equal to 10^{-6}/K or (μm/m)/K.

Volume The unit of volume, the cubic meter (m^3), is too large for many applications, while a cubic millimeter (mm^3) is too small. Therefore it is proper to use the cubic decimeter (dm^3) and the cubic centimeter (cm^3) as well. If liquids are measured, the cubic decimeter may be called a liter (L) and the cubic centimeter a milliliter (mL). The symbol for liter is actually l (lowercase), but the capital L is permissible because the former is so easily confused with a digit 1 or the letter I.

Cross-sectional properties Dimensions of specimens are obtained in millimeters. Cross-sectional *areas* may thus conveniently be computed in square millimeters, but for larger specimens square meters may be more appropriate (1 mm^2 = 10^{-6} m^2). The use of centimeters should be avoided, as the resulting powers of 10 (in contrast to the case of volumes) will not be the desired multiples of 3. The same argument pertains to *moments of inertia of area*, for which 10^{-6} m^4 (10^6 mm^4) often is a convenient unit. *Section moduli* can then be expressed in 10^{-6} m^3 (10^3 mm^3).

Viscosity The unit of *dynamic viscosity* is the Pa $\cdot$ s; formerly the poise (P) was commonly used. One P equals 10^{-1} Pa $\cdot$ s, and the often used centipoise (cP) is 1 mPa $\cdot$ s. The SI unit for *kinematic viscosity* is the m^2/s, replacing the stokes (St); 1 St equals 10^{-4} m^2/s, and a centistokes (cSt) is 1 mm^2/s. The dynamic viscosity may be obtained by multiplying the kinematic viscosity by the density of the material in kg/m^3. Multiplying the kinematic viscosity in mm^2/s by the density in Mg/m^3 gives the dynamic viscosity in mPa $\cdot$ s.

A.3 STYLE AND USAGE

If confusion is to be avoided, it is important to observe various internationally accepted rules of style. This applies particularly to the *symbols*, for which the following specific rules should be memorized:

1. Symbols are always printed in roman (upright) type, regardless of the style of the surrounding text: *2.5* kg, not *2.5 kg.*
2. Capital and lowercase letters are never interchanged (unless, as with a computer, a full font is unavailable, in which case special rules apply[†]): mm, not MM.
3. Symbols are unaltered in the plural: kg, not kgs.
4. No space is used between a prefix and a unit symbol: MPa, not M Pa.
5. Symbols are not abbreviations, and they must not be replaced by abbreviations: s, not sec.; A, not amp.
6. For products, a raised dot is used (or if necessary, a dot on line): kN · m or kN . m.
7. For quotients, one solidus (/), a fraction line, or a negative power is used: kg/(m · s) or $\dfrac{kg}{m \cdot s}$ or kg · m^{-1} · s^{-1}, not kg/m/s.

Concerning *numbers*, the most important rule is that either a period (.) or a comma (,) may be used as a decimal marker, depending on local custom. This means that neither should be used to separate groups of digits, for which purpose a space may be left (e.g., 1 234 000 J, 0.000 123 s). It is preferable in most cases to use numbers between 0.1 and 1000 and use prefixes to change the sizes of the units (e.g., 1.234 MJ, 123 μs). For numbers smaller than unity, the leading zero is never omitted: 0.123, not .123.

There should normally be a space between a number and its unit (e.g., 12.3 m/s), except for degrees Celsius (e.g., 12°C) and plane angles expressed in degrees, minutes, and seconds (e.g., 45°). When a number and unit are used in an adjectival position, however, they should be joined by a hyphen, not separated by a space (e.g., 200-mm cube).

The unit *names* are a part of the language with which they are used. In English they are not capitalized (the *symbols* derived from proper names are capitalized, but not the names: Pa, pascal), and the plural is formed in the usual way (newtons, henries), although names ending in an s sound remain unchanged in the plural (60 hertz). For products, a space or a hyphen may be used; for quotients, the word "per": newton meters (or newton-meters) and meters per second. For powers, the words "squared" or "cubed" are used following the unit, except in the case of areas and volumes: seconds squared, but cubic meters. For complicated compound units, symbols are preferred to names. For a few units, different English spellings are used: meter and metre, liter and litre, ton and tonne, it is desirable to be consistent, however.

Pronunciation, of course, is also language related, but prefixes are always accented on the first syllable (including *ki*lometers). In English, pascal rhymes with "rascal," joule sounds like "jewel," giga like "jig," and pico like "pique."

A.4 COMPUTATIONS

SI makes computations extremely simple, particularly since cumbersome conversions among various Btus, calories, ft-lb, ergs, kW · h, ft-poundals, tons (equivalent of

[†]See ISO 2955.

energy is expressed in J, kJ, MJ, etc., and it is not necessary to memorize the relations among various Btus, calories, ft-lbs, ergs, kW · h, ft-poundals, tons (equivalent of TNT), therms, and the like. There is one rule of good practice, however, that should be observed: In making computations involving multiples of units, the prefixes should always be replaced by powers of 10 before the computation is started, and one prefix should be selected at the end to replace the resulting power of 10. The prefixes are aids to the language, not to the mathematics.

For example, if a stress σ is to be computed by dividing a force of 12 kN by a cross-sectional area of 40 mm^2, the computation takes the form

$$\sigma = \frac{12 \cdot 10^3 \text{ N}}{40 \cdot 10^{-6} \text{ m}^2} = 0.3 \cdot 10^9 \frac{\text{N}}{\text{m}^2} = 300 \cdot 10^6 \text{ N/m}^2 = 300 \text{ MN/m}^2 = 300 \text{ MPa}$$

When obsolete units must be converted to SI units, the "ratio" method is quick and reliable. It works in that dimensionless ratios with a value of unity are formed. For example, let a stress σ of 36 000 psi (lbf/in^2) be converted to pascals. From a table such as the one in Appendix B, it is seen that 1 lbf = 4.448 222 N, and that 1 in = 0.025 400 m. Then

$$\sigma = 36\ 000 \ \frac{\text{lbf}}{\text{in}^2} \left[\frac{4.448\ 222 \text{ N}}{1 \text{ lbf}} \right] \left[\frac{(1 \text{ in})^2}{(0.025\ 400 \text{ m})^2} \right] = 248.21 \text{ MPa}$$

Neither the value to be converted nor any of the conversion factors should be rounded before the operation, or serious inaccuracies may occur. The result, however, should be rounded to the approximate intended accuracy of the original value. This may require some judgment. In the example above, it could be that a stress of 36 000 psi rather than 36 001 psi was given. The equivalent then is 248.21 MPa. More likely, though, the stress was to be distinguished from one of, say, 36 500 psi, in which case the answer should be rounded off to 250 MPa. The rounding of converted values is addressed in some detail in ASTM E 380.

USEFUL FACTS

B.1 CONVERSION FACTORS

To convert	from	to	multiply by[†]
Quantity:	*Non-SI unit* =	*SI unit* ×	*Conversion factor*
Acceleration	ft/s^2	m/s^2	3.048 000 E – 01
Area	in^2	mm^2	6.451 600 E + 02
	ft^2	m^2	9.290 304 E – 02
Corrosion rate			
mass per area per time	oz/ft^2/year	g/(m^2 · year)	3.051 517 E + 02
thickness per unit time	mil/year	μm/year	2.540 000 E + 01
Creep rate	in/in/h	m/(m · s)	2.777 778 E – 04
Density	lbm/in^3	kg/m^3	2.767 990 E + 04
(mass per unit volume)	lbm/ft^3	kg/m^3	1.601 846 E + 01
Energy	Btu	kJ	1.055 056 E + 00
(including heat, work,	in-lbf	J	1.129 848 E – 01
etc.)	ft-lbf	J	1.355 818 E + 00
Force	pdl	N	1.382 550 E – 01
	lbf	N	4.448 222 E + 00
	kgf	N	9.806 650 E + 00
Fracture toughness	ksi $\sqrt{\text{in}}$	MPa · m$^{1/2}$	1.098 843 E + 00
Hardness	kgf/mm^2	MPa	9.806 650 E + 00
	kgf	N	9.806 650 E + 00
Linear measure	angstrom, Å	nm	1.000 000 E – 01
(including width,	microinch	μm	2.540 000 E – 02
diameter, height,	mil	μm	2.540 000 E + 01
length, mesh size,	in	mm	2.540 000 E + 01
etc.)	ft	m	3.048 000 E – 01
Mass	oz (avdp)	g	2.834 952 E + 01
	lbm (avdp)	kg	4.535 924 E – 01
	slug	kg	1.459 388 E + 01

To convert	from	to	multiply by[†]
Quantity:	*Non-SI unit* =	*SI unit* ×	*Conversion factor*
Modulus of elasticity	lbf/in^2	GPa	6.894 757 E – 06
Moment of force, including bending moment, torque	lbf-in lbf-ft	N · m/rad N · m/rad	1.129 848 E – 01 1.355 818 E + 00
Moment of inertia of area	in^4	mm^4	4.162 314 E + 05
Momentum, linear	lbm-ft/s	kg · m/s	1.382 550 E – 01
Momentum, angular	lbm ft^2/s	kg · m^2/(rad · s)	4.214 011 E – 02
Power	Btu/min ft-lbf/min hp (550 ft-lbf/s) hp (electric)	kW W kW kW	1.758 426 E – 02 2.259 697 E – 02 7.456 999 E – 01 7.460 000 E – 01
Section modulus	in^3	mm^3	1.638 706 E + 04
Specific area	ft^2/lbm	m^2/kg	2.048 160 E – 01
Stress	lbf/in^2 ksi	MPa MPa	6.894 757 E – 03 6.894 757 E + 00
Temperature	°F °C °F	°C K K	(5/9)(°F – 32) (°C + 273.15) (5/9)(°F + 459.67)
Temperature interval	°F	°C or K	5/9
Thermal expansion, linear coefficient	in/in/°F	K^{-1}	1.800 000 E + 00
Velocity, linear	in/s ft/s in/min ft/min	mm/s m/s mm/s m/s	2.540 000 E + 01 3.048 000 E – 01 4.233 333 E – 01 5.080 000 E – 03
Velocity, angular	rev/min	rad/s	1.047 198 E – 01
Viscosity, dynamic	lbf-s/in^2 P (poise)	Pa · s Pa · s	6.894 757 E + 03 1.000 000 E – 01
Viscosity, kinematic	in^2/s St (stokes)	mm^2/s m^2/s	6.451 600 E + 02 1.000 000 E – 04
Volume	in^3 ft^3 yd^3 gal (fluid)	mm^3 m^3 m^3 dm^3, L	1.638 706 E + 04 2.831 685 E – 02 7.645 549 E – 01 3.785 412 E + 00
Water-cement ratio	gal/ft^3-sack	(mass ratio)	8.861 702 E – 02

[†]The non-SI unit equals the SI unit times the conversion factor; e.g., 1.0 ft/s^2 = 0.3048 m/s^2.

B.2 SELECTED CHEMICAL ELEMENTS

Element	Symbol	Common valance	Atomic number	Atomic mass	Density, Mg/m^3	Melting point, $^\circ C$
Aluminum	Al	+3	13	26.98	2.70	660
Beryllium	Be	+2	4	9.01	1.85	1279
Cadmium	Cd	+2	48	112.40	8.64	321
Carbon (graphite)	C	+4	6	12.01	2.25	
Chlorine	Cl	−1	17	35.46	−	−101
Chromium	Cr	+3	24	52.01	7.20	1878
Copper	Cu	+1	29	63.54	8.95	1084
Gold	Au	+1	79	197.00	19.32	1064
Hydrogen	H	−1	1	1.01	−	−259
Iron	Fe	+2	26	55.85	7.87	1538
Lead	Pb	+2	82	207.21	11.34	327
Magnesium	Mg	+2	12	24.32	1.74	650
Manganese	Mn	+2	25	54.94	7.20	1247
Mercury	Hg	+2	80	200.61	13.55	−38
Nickel	Ni	+2	28	58.71	8.90	1455
Nitrogen	N	+5	7	14.01	−	−210
Oxygen	O	−2	8	16.00	−	−219
Platinum	Pt	+2	78	195.09	21.45	1772
Potassium	K	+1	19	39.10	0.86	64
Silicon	Si	+4	14	28.09	2.42	1412
Silver	Ag	+1	47	107.88	10.49	962
Sodium	Na	+1	11	22.99	0.97	98
Sulfur	S	−2	16	32.07	2.07	119
Tin	Sn	+4	50	118.70	7.30	232
Zinc	Zn	+2	30	65.38	7.14	419
Zirconium	Zr	+4	40	91.22	6.49	1855

B.3 STANDARD SIEVE SERIES

A standard series of wire mesh sieves, used for grading aggregates, has been adopted by ISO, ANSI, and ASTM (ASTM E 11). The sieve opening sizes are determined by starting with a 1.0-mm opening and then multiplying or dividing by $\sqrt[4]{2}$ in order to obtain successively larger or smaller sizes. The table below lists the various sizes along with their equivalents in inches and "Tyler" designations.

Size, mm	Size, in[†]	Tyler no.	Size, mm	Size, in	Tyler no.
107.6	4.24		1.68	0.0661	12
90.5	$3\frac{1}{2}$		1.41	0.0555	14
76.1	3		1.19	0.0469	16
64	$2\frac{1}{2}$		1.00	0.0394	18
53.8	2.12		0.841	0.0331	20
45.3	$1\frac{3}{4}$		0.707	0.0278	25
38.1	$1\frac{1}{2}$		0.595	0.0234	30
32	$1\frac{1}{4}$		0.500	0.0197	35
26.9	1.06		0.420	0.0165	40
22.6	$\frac{7}{8}$		0.354	0.0139	45
19.0	$\frac{3}{4}$		0.297	0.0117	50
16.0	$\frac{5}{8}$		0.250	0.0098	60
13.5	0.530		0.210	0.0083	70
11.3	$\frac{7}{16}$		0.177	0.0070	80
9.51	$\frac{3}{8}$		0.149	0.0059	100
8.0	$\frac{5}{16}$		0.125	0.0049	120
6.73	0.265		0.105	0.0041	140
5.66	0.223	$3\frac{1}{2}$	0.088	0.0035	170
4.76	0.187	4	0.074	0.0029	200
4.0	0.157	5	0.063	0.0025	230
3.36	0.132	6	0.053	0.0021	270
2.83	0.111	7	0.044	0.0017	325
2.38	0.0937	8	0.037	0.0015	400
2.00	0.0787	10			

[†]Special sizes of 4, 2, 1, and $\frac{1}{2}$ in are also available.

B.4 FORMULAS FROM MECHANICS

Various formulas useful in materials testing are listed and illustrated on the following pages. They are intended to refresh the memory of those already familiar with the principles of mechanics of materials, not to substitute for study of that subject matter. The usual symbols are used in the figures and equations.

Simple stress conditions Stress variations over the cross sections of slender structural members under linearly elastic conditions are indicated.

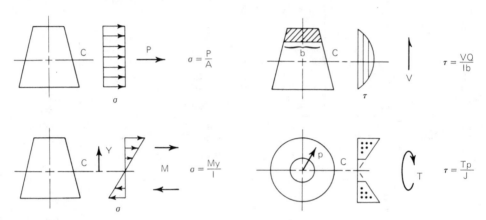

Cross sections For common solid cross sections, the area, the moment of inertia about the horizontal centroidal axis, and the polar moment of inertia about the centroid are shown.

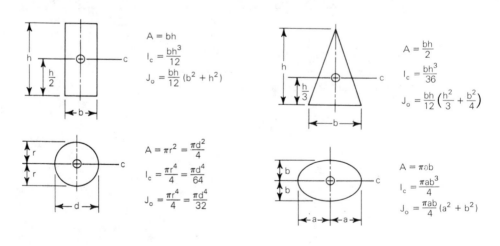

Beams The loading arrangements shown are most common in materials testing.

Simple beam with concentrated load at center:

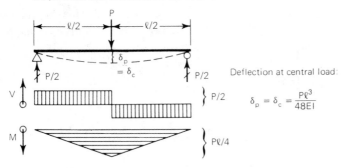

Deflection at central load:

$$\delta_p = \delta_c = \frac{P\ell^3}{48EI}$$

Simple beam with two equal loads symmetrically placed:

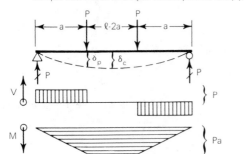

Deflection at load:

$$\delta_p = \frac{Pa^2}{6EI}(3\ell\text{-}4a)$$

Deflection at center:

$$\delta_c = \frac{Pa}{24EI}(3\ell^2\text{-}4a^2)$$

Cantilever beam with concentrated load at free end:

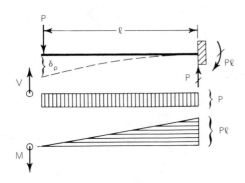

Deflection at load:

$$\delta_p = \frac{P\ell^3}{3EI}$$

Columns Euler's buckling equation applies to slender columns that fail by elastic instability. The factor k, shown for some idealized cases, accounts for end conditions.

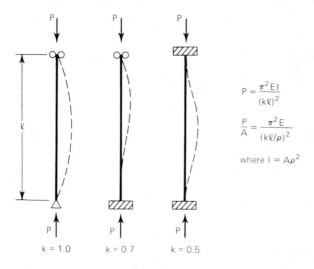

$$P = \frac{\pi^2 EI}{(k\ell)^2}$$

$$\frac{P}{A} = \frac{\pi^2 E}{(k\ell/\rho)^2}$$

where $I = A\rho^2$

$k = 1.0 \qquad k = 0.7 \qquad k = 0.5$

Mohr's circle The diagram illustrates the construction of Mohr's circle for a plane stress condition at a point.

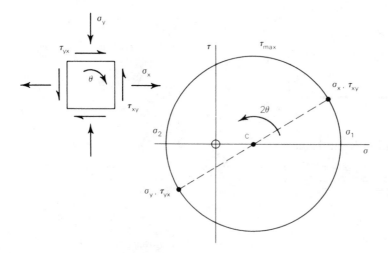

SOURCES OF INFORMATION

C.1 ADDRESSES OF ORGANIZATIONS

The organizations listed below have a major interest in engineering materials, testing, or standards. All publish literature in their areas of interest. The addresses shown were obtained in 1981.

AA Aluminum Association, 818 Connecticut Ave., N.W., Washington DC 20006

AASHTO American Association of State Highway and Transportation Officials, 444 N. Capitol, Washington DC 20001

ACI American Concrete Institute, P.O. Box 19150, Redford Station, Detroit MI 48219

AI The Asphalt Institute, Asphalt Institute Building, College Park MD 20740

AIME American Institute of Mining, Metallurgical and Petroleum Engineers, 345 E. 47th St., New York NY 10017

AISC American Institute of Steel Construction, 400 North Michigan Avenue, Chicago IL 60611

AISI American Iron and Steel Institute, 1000 16th St., N.W., Washington DC 20036

ANMC American National Metric Council, 1625 Massachusetts Ave., N.W., Washington DC 20036

ANSI American National Standards Institute, 1430 Broadway, New York NY 10018

APA American Plywood Association, 1119 A Street, Tacoma WA 98401

API American Petroleum Institute, 2101 L St., N.W., Washington DC 20037

AREA American Railway Engineering Association, 59 E. Van Buren St., Chicago IL 60605

ASCE American Society of Civil Engineers, 345 East 47th St., New York NY 10017

ASME American Society of Mechanical Engineers, 345 East 47th St., New York NY 10017

ASTM American Society for Testing and Materials, 1916 Race St., Philadelphia PA 19103

AWPI American Wood Preservers Institute, 1651 Old Meadow Road, McLean VA 22101

BIA Brick Institute of America, 1750 Old Meadow Road, McLean VA 22102

BIPM International Bureau of Weights and Measures, Pavillon de Breteuil, F-92310 Sèvres, France

CSA Canadian Standards Association, 178 Rexdale Blvd., Rexdale, Ont. M9W 1R3, Canada

ISO International Organization for Standardization, 1 rue de Varembe, CH-1211 Geneva, Switzerland

NBS National Bureau of Standards, Washington DC 20234

NFPA	National Forest Products Association, 1619 Massachusetts Ave. N.W., Washington DC 20036
PCA	Portland Cement Association, Old Orchard Road, Skokie IL 60076
PIA	Plastics Institute of America, Stevens Institute of Technology, Hoboken NJ 07030
SAE	Society of Automotive Engineers, 400 Commonwealth Dr., Warrendale PA 15096
SESA	Society for Experimental Stress Analysis, 14 Fairfield Dr., Brookfield Center CT 06805
SPI	The Society of the Plastics Industry, 355 Lexington Ave., New York NY 10017
USMB	United States Metric Board, 1815 N. Lynn, Arlington VA 22209

C.2 SELECTED ASTM STANDARDS

The following is an alpha-numeric list of ASTM specifications, methods, and definitions extracted from the index to the *1980 Annual Book of ASTM Standards*. The designations of the year of adoption or latest revision have been omitted in order to minimize obsolescence: the current edition of the source should be consulted. An asterisk preceding a designation indicates that the standard has been approved as an American National Standard by ANSI. Decimal numbers following a title show sections of this book in which the standard is referenced (P means "Problem").

A. Ferrous Metals
- *A 36* Structural Steel, Spec. for, 8.2, 11.7, 17.4
- *A 48* Gray Iron Castings, Spec. for, 8.2, 11.3, 17.3
- *A 327* Impact Testing of Cast Irons, 13.2
- *A 340* Magnetic Testing, Def. of Terms, Symbols, and Conversion Factors Relating to, 16.3
- *A 370* Mechanical Testing of Steel Products, Methods and Def. for, 17.7
- *A 615M* Deformed and Plain Billet-Steel Bars for Concrete Reinforcement, Spec. for, 8.2, 11.7
- A 644 Iron Castings, Def. of Terms Relating to

B. Nonferrous Metals
- *B 106* Flexivity of Thermostat Metals, Test for, 11.7
- *B 593* Bending Fatigue Tests for Copper-Alloy Spring Materials, Rec. Practice for Conducting, 14.3

C. Cementitious, Ceramic, Concrete, and Masonry Materials
- *C 31* Making and Curing Concrete Test Specimens in the Field, 5.4, 9.2, 18.3, 18.5
- *C 33* Concrete Aggregates, Spec. for, 18.2
- *C 39* Compressive Strength of Cylindrical Concrete Specimens, Test for, 9.3, 9.4, 9.P, 18.3, 18.5
- *C 42* Drilled Cores and Sawed Beams of Concrete, Obtaining and Testing, 9.6
- *C 43* Structural Clay Products, Def. of Terms Relating to
- *C 62* Building Brick (Solid Masonry Units Made from Clay or Shale), Spec. for
- *C 67* Brick and Structural Clay Tile, Sampling and Testing, 5.4, 9.2, 11.3
- *C 78* Flexural Strength of Concrete (Using Simple Beam with Third-Point Loading), Test for, 11.3
- *C 88* Soundness of Aggregates by Use of Sodium Sulfate or Magnesium Sulfate, Test for, 19.3
- *C 109* Compressive Strength of Hydraulic Cement Mortars (Using 2-in. or 50-mm Cube Specimens), Test for, 9.2, 18.2
- *C 115* Fineness of Portland Cement by the Turbidimeter, Test for, 18.2
- *C 117* Materials Finer Than 76-μm (No. 200) Sieve in Mineral Aggregates by Washing, Test for, 19.3
- *C 125* Concrete and Concrete Aggregates, Def. of Terms Relating to
- *C 127* Specific Gravity and Absorption of Coarse Aggregate, Test for, 19.3
- *C 128* Specific Gravity and Absorption of Fine Aggregate, Test for, 19.3

*C 131 Resistance to Abrasion of Small Size Coarse Aggregate by Use of the Los Angeles Machine, Test for, 12.2, 19.3

*C 133 Cold Crushing Strength and Modulus of Rupture of Refractory Brick and Shapes, Tests for, 9.2

*C 136 Sieve or Screen Analysis of Fine and Coarse Aggregates, Test for, 3.P, 19.3

*C 138 Unit Weight, Yield, and Air Content (Gravimetric) of Concrete, Test for, 18.6

*C 140 Concrete Masonry Units, Sampling and Testing, 5.4

*C 143 Slump of Portland Cement Concrete, Test for, 18.3, 18.4, 18.P

*C 151 Autoclave Expansion of Portland Cement, Test for, 18.2

*C 170 Compressive Strength of Natural Building Stone, Test for, 9.2

*C 173 Air Content of Freshly Mixed Concrete by the Volumetric Method, Test for, 18.6

*C 183 Hydraulic Cement, Sampling, 5.4

*C 188 Density of Hydraulic Cement, Test for, 18.2

*C 190 Tensile Strength of Hydraulic Cement Mortars, Test for, 8.2, 18.2

*C 191 Time of Setting of Hydraulic Cement by Vicat Needle, Test for, 18.2

*C 192 Making and Curing Concrete Test Specimens in the Laboratory, 9.3

*C 219 Hydraulic Cement, Def. of Terms Relating to

*C 231 Air Content of Freshly Mixed Concrete by the Pressure Method, Test for, 18.6

*C 289 Potential Reactivity of Aggregates (Chemical Method), Test for, 18.2

*C 293 Flexural Strength of Concrete (Using Simple Beam with Center-Point Loading), Test for, 11.3

*C 359 Early Stiffening of Portland Cement (Mortar Method), Test for, 18.2

*C 360 Ball Penetration in Fresh Portland Cement Concrete, Test for, 18.4

*C 441 Effectiveness of Mineral Admixtures in Preventing Excessive Expansion of Concrete Due to the Alkali-Aggregate Reaction, Test for, 18.2

*C 451 Early Stiffening of Portland Cement (Paste Method) Exposed to Sulfate, Test for, 18.2

*C 512 Creep of Concrete in Compression, Test for, 15.2

*C 822 Concrete Pipe and Related Products, Def. of Terms Relating to

 C 896 Clay Products, Def. of Terms Relating to

D. *Miscellaneous Materials*

*D 5 Penetration of Bituminous Materials, Test for, 19.3

*D 6 Loss on Heating of Oil and Asphaltic Compounds, Test for, 19.3

*D 8 Materials for Roads and Pavements, Def. of Terms Relating to, 19.1

*D 9 Timber, Def. of Terms Relating to, 20.1

*D 16 Paint, Varnish, Lacquer, and Related Products, Def. of Terms Relating to

*D 36 Softening Point of Bitumen (Ring-and-Ball Apparatus), Test for, 19.3

*D 70 Specific Gravity of Semi-Solid Bituminous Materials, Test for, 19.3

*D 75 Aggregates, Sampling, 3.P, 5.4

*D 92 Flash and Fire Points by Cleveland Open Cup, Test for, 19.3

*D 93 Flash Point by Pensky-Martens Closed Tester, Test for, 19.3

*D 95 Water in Petroleum Products and Bituminous Materials by Distillation, Test for, 19.3

*D 113 Ductility of Bituminous Materials, Test for, 19.3

*D 139 Bituminous Materials, Float Test for, 19.3

*D 143 Small Clear Specimens of Timber, Testing, 5.4, 5.P, 8.2, 9.2, 9.5, 10.6, 11.3, 11.6, 11.P, 12.3, 13.2 13.7, 20.4, 20.P

*D 198 Static Tests of Timbers in Structural Sizes, 9.2, 11.3, 20.4

*D 244 Emulsified Asphalts, Testing, 19.3

*D 245 Structural Grades and Related Allowable Properties for Visually Graded Lumber, Establishing, 20.1

*D 256 Impact Resistance of Plastics and Electrical Insulating Materials, Tests for, 13.2, 21.5

*D 402 Distillation of Cut-Back Asphaltic (Bituminous) Products, Test for, 19.3

 D 530 Hard Rubber Products, Testing, 12.2

*D 570 Water Absorption of Plastics, Test for, 21.5

 D 575 Rubber Properties in Compression, Tests for, 9.2, 9.4

*D 638 Tensile Properties of Plastics, Test for, 21.5, 21.P
*D 651 Tensile Strength of Molded Electrical Insulating Materials, Test for, 8.2, 21.P
*D 671 Flexural Fatigue of Plastics by Constant, Amplitude-of-Force, Test for, 14.2, 14.3
*D 695 Compressive Properties of Rigid Plastics, Test for, 21.5
*D 696 Coefficient of Linear Thermal Expansion of Plastics, Test for, 21.5
*D 747 Stiffness of Plastics by Means of a Cantilever Beam, Test for, 11.7
*D 785 Rockwell Hardness of Plastics and Electrical Insulating Materials, Test for, 21.5
*D 790 Flexural Properties of Plastics and Electrical Insulating Materials, Tests for, 21.5
*D 792 Specific Gravity and Density of Plastics by Displacement, Tests for, 21.5, 21.P
*D 883 Plastics, Def. of Terms Relating to, 21.1
*D 897 Tensile Properties of Adhesive Bonds, Test for, 21.4
*D 907 Adhesives, Def. of Terms Relating to
*D 946 Asphalt Cement for Use in Pavement Construction, Spec. for, 19.1
*D 977 Emulsified Asphalt, Spec. for, 19.2, 19.4
*D 1002 Strength Properties of Adhesives in Shear by Tension Loading (Metal-to-Metal), Test for, 21.4
*D 1037 Wood-Base Fiber and Particle Panel Materials, Evaluating the Properties of, 5.P
*D 1038 Veneer and Plywood, Def. of Terms Relating to, 20.6
*D 1043 Stiffness Properties of Plastics as a Function of Temperature of Means of a Torsion Test, Test for, 10.2, 10.3
*D 1075 Effect of Water of Cohesion of Compacted Bituminous Mixtures, Test for, 19.3
*D 1165 Domestic Hardwoods and Softwoods, Nomenclature of, 20.1
 D 1310 Flast Point of Liquids by Tag Open-Cup Apparatus, Test for, 19.3
*D 1554 Wood-Base Fiber and Particle Panel Materials, Def. of Terms Relating to
*D 1559 Resistance to Plastic Flow of Bituminous Mixtures Using Marshall Apparatus, Test for, 19.6, 19.P
*D 1560 Resistance to Deformation and Cohesion of Bituminous Mixtures by means of Hveem Apparatus, Tests for, 19.6
*D 1561 Compaction of Test Specimens of Bituminous Mixtures by Means of California Kneading Compactor, 19.6
*D 1600 Plastics, Abbreviations of Terms Relating to, 21.1
*D 1664 Coating and Stripping of Bitumen-Aggregate Mixtures, Test for, 19.3
*D 1754 Effect of Heat and Air on Asphaltic Materials (Thin-Film Oven Test), Test for, 19.3
*D 1761 Mechanical Fasteners in Wood, Testing, 5.P
*D 1860 Moisture and Creosote-Type Preservative in Wood, Test for, 20.5
 D 2026 Cutback Asphalt (Slow-Curing Type), Spec. for, 19.2, 19.4
*D 2027 Cutback Asphalt (Medium-Curing Type), Spec. for, 19.2, 19.4
*D 2028 Cutback Asphalt (Rapid-Curing Type), Spec. for, 19.2, 19.4
*D 2161 Conversion of Kinematic Viscosity to Saybolt Universal Viscosity or to Saybolt Furol Viscosity, 19.3
*D 2170 Kinematic Viscosity of Asphalts (Bitumens), Test for, 19.3
*D 2171 Viscosity of Asphalts by Vacuum Capillary Viscometer, Test for, 19.3
*D 2395 Specific Gravity of Wood and Wood-Base Materials, Tests for, 20.3
 D 2397 Cationic Emulsified Asphalt, Spec. for, 19.2, 19.4
*D 2398 Softening Point of Bitumen in Ethylene Glycol (Ring-and Ball), Test for, 19.3
*D 2555 Clear-Wood Strength Values, Establishing, 20.4, 20.P
*D 2718 Plywood in Rolling Shear (Shear in Plane of Plies), Testing, 20.6
*D 2719 Plywood in Shear Through-the-Thickness, Testing, 20.6
*D 2872 Effect of Heat and Air on a Moving Film of Asphalt (Rolling Thin-Film Oven Test), Test for, 19.3
 D 2915 Allowable Properties for Grades of Structural Lumber, Evaluating, 3.1
*D 2990 Tensile, Compressive, and Flexural Creep and Creep Rupture of Plastics, Tests for, 15.2
 D 3029 Impact Resistance of Rigid Plastic Sheeting or Parts by Means of a Tup (Falling Weight), Test for, 13.2
*D 3043 Plywood in Flexure, Testing, 20.6

*D 3044 Shear Modulus of Plywood, Test for, 20.6
*D 3142 Specific Gravity or API Gravity of Liquid Asphalts by Hydrometer Method, Test for, 19.3
*D 3381 Viscosity-Graded Asphalt Cement for Use in Pavement Construction, Spec. for, 19.2, 19.4
*D 3499 Toughness of Plywood, Test for, 20.6
*D 3500 Plywood in Tension, Testing, 20.6
*D 3501 Plywood in Compression, Testing, 20.6
*D 3515 Hot-Mixed, Hot-Laid Bituminous Paving Mixtures, Spec. for, 19.6

E. *Miscellaneous Subjects*
  *E 4   Load Verification of Testing Machines, 6.6, 6.P
  *E 6   Methods of Mechanical Testing, Def. of Terms Relating to, 2.3
  *E 8   Tension Testing of Metallic Materials, 8.2, 8.3, 8.4, 8.P
  *E 9   Compression Testing of Metallic Materials at Room Temperature, 9.2, 9.4
  *E 10   Brinell Hardness of Metallic Materials, Test for, 12.3, 12.6, 12.8, 12.P
  *E 11   Wire-Cloth Sieves for Testing Purposes, Spec. for, 3.P, 18.2
   E 12   Density and Specific Gravity of Solids, Liquids, and Gases, Def. of Terms Relating to, 2.1
  *E 18   Rockwell Hardness and Rockwell Superficial Hardness of Metallic Materials, Tests for, 12.3, 12.4, 12.7, 12.P
   E 23   Notched Bar Impact Testing of Metallic Materials, 13.2, 13.4, 13.P
   E 29   Indicating Which Places of Figures Are to Be Considered Significant in Specified Limiting Value, Rec. Practice for
   E 44   Heat Treatment of Metals, Def. of Terms Relating to
  *E 74   Force-Measuring Instruments for Verifying the Load Indication of Testing Machines, Calibration of, 6.6
  *E 92   Vickers Hardness of Metallic Materials, Test for, 12.3
  *E 94   Radiographic Testing, Rec. Practice for, 16.2
   E 102 Saybolt Furol Viscosity of Bituminous Materials at High Temperatures, Test for, 19.3
  *E 139 Creep, Creep-Rupture, and Stress-Rupture Tests of Metallic Materials, Rec. Practice for Conducting, 15.2, 15.P
  *E 140 Standard Hardness Conversion Tables for Metals (Relationship Between Brinell Hardness, Vickers Hardness, Rockwell Hardness, Rockwell Superficial Hardness, and Knoop Hardness), 12.9
  *E 150 Creep and Creep-Rupture Tension Tests of Metallic Materials Under Conditions of Rapid Heating and Short Times, Rec. Practice for Conducting, 15.2, 15.P
  *E 165 Liquid Penetrant Inspection Method, Rec. Practice for, 16.8, 20.5
  *E 178 Outlying Observations, Rec. Practice for Dealing with, 3.2
  *E 206 Fatigue Testing and the Statistical Analysis of Fatigue Data, Def. of Terms Relating to, 3.1, 14.2
  *E 268 Electromagnetic Testing, Def. of Terms Relating to, 16.3
  *E 269 Magnetic Particle Inspection, Def. of Terms Relating to, 16.4
  *E 284 Appearance of Materials, Def. of Terms Relating to
   E 344 Temperature Measurement, Def. of Terms Relating to
  *E 380 Metric Practice, 1.P, A.1
  *E 399 Plane-Strain Fracture Toughness of Metallic Materials, Test for, 14.3
  *E 448 Scleroscope Hardness Testing of Metallic Materials, Rec. Practice for, 12.3, 12.P
   E 456 Statistical Methods, Def. of Terms Relating to, 3.1
  *E 466 Constant Amplitude Axial Fatigue Tests of Metallic Materials, Rec. Practice for, 14.2, 14.3, 14.5, 14.P
   E 500 Ultrasonic Testing, Def. of Terms Relating to, 16.6
   E 513 Constant-Amplitude Low-Cycle Fatigue Testing, Def. of Terms Relating to, 14.2
   E 529 Flexural Tests on Beams and Girders for Building Construction, 7.P
  *E 558 Torsion Testing of Wire, 10.2

E 586 Gamma and X-Radiography, Def. of Terms Relating to, 16.2
E 606 Constant-Amplitude Low-Cycle Fatigue Testing, Rec. Practice for, 14.2
*E 616 Fracture Testing, Def. of Terms Relating to, 14.1
*E 621 Use of Metric (SI) Units in Building Design and Construction, Practice for, A.1
*E 631 Building Construction, Terminology Used in
E 709 Magnetic Particle Examination, Practice for, 16.4

F. Materials For Specific Applications
*F 412 Plastic Piping Systems, Def. of Terms Relating to
*F 547 Nails for Use with Wood and Wood-Base Materials, Def. of Terms Relating to, 5.P

G. Corrosion, Deteriorations, and Degradation of Materials
G 15 Corrosion and Corrosion Testing, Def. of Terms Relating to

C.3 TEXT REFERENCES

The following sources are cited in the text. The list is not, and is not intended to be, a complete bibliography on the subject.

1. Anderegg, F. O., R. Weller, and B. Fried: "Tension Specimen Shape and Apparent Strength," *Proc. ASTM*, vol. 39 (1939), pp. 1261–1269.
2. Anderson, Paul: "Experiments with Concrete in Torsion," *Trans. ASCE*, vol. 100 (1935), pp. 949–983. See especially discussion by Gilkey.
3. Armstrong, T. N., and A. P. Gagnebin: "Impact Properties of Some Low Alloy Nickel Steels at Temperature Down to –220 Degrees Farh.," *Trans. Am. Soc. Metals*, March 1940.
4. ASTM Committee A-3 on Cast Iron, Subcommittee VI, "Report of Subcommittee on Impact Testing of Cast Iron," *Proc. ASTM*, vol. 33, pt. I (1933), pp. 87–129. Also vol. 49 (1949), pp. 100–109.
5. Barba, J.: "Résistance à la traction et allongements des métaux après rupture," *Mem. soc. ing. civils France*, 1880, 1re partie, p. 682.
6. Baron, F., and E. W. Larson, Jr.: *Static and Fatigue Tests of Riveted and Bolted Joints Having Different Lengths of Grip*, Department of Civil Engineering, Northwestern University, with Cooperation of Research Council on Riveted and Bolted Structural Joints of Engineering Foundation, Project V, January 1952.
7. Batson, R. G., and J. H. Hyde: *Mechanical Testing*, Vol. I: *Testing of Materials of Construction*, Dutton, New York (Chapman & Hall, London), 1922.
8. Blanks, R. F., and C. C. McNamara: "Mass Concrete Tests in Large Cylinders," *Proc. Am. Concrete Inst.*, vol. 31 (1935), pp. 280–302, and discussion vol. 32 (1936), pp. 234–262.
9. Breuner, A.: "Magnetic Method for Measuring the Thickness of Nickel Coatings on Nonmagnetic Base Metals," *J. Research*, Natl. Bur. Standards, vol. 18 (1937), p. 565 (R.P. 994).
10. Breuner, A.: "Magnetic Method for Measuring the Thickness of Nonmagnetic Coatings on Iron and Steel," *J. Research*, Natl. Bur. Standards, vol. 20 (1938), p. 357 (R.P. 1081).
11. Breunich, T. R.: "Fatigue Testing: Its Machines and Methods," *Product Eng.*, vol. 24, no. 2 (February 1953), pp. 128–134; and vol. 24, no. 3 (March 1953), pp. 148–154.
12. Burgess, G. K.: "The Bureau of Standards," *Civil Eng.*, vol. 1, no. 6 (March 1931), pp. 491–494.
13. Clark, C. L., and A. E. White: "Creep Characteristics of Metals," *Trans. Am. Soc. Metals* (December 1936), pp. 831–868.
14. Clark, D. S., and P. E. Duwez: "The Influence of Strain Rate on Some Tensile Properties of Steel," *Proc. ASTM*, vol. 50 (1950), pp. 560–575.
15. Davis, E. A.: "The Effect of the Speed of Stretching and the Rate of Loading on the Yielding of Mild Steel," *J. Appl. Mech.*, vol. 5, no. 4 (December 1938), pp. A137–A140.
16. Davis, H. E., E. R. Parker, and A. Boodberg: "A Study of the Transition from Shear to Cleavage Fracture in Mild Steel," *Proc. ASTM*, vol. 47 (1947), pp. 483–499.

17. *Design and Control of Concrete Mixtures*, 10th ed., Portland Cement Assn., Chicago, Ill., 1952.

18. Doane, F. B., and C. E. Betz: *Principles of Magnaflux*, Magnaflux Corp., Chicago, 1941.

19. Dolan, T. J., and C. S. Yen: "Some Aspects of the Effect of Metallurgical Structure on Fatigue Strength and Notch Sensitivity of Steel," *Proc. ASTM*, vol. 48 (1948), pp. 664–695. See especially pp. 685–688. See also vol. 50 (1950), pp. 587–618.

20. Driscoll, D. E.: "The Charpy Impact Machine and Procedure for Inspection and Testing Charpy V-notch Impact Specimens," *ASTM Bull.* 191 (July 1953), pp. 60–64.

21. Duwez, P. E., and D. S. Clark: "An Experimental Study of the Propagation of Plastic Deformation under Conditions of Longitudinal Impact," *Proc. ASTM*, vol. 47 (1947), pp. 502–522. See also discussion by L. H. Donnell and LeVan Griffis, pp. 523–532; Von Kármán, T., "On the Propagation of Plastic Strains in Solids," *Sixth International Congress for Applied Mechanics*, 1946, Paris.

22. "Effect of Load Rate on Compressive Strength of Mortar Cubes," in Report of ASTM Committee C-1, *Proc. ASTM*, vol. 34, pt. II (1934), pp. 336–338.

23. "Electric Thickness Gage," *Instruments*, vol. 8 (1935), p. 341.

24. De Forest, A. V., and G. Ellis: "Brittle Lacquers as an Aid to Stress Analysis," *J. Aeronaut. Sci.*, vol. 7, no. 5 (March 1940), pp. 205–208. See also de Forest, A. V., G. Ellis, and F. B. Stern, Jr.: "Brittle Coatings for Quantitative Strain Measurements," *J. Appl. Mech.*, vol. 64 (December 1942), pp. A184–A188.

25. Frankel, H. E., J. A. Bennett, and W. A. Pennington: "Fatigue Properties of High-strength Steels," Natl. Bur. Standards *Tech. News Bull.*, vol. 43, no. 4 (1959), p. 74.

26. Gensamer, M.: *Strength of Metals under Combined Stresses*, American Society for Metals, Cleveland, Ohio, 1941.

27. Gensamer, M.: "Fundamentals of Fractures in Metals," in Symposium on Effect of Temperature on the Brittle Behavior of Metals, *ASTM Spec. Tech. Pub.* 158 (1953), p. 164.

28. Gibbons, C. H.: "Load-weighing and Load-indicating Systems," *ASTM Bull.* 100 (October 1939), pp. 7–13.

29. Goldbeck, A. T.: "Tensile and Flexural Strengths of Concrete," in *Significance of Tests of Concrete . . .*, 2d ed., pp. 9–14, American Society for Testing Materials, Philadelphia, 1943.

30. Gonnerman, H. F.: "Effect of End Condition of Cylinder on Compressive Strength of Concrete," *Proc. ASTM*, vol. 24, pt. II (1924), pp. 1036–1065.

31. Gough, H. J.: "Crystalline Structure in Relation to Failure of Metals—Especially by Fatigue," *Proc. ASTM*, vol. 33, pt. II (1933), pp. 3–114.

32. Habart, H., and W. J. Herge: "Sub-size Charpy Relationships at Sub-zero Temperatures," *Proc. ASTM*, vol. 39, pt. II (1939), pp. 649–658.

33. Herbert, E. G.: "Work-hardening Properties of Metals," *Trans. ASME*, vol. 48 (1926), pp. 705–745.

34. Hetenyi, M.: *Handbook for Experimental Stress Analysis*, Wiley, New York, 1950.

35. Heyer, R. H.: "Analysis of the Brinell Hardness Test," *Proc. ASTM*, vol. 37, pt. II (1937), pp. 119–145.

36. Heyer, R. H., and Y. E. Lysaght: "Survey of Investigations of Specimen Thickness on Rockwell Tests," *ASTM Bull.* 193 (October 1953), pp. 32–39.

37. Horger, O. J., and H. R. Neifert: "Fatigue Strength of Machine Forgings 6 to 7 in. in Diameter," *Proc. ASTM*, vol. 39 (1939), pp. 723–740.

38. Hornibrook, F. B.: "Application of Sonic Method to Freezing and Thawing Studies of Concrete," *ASTM Bull.* 101 (December 1939), pp. 5–8.

39. Jones, P. G., and F. E. Richart: "The Effect of Testing Speed on Strength and Elastic Properties of Concrete," *Proc. ASTM*, vol. 36, pt. II (1936), pp. 380–392.

40. Jones, P. G., and H. F. Moore: "An Investigation of the Effect of Rate of Strain on the Results of Tension Tests of Metals," *Proc. ASTM*, vol. 40 (1940), pp. 610–624.

41. Jones, P. G.: "On the Transition from a Ductile to a Brittle Type of Fracture in Several Low-alloy Steels," *Proc. ASTM*, vol. 43 (1943), pp. 547–555.

42. Jones, R.: *Non-destructive Testing of Concrete*, Cambridge University Press, 1962.

43. Kahn, N. A., E. A. Imbembo, and F. Ginsberg: "Effect of Variations in Notch Acuity on the

Behavior of Steel in the Charpy Notched-bar Test," *Proc. ASTM*, vol. 50 (1950), pp. 619–648. See also *ASTM Bull.* 146 (May 1947), p. 66.

44. Kellerman, W. F.: "Effect of Size of Specimen, Size of Aggregate, and Method of Loading upon the Uniformity of Flexural Strength Tests," *Public Roads*, vol. 13, no. 11 (January 1933), pp. 177–184.

45. Knerr, H. C.: "Electrical Detection of Flaws in Metal," *Metals & Alloys*, vol. 11, no. 10 (October 1940).

46. Knoop, F., C. G. Peters, and W. B. Emerson: "A Sensitive Pyramidal-diamond Tool for Indentation Tests," *Natl. Bur. Standards Research Paper* 1220 (July 1939). See also *Natl. Bureau Standards Letter Circ.* 819.

47. Kommers, J. B.: "The Effect of Overstressing and Understressing on Fatigue," *Proc. ASTM*, vol. 38, pt. II (1938), pp. 249–268.

48. Lessells, J. M., and W. M. Murray: "The Effect of Shot Blasting and Its Bearing on Fatigue," *Proc. ASTM*, vol. 41 (1941), pp. 659–673.

49. Lester, H. H.: "Some Aspects of Radiographic Sensitivity in Testing with X-Rays," *ASTM Bull.* 100 (October 1939), pp. 33–40.

50. Luerssen, G. V., and O. V. Greene: "The Torsion Impact Test," *Proc. ASTM*, vol. 33, pt. II (1933), pp. 315–333.

51. Luerssen, G. V., and O. V. Greene: "The Torsion Impact Properties of Tool Steel," *Trans. Am. Soc. Metals*, vol. 22 (1934), p. 311.

52. Luerssen, G. V., and O. V. Greene: "Interpretation of Torsion Impact Properties of Carbon Tool Steel," *Trans. Am. Soc. Metals*, vol. 23 (1935), p. 861.

53. Lysaght, V. E.: *Indentation Hardness Testing*, Reinhold, New York, 1949.

54. Lyse, I., and C. C. Keyser: "Effect of Size and Shape of Test Specimen upon the Observed Physical Properties of Structural Steel," *Proc. ASTM*, vol. 34, pt. II (1934), pp. 202–215.

55. MacGregor, C. W.: "The Tension Test," *Proc. ASTM*, vol. 40 (1940), pp. 508–534. Bibliography.

56. MacKenzie, J. T.: "Tests of Cast Iron Specimens of Various Diameters," *Proc. ASTM*, vol. 31, pt. I (1931), pp. 160–166.

57. MacKenzie, J. T., and C. K. Donoho: "A Study of the Effect of Span on the Transverse Test Results for Cast Iron," *Proc. ASTM*, vol. 37, pt. II (1937), pp. 71–87.

58. Mann, H. C.: "The Relation between the Tension Static and Dynamic Tests," *Proc. ASTM*, vol. 35, pt. II (1935), pp. 323–340.

59. Mann, H. C.: "High-velocity Tension-impact Tests," *Proc. ASTM*, vol. 36, pt. II (1936), pp. 85–109.

60. Markwardt, L. J., and T. R. C. Wilson: "Strength and Related Properties of Woods Grown in the United States," *U.S. Dept. Agr. Tech. Bull.* 479 (September 1935).

61. McCune, C. A.: "Magnaflux Inspection of Railroad Steel Parts," *Proc. Southern and Southwestern Railway Club*, July, 1940.

62. McVetty, P. G.: "Working Stresses for High-temperature Service," *Mech. Eng.*, vol. 56, no. 4 (March 1934), pp. 149–154. Also *Proc. ASTM*, vol. 43 (1943), pp. 707–727.

63. "Methods of Impact Testing of Cast Iron," *ASTM Standard* A327-54.

64. *Metals Handbook—1961*, American Society for Metals, Cleveland, Ohio, 1961.

65. Mitchell, N. B., Jr.: "The Indirect Tension Test for Concrete," *Materials Research and Standards*, vol. 1, no. 10 (October 1961), pp. 780–788. Also vol. 2, no. 3 (March 1962), p. 179.

66. Moore, H. F.: "Tension Tests of Steel with Test Specimens of Various Size and Form," *Proc. ASTM*, vol. 18, pt. I (1918), pp. 403–421.

67. Moore, H. F., and J. B. Kommers: *Fatigue of Metals*, McGraw-Hill, New York, 1927.

68. Moore, H. F.: "Stress, Strain and Structural Damage," *Proc. ASTM*, vol. 39 (1939), pp. 549–570.

69. Moore, H. F., and M. B. Moore: *Textbook of the Materials of Engineering*, 8th ed., McGraw-Hill, New York, 1953.

70. Morrison, J. L. M.: "The Influence of Rate of Strain in Tension Tests," *Engineer*, vol. 158, no. 4102 (Aug. 24, 1934), pp. 183–185.

71. Muhlenbruch, C. W.: *Testing of Engineering Materials*, Van Nostrand, New York, 1944.

72. Norton, F. H.: *The Creep of Steel at High Temperatures*, McGraw-Hill, New York, 1929.
73. Obert, L.: "Sonic Method of Determining the Modulus of Elasticity of Building Materials under Pressure," *Proc. ASTM*, vol. 39 (1939), pp. 987–998.
74. Parker, E. R., H. E. Davis, and A. E. Flanigan: "A Study of the Tension Test," *Proc. ASTM*, vol. 46 (1946), pp. 1159–1174.
75. Petrenko, S. N., W. Ramberg, and B. Wilson: "Determination of the Brinell Number of Metals," *J. Research Natl. Bur. Standards*, vol. 17, no. 1 (July 1936), pp. 59–96 (R. P. 903).
76. "Proposed Method for Torsion Tests to Determine the Mechanical Properties of Metallic Materials under Shearing Stress," in Report of ASTM Committee E-1, *Proc. ASTM*, vol. 25, pt. I (1925), pp. 430–436.
77. Ramberg, W., and W. R. Osgood: "Description of Stress-Strain Curves by Three Parameters," *NACA, TN* 902 (July 1943).
78. Roberts, I.: "Prediction of Relaxation of Metals from Creep Data," *Proc. ASTM*, vol. 51 (1951), p. 811.
79. Seely, F. B., and W. J. Putnam: "Relation between the Elastic Strengths of Steel in Tension, Compression and Shear," *Univ. Ill. Eng. Expt. Sta. Bull.* 115 (1919).
80. Seely, F. B.: "The Strength Features of the Tension Test," *Proc. ASTM*, vol. 40 (1940), pp. 535–550. Bibliography.
81. Shanley, F. R.: *Strength of Materials*, McGraw-Hill, New York, 1957, chaps, 24, 25.
82. Simmons, W. F., and H. C. Cross: "Elevated-temperature Properties of Stainless Steels," *ASTM Spec. Tech. Pub.* 124 (1952).
83. Sisco, F. T.: *Modern Metallurgy for Engineers*, Pitman, New York, 1948.
84. Soderberg, C. R.: "Working Stresses," *Trans. ASME*, vol. 55 (1933). APM 55–16.
85. St. John, A., and H. R. Isenburger: *Industrial Radiography*, Wiley, New York, 1934.
86. "Stresses in Overstrained Materials," *Engineer*, vol. 140, no. 3635 (Sept. 13, 1935), pp. 291–292.
87. Swain, G. F.: *Structural Engineering–Strength of Materials*, McGraw-Hill, New York, 1924.
88. *Symposium on High Speed Testing*, Interscience Publishers, New York, 1958, p. 12.
89. "Symposium on Impact Testing," *Proc. ASTM*, vol. 38, pt. II (1938), pp. 21–156.
90. "Symposium on Plastics," *ASTM Spec. Tech. Pub.* 59 (1944), 200 pp.
91. Templin, R. L., and R. L. Moore: "Specimens for Torsion Tests of Metals," *Proc. ASTM*, vol. 30, pt. II (1930), pp. 534–543.
92. Templin, R. L.: "Some Factors Affecting Strain Measurements in Tests of Metals," *Proc. ASTM*, vol. 34, pt. II (1934), pp. 182–201.
93. Templin, R. L.: "Fatigue Machines for Testing Structural Units," *Proc. ASTM*, vol. 39 (1939), pp. 711–722.
94. *The Testing of Materials, Natl. Bur. Standards Circ.* 45 (1913).
95. Timoshenko, S.: *Strength of Materials*, 2 vols., 3d ed., Van Nostrand, New York, 1955. See part II, chap. 10, "Mechanical Properties."
96. Troxell, G. E.: "The Effect of Capping Methods and End Conditions before Capping upon the Compressive Strength of Concrete Test Cylinders," *Proc. ASTM*, vol. 41 (1941), pp. 1038–1052.
97. Upton, G. B.: *Materials of Construction*, Wiley, New York, 1916.
98. Vose, R. W.: "An Application of the Interferometer Strain Gage in Photoelasticity," *J. Appl. Mech.*, vol. 2, no. 3 (September 1935), pp. A99–A102.
99. Warwick, C. L.: "The Work in the Field of Standardization of the American Society for Testing and Materials," *Ann. Am. Acad. Political and Social Sci.*, vol. 137 (May 1928).
100. Williams, S. R.: *Hardness and Hardness Measurements*, American Society for Metals, Cleveland, Ohio, 1942.
101. Wilson, W. M., and F. P. Thomas: "Fatigue Tests of Riveted Joints," *Univ. Ill. Eng. Expt. Sta. Bull.* 302 (1938), pp. 36–41.
102. Wilson, W. M., and F. P. Thomas: "Fatigue Tests of Riveted Joints," *Civil Eng.*, vol. 8, no. 8 (August 1938), pp. 513–156.
103. Wilson, W. M., et al.: "Fatigue Tests of Welded Joints in Structural Steel Plates," *Univ. Ill. Eng. Expt. Sta. Bull.* 327 (1941).

104. Withey, M. O., and G. W. Washa: *Johnson's Materials of Construction*, 9th ed., Wiley, New York, 1954.
105. *Wood Handbook*, Forest Products Laboratory, U.S. Department of Agriculture 1955 and 1974.
106. Wright, P. J. F.: "Effect of the Method of Test on the Flexural Strength of Concrete," *Mag. Concrete Research (London)*, no. 11 (October 1952), pp. 67–76.
107. Zuschlag, T.: "Magnetic Analysis Applied to the Inspection of Bar Stock and Pipe," *ASTM Bull.* 99 (August 1939), pp. 35–40.
108. *Symposium on Behavior of Metals at Low Temperatures*, American Society for Metals, Cleveland, Ohio, 1952.
109. *Symposium on Short-time High-temperature Testing*, American Society for Metals, Cleveland, Ohio, 1957.
110. Draffin, J. O., and W. L. Collins: "Effect of Size and Type of Specimen on the Torsional Properties of Cast Iron," *Proc. ASTM*, vol. 38, pt. II (1938), pp. 235–248.
111. Draffin, J. O., W. L. Collins, and C. H. Casberg: "Mechanical Properties of Gray Cast Iron in Torsion," *Proc. ASTM*, vol. 40 (1940), pp. 840–848.
112. *Building Code Requirements for Reinforced Concrete*, American Concrete Institute, Detroit, 1977.
113. Abraham, H.: *Asphalt and Allied Substances*, 6th ed., Van Nostrand, Princeton, N. J., 1960–63 (5 vols.).
114. Goetz, W. H., and L. E. Wood: "Bituminous Materials and Mixtures," sec. 18 in K. B. Woods (ed.), *Highway Engineering Handbook*, McGraw-Hill, New York, 1960.
115. Seely, F. B., and J. O. Smith: *Resistance of Materials*, Wiley, New York, 1956.
116. LaTour, H., and R. S. Sutton: "Improved Scale System for Stiffness Testing Machine," *ASTM Bull.* 196 (February 1954), pp. 40–42.
117. Shewhart, W. A.: *Economic Control of Quality of Manufactured Product*, Van Nostrand, New York, 1931.
118. *Mix Design Methods for Asphalt Concrete and Other Hot-Mix Types*, The Asphalt Institute, Publ. MS-2, 1979.
119. *A. Basic Asphalt Emulsion Manual*, The Asphalt Institute, Publ. MS-19, March 1979.
120. *Mix Design Methods for Liquid Asphalt Mixtures*, The Asphalt Institute, Publ. Misc-74-2, February 1974.
121. *Interim Guide to Full-Depth Asphalt Paving*, The Asphalt Institute, Pacific Coast Div., Publ. PDC-1, January 1976.
122. *The Grading and Behavior of Paving Asphalts*, The Asphalt Institute, Pacific Coast Div., Publ. PCD-8, March 1979.
123. *Bituminous Emulsions for Highway Pavements*, NCHRP Synthesis of Highway Practice Report no. 30, Natl. Res. Council, Transp. Res. Board, 1975.
124. Mertens, E. V., and J. R. Wright: "Cationic Emulsions: How They Differ from Conventional Emulsions in Theory and Practice," *Proc. Hwy. Res. Bd.*, vol. 38 (1959), pp. 385–397.
125. Ricketts, W. C., et al.: "An Evaluation of the Specific Gravity of Aggregates for Use in Bituminous Mixtures," *Proc. ASTM*, vol. 54 (1954), pp. 1246–1257.
126. *Symposium on Specific Gravity of Bituminous Coated Aggregates*, ASTM Spec. Tech. Publ. No. 191.
127. McLeod, N. W.: "Selecting Aggregate Specific Gravity for Bituminous Paving Mixtures," *Proc. Hwy. Res. Bd.*, vol. 36, 1957.
128. *Factors Involved in the Design of Asphaltic Pavement Surfaces*, NCHRP Report no. 39, Natl. Res. Council, Hwy. Res. Bd., 1967.
129. *Emulsion Mix. Design, Stabilization and Compaction*, Transp. Res. Record 754, Natl. Res. Council, Transp. Res. Bd., 1980.
130. *Mix Design Methods for Base and Surface Courses Using Emulsified Asphalt—A State-of-the-Art Report*, Report No. FHWA-RD-78-113, USDOT, Fed. Hwy. Admin., October 1978.
131. Timoshenko, S., and J. N. Goodier: *Theory of Elasticity*, 2d ed., McGraw-Hill, New York, 1951.
132. Frocht, M. M.: *Photoelasticity*, vol. 2, Wiley, New York, 1948.

133. Kennedy, T. W.: "Characterization of Asphalt Paving Materials Using the Indirect Tension Test," *Proc. Assn. Asphalt Paving Technologists*, vol. 46, 1977, pp. 132–150.
134. Schmidt, R. J.: "A Practical Method for Measuring the Resilient Modulus of Asphalt-Treated Mixes," *Transp. Res. Record 404*, Natl. Res. Council, Transp. Res. Bd., 1972, pp. 22–32.
135. Mamlouk, M. S., and L. E. Wood: "Characterization of Asphalt Emulsion Treated Bases," *Proc. ASCE, Transp. Eng. Jour.*, vol. 107, TE 2 (March 1981), pp. 183–196.
136. Mamlouk, M. S., and L. E. Wood: "Evaluation of the Use of Indirect Tensile Test Results for Characterization of Emulsion Treated Bases," *Transp. Res. Record 733*, Natl. Res. Council, Transp. Res. Board, 1979, pp. 99–105.
137. *A Brief Introduction to Asphalt and Some of Its Uses*, The Asphalt Institute, Publ. MS-5, 1977.
138. *Asphalts–Paving, Cutback and Emulsified*, The Asphalt Institute, Pacific Coast Div., Publ. PCD-7, 1978.
139. Cechetini, J. A.: "A Modified CKE Test," *Proc. Assn. Asphalt Paving Technologists*, vol. 40 (1971), pp. 509–526.

INDEX

Abbreviations for plastics, (table)
    414-415
Abrasion resistance, 198
  of aggregates, 204, 376
  of wire rope, 326
ABS plastics, 419
Absorption:
  by aggregates, 334, 372
  by plastics, 421
Abstracts in report preparation, 63
Accelerated soundness test, 376
Accessories for testing machines, 82,
    91-93
Accidental errors, 75-76
Accuracy:
  adequacy of, 72
  control of, 75
  definition of, 10
Acetal, 419
Acrylic, 413, 419
Addition polymerization, 415
Additives to plastics, 413-416
Addresses of organizations, 443-444
Adhesives, 420-421
  for plywood, 406
Admixtures to concrete, 333
Agencies, standardizing, 12-15
Aggregate particles, shape of, 375

Aggregates:
  abrasion resistance of, 204, 376
  absorption by, 334, 372
  for asphaltic mixtures, 370-378
  cleanness of, 376-377
  for concrete, 330-331, 333-336
  dense-graded, 374
  density of, 334-336, 371-373
  grading of, 331, 373-374
  mineral composition of, 377
  moisture content of, 334-335
  soundness of, 376
  stripping of, 377-378
  toughness of, 376
Air entrainment, 342, 347-348
Air-blown asphalt, 355
Alkyd, 420
Allowance in gages, 103
Alloying, 305
  of cast iron, 311
  of steel, 313-316
Alloys:
  nonferrous (see Nonferrous alloys)
  solutions, 305
Allylic, 420
Alumina, 307, 310
Aluminum, 307
  designations for, 318

Aluminum (*Cont.*):
  production of, 310–311
  properties of, 318, (table) 321
  in steels, 315
  (*See also* Metals)
American National Standards Institute
    (ANSI), 14–15
American Society for Testing and
    Materials (ASTM), 13–14, (table)
    444–448
Amine, 420
Analysis of data, 37–55
  control charts, 51–52
  correlation, 50–51
  distribution characteristics, 44–48
  grouped frequency, 39–44
  management of data, 37–38
    reduction, 37
    summary, 37
  sampling and errors, 48–50
  statistical summaries, 53
  variations, 38–39
    ungrouped, 38
Angle:
  plane, units of, 432
  of twist, 173
Anionic emulsified asphalt, 355
Annealing, 309–311
Antioxidants, 416
Anvil, 226
Apparent density, 371
Apparatus, 74
  for Brinell hardness test, 205–206
  for compression tests, 154–156
  for creep tests, 266–268
  for fatigue tests, 251–253
  for flexure tests, 184–186
  for impact tests, 226–230
  for photoelastic analysis, 295–299
  for radiography, 276–277
  for Rockwell hardness test, 210, (fig.)
    211
  for scleroscope hardness test, 215–216
  selection of, 75
  Sonizon, 290–291

Apparatus (*Cont.*):
  Sperry, 285–286
  for tension tests, 131–135
  for torsion tests, 171–172
  for ultrasonic testing, 288–291
Appendixes (to reports), 57
Arithmetic mean, 44
Arrangement of report, 56
Asphalt, 351–390
  air-blown, 355
  anionic emulsified, 355
  cationic emulsified, 355
  cement, 354, 363–364
  composition of, 357–358
  consistency of, 358–363
  curing of, 355
  cutback, 354–355, (table) 365–366,
    369
  density of, 356–357
  diluents for, 355
  distillation process in manufacture of,
    353, 357–358
  emulsified (*see* Emulsified asphalts)
  grades of, 363–370
  liquid, 354–355
    (*See also* Cutback asphalt; Emulsified
    asphalts)
  mixtures (*see* Asphalt mixtures)
  rock, 352
  softening point of, 359–360
  solvents for, 355
  temperature susceptibility of, 358–359
  viscosity of, 358–363
Asphalt Institute vacuum viscometer,
    360–361
Asphalt mixtures, 378–389
  concrete, 378–379
  consolidation in, 379
  design of, 378–379
  ductility of, 362
  durability of, 379
  mix-design process, 379–380
  selection of, 380–382, 387–389
  stability of, 378–379, 384–389
  structure of, internal, 378

Asphalt mixtures (*Cont.*):
  trial, 381–387
  volumetric relations in, 383
ASTM (American Society for Testing and
      Materials), 13–14, (table) 444–448
ASTM standards, 13–14
Autocollimator, 106
Auxilliary measurements, 76
Average, 44
Average deviation, 45
Average error, 49

Baekeland, Leo Hendrik, 412
Bakelite, 412
Ball penetration test, 340–341
Barba's law, 141
Base units (SI), 429, (table) 430
Beams (*see* Bending)
Bend tests, 183, 191–193
Bending, 179–182
  tests (*see* Bending tests)
Bending moment diagrams, 179–182, 441
Bending specimens of wood, 184
Bending stresses, 179–182
Bending tests, 70, 179–193
  apparatus for, 184–186
  observations in, 187–189
    failure, 187–188
    yielding, 189
  procedure, 186–187
  scope and applicability, 182–183
  specimens for, 183–184
    blocks, 184
    cast iron, 184
  variables in, 189–191
    stability of beams, 190
    test speed, 191
Benzene, 415
Beryllium, 307, 318
Bessemer converter, 308
Beta rays, 275–279
Birefringence, 295–300
Bitumen, 351–352
Blast furnace, 308

Bleeding, 342
Blocks for flexure tests, 184
Bourdon tube, 81, 83
Brass, 318, (table) 320
Bridge, Wheatstone, 112
Brinell hardness test, 199, 204–210, 259
  apparatus for, 205–206
  observations in, 207–208
  specimens for, 204–205
  variables in, 208–210
Brittle coatings, 116–120
  calibration of, 117
Brittleness, 28, 241
Bronze, 318, (table) 320
Buckling, 31
Bulk density, 335–338, 371
Bulk modulus, 20

Calibration:
  of brittle coating, 117
  of testing machine, 93–97
Calibration ring, 93–94, (fig.) 96
Calipers, 99–100
Cambium layer, 393
Cannon-Manning vacuum viscometer,
      360–361
Capacity of testing machines, 86
Capillary tube viscometer, 360
Capping, 155, (table) 161
Carbon content:
  of cast iron, 311
  of ferrous metals, 306
  of steel, 313
Carbon disulfide, 351, 357
Carrier wave, 115
Cast iron, 306, 311–312
  alloying of, 311
  bending specimens of, 184
  carbon content of, 311
  flexural failure of, 187–188
  impact tests for, 225
  properties of, 311–312
  tension specimen of, (fig.) 130
  (*See also* Metals)

Casting, 418
Cathode-ray tube (CRT), 115
Cationic emulsified asphalt, 355
Cell (interval), 41
Cell (wood), 392
Cells, galvanic, 322
Cellulose, 392, 415
Cellulosic, 420
Cement:
  asphalt, 354, 363–364
  fineness of, 332
  portland, 330–332
  set of, 332
  soundness of, 332
Cement mixing test (asphalt), 362
Central tendency, 44–45
Centrifuge kerosene equivalent (CKE)
  test, 375
Characteristics of a material,
  determination of by nondestructive
  tests, 275
Charpy test, 224, 231–233
Charpy values, effects of steel on, (tables)
  238, (tables) 240
Chemical corrosion, 319, 322
Chromium, 311, 313
Circular polarization, 298–299
CKE (centrifuge kerosene equivalent)
  test, 375
Class frequency, 41
Class or step interval, 41
Classification:
  of tests, 6–9
  of variables, 69
Cleanness of aggregates, 376–377
Clearance in gages, 103
Cleavage fracture, 31
Cleveland open cup (COP) test, 357
Clinker, 331
Clip gage, 111
Closed-loop system, 82, 91, (fig.) 93
Coatings:
  brittle (*see* Brittle coatings)
  seal, 353

Coefficient:
  of compressibility, 20
  of correlation, 50–51
  of thermal expansion, 311n., 433
  of variation, 46
Cohesiometer, 384–385
Coincidence, edge, 99, (fig.) 100
Cold-bend tests, 191–192
Cold-working, (cold-forming), 308
Collar deformeter, 104
Collar extensometer (compressometer),
  104
Column headings, 60
Column loads (Euler), 442
Columns, stability of, 442
Commercial testing, 7
Common sense in testing, 77–78
Compact tension specimen (CTS), 250
Comparitor, optical, 106, (fig.) 107
Complex stress condition, 70, (fig.) 166
Composition of asphalts, 357–358
Compressibility, coefficient of, 20
Compression molding, 417
Compression tests, 151–163
  apparatus for, 154–156
  by brittle coating, 119–120
  eccentricity in, 154–156, 160–161
  failure in, 157–159
  fracture in, 157–159
  nature of, 69
  observations in, 157–159
  procedure for, 156–157
  specimens for, 152–154
    metals, 152, (fig.) 153, 155
    wood, 153
  speed in, 156–157, 162
  triaxial, 164n.
  variables in, 159–162
Compressive failure of wood, 158, (fig.)
  159
Compressive strength, 30
Compressometers, 104–106, 156
Computations in SI, 434–435
Computer (team member), 77

Concentration cells, 322
Concentration yards, 391
Concrete:
  asphalt (*see* Asphalt; Asphalt mixes)
  portland cement, 330–350
    bending specimens of, 184
    composition of, 330–331
      aggregates, 330–331
      corrosion resistance, 330
    compression specimens of, 152–153,
      155
    flexural failure of, 187–188
    manufacture of, 336–340
      design of, 336–339
      mixing, placing, and curing, 339–340
    materials for making, 331–336
      admixtures, 333
      aggregates, 333–336
      water, 333
    properties of, 340–342, 346–349
      air entrainment, 342
      bleeding, 342
      consistency, 340–342
      density, 346
      durability and impermeability,
        346–348
      volume changes, 348–349
    splitting tension test, 134–135
    strength of, 342–346
      creep, 345–346
      stiffness, 344–345
      stress-strain diagram, (fig.) 345
    test hammer method for, 198, 292
Concrete mixes, 336–342
  consistency of, 340–342
  design of, 336–339
  (*See also* Asphalt mixes)
Concrete specimens, 152–153
Condensation polymerization, 413
Conduct of tests, 76–78
Conifers, 391
Consistency:
  of asphalts, 358–363
  of concrete mixes, 340–342

Consolidation in asphalt mixes, 379
Constantan wire, 108
Control charts, 51–52
Control of material properties, 3–16
Control systems, 91
Control unit, 91
Conversion factors, (table) 436–438
Conversions in SI units, 435
Coolants, 234
Coolidge X-ray tube, 276–277
COP (Cleveland open cup) test, 375
Copper, 306
  corrosion of, 319, 322
  properties of, 318, (table) 320
  in steel, 314
Correlation:
  coefficient of, 50–51
  of fatigue limits, 259
  of hardness, (table) 217, 218
Correspondence, covering, 65–66
Corrosion, 319, 322–323
  chemical reaction, 319, 322
  electrolytic, 322
  of iron, 319, 322
  protection from, 322–323
  of steel, 319
Corrosion fatigue, 247
Corrosion protection, 322–323
Corrosion resistance of concrete, 330
Corrosive environment, 71
Cover page, 63
Covering correspondence, 65–66
Crack formation, 245
Creep, 261–263
  of concrete, 345–346
  nature of, 27, 31, 262
  occurrence of, 261
  stages of, 262–263
  of wood, 402
Creep data, extrapolation of, 265, (fig.)
    269, 270
Creep diagrams, (fig.) 269, 270–271
Creep resistance:
  effect of heat treatment on, 272

Creep resistance (*Cont.*):
  of steel, (fig.) 263, (fig.) 270, 271–272
Creep tests, 261–273
  apparatus for, 266–268
  eccentricity in, 269–270
  furnaces for, 266–268
  procedure and observations, 268–271
  scope and applicability of, 264–266
  temperature in, 261–262, 271
  variables in, 265, 271–272
Creosote, 405
Cresol, 415
Crossheads, 86
Cross-sectional properties:
  formulas for, 440
  units for, 187, 433
CRT (cathode-ray tube), 115
Cryolite, 310
Crystal probes, 288–291
Crystallization, 245
Crystals, 31, 244
CTS (compact tension specimen), 250
Cubical dilation (dilatation), 19
Cumulative frequency, 41, 43, 44
Curing:
  of asphalt, 355
  of concrete, 340
Curves (*see* Diagrams)
Cutback asphalt, 354–355, (table)
  365–366, 369

Data:
  analysis of (*see* Analysis of data)
  creep, extrapolation of, 265, (fig.)
    269, 270
Date, ISO standard, 65*n.*
Deciduous trees, 391
Defects:
  detection of using nondestructive tests,
    274
  electrical methods, 285–286
  electromagnetic methods, 279–282
  magnetic-particle method, 282–285
  radiographic methods, 275–279
  ultrasonic methods, 288–290

Defects (*Cont.*):
  in wood, 395, 397
Deflectometer, 104, 185–186
Deformation, definition of, 18
Deformeters, 103–107
  collar, 104
Demulsibility test, 362
Dense-graded aggregate, 374
Density:
  of aggregates, 334–336, 371–373
  apparent, 371
  of asphalts, 356–357
  bulk, 335–338, 371
  of concrete, 346
  definition of, 17, 431
  effective, 372
  of nonferrous alloys, 319–321
  relative, 17, 413
  of steel, 316
  units of, 431
  of wood, 396–398, (tables) 403–404
Derived units (SI), 429, (tables) 430
Descriptive methods, 38
Design:
  of asphalt mixes, 378–389
  of concrete mixes, 336–339
  of tests, 71–72
Destructive vs. nondestructive testing, 8
Detrusion, 18, 166
Detrusion indicator, 104, 172–173
Deviation, 45–46
  average, 45
  mean, 45
  standard, 46
Diagonal tension, 168
Diagrams, 61
  creep, (fig.) 269, 270–271
  Haigh-Soderberg, 257, (fig.) 258
  Schenk-Peterson, 257, (fig.) 258
  shear and moment, 180, 441
  S-N (fatigue), 246–247, 254
  stress-strain (*see* Stress-strain diagrams)
Dial micrometer, 104–105
Diametral resilient modulus, 387
Diffuse-porous woods, 393

Digital indicator, 93
Dilatation (cubical dilation), 19
Diluents for asphalt, 355
Direct shear, 164
Direct shear test (*see* Shear test, direct)
Discrepancies in measurements, 75
Dispersion, 45–47
Displacement, determination of, 103–107
Distillation process in asphalt
    manufacture, 353, 357–358
Distribution, frequency, 38–40
Distribution curve, normal, 42–43
Double shear, 164, 170
Drop-weight machines, 227, 230
Ductile metals, stress-strain diagram for,
    139
Ductility, 28, 176, 241
  of asphaltic mixtures, 362
Ducts, resin, 392–393
Durability:
  of asphalt mixtures, 379
  of concrete, 346–348
  of metals, 319, 322–323
  of wire rope, 327
  of wood, 395, 402, 405
Durometers, 201–202
Dyes, 416
Dynamic testing machines, 90–91, 251
Dynamic tests:
  definition of, 70
  instrumentation for, 114–115
Dynamic (absolute) viscosity, 360–361,
    433
Dynamic-hardness tests, 198, 202–203
Dynamometers, 84–85

Earlywood, 393, 397
Eberbach tester, 201
Eccentricity:
  in compression tests, 154–156, 160–161
  in creep tests, 269–270
  in tension tests, 144
Edge coincidence, 99, (fig.) 100

Effective density, 372
Efficiency of rope, 325–326
Elastic action, 24–25
Elastic limit, 25
Elasticity, 24–27, 69
  elastic action, 24–25
  modulus of, 19, 183
  yielding, 25–27
Electric furnace, 308
Electric load measurement, 85
Electric resistance strain gages, 107–115,
    268
Electrical analyses, 285–288
Electrical determination of moisture
    content, 287–288, 396–397
Electrical insulating materials, molded,
    tension specimen of, (fig.) 131
Electrolytic corrosion, 322
Electromagnetism, 279–282
Electrometallurgy, 308
Electromotive force series, 322
Embrittlement and impact tests, 224
Emery capsule, 81, 84
Emulsified asphalts, 354–355, (table)
    367–368, 369–370
  inverted, 355
  stability of, 362
Endurance (*see* Fatigue)
Energy, units of, 432
Energy capacity, 32–34, 69
Engineering materials, 4–8
Entrainment, air, 342, 347–348
Environment, corrosive, 71
Epoxy, 420
Equicohesive temperature, 262–263
Errors:
  probable, 49–50
  types of, 40, 49–50, 75–76
    accidental, 75–76
    average, 49
    instrument, 75
    natural, 75
    relative, 49
    standard error of the mean, 49
    systematic, 75

Ethylene, 415
Eutectiod, 313
Experimental stress analysis, 7, 275
Experiments vs. tests, 7
Extensometers, 104–106, 136, 268
  Marten's, 105, (fig.) 106
  mirror, 105–106
Extruding, 416–417

Fabrication of plastics, 418–419
Failure, 30–32
  in bending, 187–189
  in compression, 157–159
  criteria of, 5
  in fatigue, 245
  flexural, of wood beams, 188
  under impact, 241
  of materials, 30–32
    buckling, 31–32
    separation, 31
    slip or flow, 31
  in shear, 174–175
  in tension, 139–140
  in torsion, 174, (fig.) 175
Fatigue, 244–248
  corrosion, 247
  failure in, 245
Fatigue limits, 30, 246–247, 255
  correlation of, 259
  of metals, 255
Fatigue strength, 30, 246–247
Fatigue tests, 70, 244–260
  apparatus for, 251–253
  low-cycle, 248, 255
  procedure and observations, 253–255
  scope and applicability of, 247–248
  specimens for, 249–250
  variables in, 255–259
Ferrous metals, 305–306
  carbon content of, 306
  corrosion of, 319
Fiber saturation point, 287, 395
Fibers, wood, 392–393
Field tests, 7–8

Figures (in reports), 60–63
File test, 203
Fillers for plastics, 416
Filtered-particle inspection, 294
Fineness of cement, 332
Fire resistance of wood, 405
Fixtures, special, 92
Flake board, 392
Flame retardants, 416
Flash point, 357
Flexural failure:
  of cast iron, 187–188
  of portland cement concrete, 187–188
  of wood beams, 188
Flexure, 180
  (*See also* Bending)
Flexure tests, 183, 191–193
Float test, 360
Flow, 31
Fluorescent penetrants, 294
Flywheel machines, 227
Foaming in plastics, 418
Force, units of, 432
Fraction defective, 41
Fractional frequency, 41
Fracture:
  cleavage, 31
  in compression, 157–159
  progressive, 244
  in tension, 136, 146
  in torsion, 174–175
  of wood, 158, (fig.) 159
Fracture mechanics, 245
Fracture toughness, units of, 433
Frequency, 38–44
  class, 41
  cumulative, 41, 43, 44
  fractional, 41
  grouped, 39–44
  relative, 41
  relative cumulative, 41
Frequency diagrams, 41–44
Frequency distribution, 38–40
  skewed, 47–48
  ungrouped, 38–39

Frequency-distribution series, 39–40
Frequency histogram, 42
Frequency polygon, 42
Friction pointer, 226, 229, 236
Frictional resistance, 166
Fulcrums, 88
Function generation in hydraulic
    machines, 91
Fungi, 405
Furnaces:
  for creep tests, 266–268
  for steel production, 308
    blast, 308
    electric, 308

Gage factor, 107–108
Gage length, 141
Gages:
  clip, 111
  electric resistance strain, 107–115, 268
  indicating, 103
  inspection, 102–103
    allowance, 103
    clearance, 103
    indicating, 103
    reference, 102
    simple, 103
    tolerance in, 103
  SR-4, 108
Galena, 307
Galvanic cells, 322
Galvanic series, 232
Galvanometer, 113–114
Gamma rays, 275–279
Glulam, 392
Goodman-Johnson diagram, 257, (fig.)
    258
Gradation of particles, 335, 373–374
Grades of asphalts, 363–370
Grading:
  of aggregates, 331, 373–374
  of lumber, 392
  of plywood, 406
Grain in wood, 395

Graph paper, 61–62
Graphics, 60–63
Graphs, 61
Graticule, 115
Grips, 91, 131–133, 251
Grouped frequency, 39–44
Grouping of data, 39–41
Growth rings in wood, 393
Guillery machine, 227

Haigh machine, 252
Haigh-Soderberg diagram, 257, (fig.) 258
Hardening of steel, 309–310
Hardness, 30, 195–196
  correlations, (table) 217, 218
  of iron, (table) 312
  as measure of strength, 68
  of nonferrous alloys, (tables) 320–321
  of steel, (table) 217, (table) 317
  of wood, 202
Hardness tests, 195–220, 292
  Brinell (see Brinell hardness test)
  dynamics, 198, 202–203
  indentation, 198–202
  Monotron, 200
  Rockwell (see Rockwell hardness test)
  scleroscope (see Scleroscope hardness
    test)
  scope and applicability of, 196–198
  scratch- and wear-, 198, 203, (table)
    204
  Tukon, 200
  units in, 196
Hardness values, units of, 432–433
Hardwood, 391
Hatt-Turner test, 225, 231, 233–234
Headings in report preparation, 63
Heartwood, 395
Heat treatment, 255, 308–311
  effect on creep resistance, 272
Heating rate, effect on steel, 148–149
Helical springs, 168
Hemicellulose, 392
Herbert tester, 198, 203

High-heating-rate test, 148
High-temperature tests, 147–149, 265
 (*See also* Creep tests)
Histogram, 42
Hooke's law, 62, 183
Hot-bend tests, 192
Hveem procedure, 384–385
Hydraulic capsule, 84
Hydraulic cylinder, 84
Hydraulic load measurement, 83–84,
 88–89
Hydraulic machines, 80, 81, 88–89, 91
 function generation in, 91
 load measurement in, 83–84, 88–89
Hydrocarbons, 353
Hydrometallurgy, 308
Hygroscopicity, 395
Hysteresis, 25, 33

Impact, 221–224
Impact specimens:
 of metals, (fig.) 232
 of wood, (fig.) 232
Impact tests, 70, 221–243
 apparatus for, 227–230
  velocity of testers, 229
 observations in, 235–236
 procedure for, 233–235
 temperature, 234–235
 scope and applicability of, 223–226
  for metals, 225–226
  for wood, 225
 specimens for, 231, (figs.) 232, 237–238
 temperature in 234–235, 238–241
 variables in, 237–241
  failure, 240–241
  specimens, 237–238
  steel, (tables) 238, (figs.) 240
  temperature, 238–241
  velocity of testers, 237
Indentation hardness, 198–202
Indicating gage, 103
Indicator:
 detrusion, 104, 172–173
 digital, 93

Indicator (*Cont.*):
 null, 113–114
 strain, 113–114
Indirect tensile test, 386–387
Inference methods, 38
Injection molding, 417
Inspection:
 filtered-particle, 294
 magnetic-particle, 282–285
 of materials, 8–9
 vs. testing, 8–9
 visual, 293–295
Inspection gages, 102–103
Instrument errors, 75
Instrument sensitivity, 75
Instrumentation, strain-gage, 111–115
Instruments (*see specific instruments*)
Interferometer, 101–102
 Vose's, (fig.) 105
International Organization for
  Standardization (ISO), 15
International System of Units (*see* SI)
Interpretation of test results, 4
Interval, class or step, 41
Inverted emulsified asphalt, 355
Iron, 305–306
 corrosion of, 319, 322
 hardness of, (table) 312
 production of, 308
ISO (International Organization for
  Standardization), 15
Isoelastic spring, 88
Isoelastic wire, 108
Izod test, 224, 231–233

Johnson shear tool, 168

Kelvin, Lord William Thomson, 107
Kilns, 392
Kinematic (relative) viscosity, 360–362,
 433
Knife-edges, 87–88
Knoop indenter, 200

Lateral strainometers, 104–105
Latewood, 393, 397
Lay (of rope), 324–325
L/d ratio, effect on steel, 142
Lead, 307
Least reading of an instrument, 75
Levers, use of, 83
Lignin, 392
Limnorias, 405
Linear measurements, 99–107
Linearly variable differential transformer (LVDT), 89–90, 106–107
Liquid asphalt, 354–355
  (*See also* Cutback asphalt; Emulsified asphalt)
Load measurement:
  in hydraulic machines, 83–84, 88–89
  mechanical, 84–85
  methods and devices for, 82–85
    electric, 85
  in screw-gear machines, 86
  in torsion machines, 171–172
Load unit, 91
Loading methods, 69, 186, 189
  two point, 186
Loading range, 96
Loading rate, types of, 70
Loading types, 69
Logsheet, (fig.) 138
Logarithmic paper, 61
Long-time tests, 70
  (*See also* Creep tests)
Los Angeles abrasion machine, 204, 376
Low-cycle fatigue tests, 248, 255
Low-temperature tests, 147, 234
Lumber, grading of, 392
LVDT (linearly variable differential transformer), 89–90, 106–107

Macadam, 353
Machinability, 198
Machines:
  drop-weight, 227, 230
  flywheel, 227
  guillery, 227

Machines (*Cont.*):
  Haigh, 252
  hydraulic (*see* Hydraulic machines)
  Los Angeles abrasion, 204, 376
  Mann-Haskell, 227
  Moore rotating-beam, 251
  screw-gear (*see* Screw-gear machines)
  single-blow pendulum, 226–227, (figs.) 228, 229
  Sonntag, 253
  testing (*see* Testing machines)
  torsion, load measurement in, 171–172
Magnaflux, 282*n.*
Magnatite, 307
Magnesium, 307, 319, (table) 321
Magnetic-particle inspection, 282–285
Magnetism, 279–282
Manganese, 314
Mann-Haskell machine, 227
Marine borers, 405
Marshall procedure, 384–386
Marten's extensometer, 105, (fig.) 106
Mass:
  units of, 431
  vs. weight, 17, 196, 431
Material properties, control of, 3–16
Materials:
  engineering, 4, 8
  inspection of, 8–9
  selection of, 6
  specification of (*see* Specification of Materials)
  variability of, 4
Materials research, 7
Mathematics (in reports), 58–59
Mean, 44
  arithmetic, 44
  standard error of the, 49
Mean deviation, 45
Measurements:
  auxiliary, 76
  of displacement, 103–107
  linear, 99–107
  of load (*see* Load measurement)
  in mechanical testing, 74–76
  of strain, 107–120, 295–300

Measurements (*Cont.*):
of surface roughness (*see* Surface roughness, measurement of)
theory of, 4
Mechanical behavior, 17–36
definitions, 17–20
Mechanical load measurement, 84–85
Mechanical machines, 80, 81, 86–88
Mechanical nondestructive testing, 291–293
Mechanical properties, 68–69
(*See also* Properties)
Mechanical tests, 5, 68–71
measurements in, 74–75
purpose of, 68–69
types of, 69–71
Mechanics, fracture, 245
Mechanics formulas, (table) 440–442
Median, 44–45
Melamine, 420
Metallic specimens, 72–73, 129–130, 152–153, 249–250
Metal, 305–329
composition and characteristics, 305–307
compression specimens of, 152, (fig.) 153, 155
corrosion of, 319, 322–323
direct shear test for, 168–169
ductile, stress-strain diagram for, 139
durability of, 319, 322–323
fatigue limits of, 255
fatigue specimens of, 249–250
ferrous (*see* Ferrous metals)
impact specimens of, (fig.) 232
impact tests for, 225–226, 231
nonferrous, 306–307
nonferrous alloys (*see* Nonferrous alloys)
production methods, 307–311
S-N diagrams for, (fig.) 254
steel, 313–317
strength of as function of temperature, 147
structure of, 244
tensile fractures of, 139–140

Metal (*Cont.*):
tension specimens of, (figs.) 129–130
tension test for, 126
wire rope (*see* Wire rope)
(*See also specific metals*)
Meter (unit), 102*n*.
Microhardness testers, 200–201
Micrometer (instrument), 100–101
dial, 104–105
Microscope, micrometer, 101
Mineral aggregates (*see* Aggregates)
Mineral composition of aggregates, 377
Mineral dressing, 307
Mineral filler, 373
Mirror extensometers, 105–106
Mirror scale, 99, (fig.) 100
Mixes:
asphalt (*see* Asphalt mixes)
concrete (*see* Concrete mixes)
Mixing of concrete, 339
(*See also* Concrete mixes)
Mode, 44, 45
Models for photoelasticity, 295, 299
Modulus:
bulk, 20
diametral resilient, 387
of elasticity, 19, 183
of resilience, 32
of rigidity, 19
of rupture, 30, 174, 181, 188
secant, 19, 344
tangent, 19, 344–345
of toughness, 33
of wire rope, 326
Young's, 19
Mohr circle, 157–158, 166, 442
Mohs' scale, 198, 203, (table) 204
Moisture conditions, 71
Moisture content:
of aggregates, 334–335
electrical determination of, 287–288, 396–397
in wood, 287–288
of wood, 395–396, 398
Molded electrical insulating materials, tension specimen of, (fig.) 131

Molding, 417
  compression, 417
  injection, 417
Molybdenum, 308, 312
Moment, units of, 432
Moment diagrams, 180, 441
Monel metal, 319
Monotron hardness test, 200
Moore rotating-beam machine, 251

Narration in reports, 57–58
Naphthalene, 415
National Bureau of Standards (NBS), 15
Natural errors, 75
Natural strain, 18, 22
Neutral axis, 180–182
Neutral surface, 181
Nick-bend test, 192
Nickel, 306, 311, 315
Nondestructive tests, 274–301
  vs. destructive tests, 8
  determination of material characteristics
      by, 275
  electrical methods, 285–288
  electromagnetic methods, 279–282
  and location of defects, 274
  magnetic-particle method, 282–285
  mechanical methods, 291–293
  photoelasticity, 295–300
  purposes of, 8, 274–275
  radiographic methods, 275–279
  sonic methods, 288–291
  visual methods, 293–295
Nondimensionalization of stress-strain
      diagram, 23–24
Nonferrous alloys, 318–321
  density of, 319–321
  hardness of, (tables) 320–321
  stiffness of, 320–321
  strength of, 320–321
  thermal expansion of, 319
Nonferrous metals, 306–307
Normal distribution curve, 42–43
Notch sensitivity, 225, 237–238, (fig.)
      240

Notched-bar tests, 222
  (See also Impact tests)
Null indicator, 113–114
Nylon, 419

Objectives of testing, 6–7
Observations:
  in bending tests, 187–189
  in Brinell hardness test, 207–208
  in compression tests, 157–159
  in creep tests, (fig.) 269, 270–271
  in fatigue tests, 254–255
  in impact tests, 235–236
  in Rockwell hardness test, 211–214
  in scleroscope hardness test, 216–218
  in shear tests, 174–176
  in tension tests, 137–140
Observer (team member), 77
Offset, 26
Ogiv (or ogee) curve, 44
Oil-whiting test, 294
Open-graded aggregate, 374
Open-hearth furnace, 308
Operator (team member), 77
Optical comparator, 106, (fig.) 107
Optical deformeters, 105–106
Ore, 307
Organization of test team, 76
Organizations, addresses of, 443–444
Orientation of wood grain, 398–402
Oscilloscopes, 93, 114, 115–116
  triggering in, 115–116
Osgood, W. R., 23
Oven-drying, 396
Oxford machine, 227
Oxidized asphalt, 355
Oxygen, 315

Pagination, 63
Paper:
  graph, 61
  logarithmic, 61–62
Parallax, 99

Particle board, 391–392
Passivation, 322–323
Pavements, 353
  surface of, 379
Pendulum, 83
Pendulum impact tests, 224, 235–236
Pendulum machines, 226, (figs.) 228, 229
Penetrant tests, 294–295
  filtered-particle inspection, 294
  fluorescent-penetrant inspection, 294
  oil-whiting test, 294
Penetration test, 359
Pensky-Martens test, 357
Pentachlorophenol, 405
Perforation plates, 393
Performance, prediction of, 9–10
Performance unit, 91
Permeability of concrete, 346–348
Personal errors, 75
Petroleum fractions, 353–354
Phenol, 415
Phenolic, 420
Phosphorous, 315
Photoelastic analysis, apparatus for,
    295–299
Photoelasticity, 295–300
  models for, 295, 299
PhotoStress, 299–300
Piezoelectric effect, 288
Pig iron, 308
Pigments, 416
Pitch, 352
Placing of concrete, 339–340
Plane angle, units of, 432
Plane bending, 181
Plane polarization, 296–298
Plastic action, 27–28
Plastic moment, 189
Plastic specimens, 249
Plasticity, 27–30, 69
  maximum strength, 28–30
  plastic action, 27–28
Plasticity test, 377
Plasticizers, 416
Plastics, 412–426
  abbreviations for, (table) 414–415

Plastics (*Cont.*):
  absorption by, 421
  additives to, 413–416
  fabrication of, 418–419
  fatigue specimen of, 249
  fillers for, 416
  foaming in, 418
  impact tests for, 226
  production methods, 415–416
  reinforcements for, 416, 419
  stabilizers for, 416
  strength of, 422–425
  strength-time data for, 145
  thermal expansion of, 422
  thermosetting, 415, 420, (table) 424
  types and characteristics, 412–415
Platen, 86
Plugs for torsion specimens, 171
Plywood, 391–392, 405–410
  adhesives for, 406
  fabrication of, 407
  grading of, 406
  shrinkage of, 409
  specifications for, 407–409
  stiffness of, 409–410
  strength of, 409–410
  wood plies, 406–407
Poisson's ratio, 18
Polariscopes, 295–300
Polarization of light, 295–299
  circular, 298–299
Polycarbonate, 420
Polyester, 420
Polyethylene, 416, 419
Polymerization, 413
  addition, 413
  condensation, 413
Polymers, 412
Polypropylene, 419
Polystyrene, 420
Polysulfide, 420
Polyurethane, 420
Polyvinyl chloride (PVC), 416, 420
Population, 38
Pores in aggregates, 371
Portland cement, 330–332

Portland cement concrete (*see* Concrete, portland cement)

Portland cement mortar:
  briquet grip for, (fig.) 134
  compression specimens of, 153
  stress distribution in tensile specimen of, 143
  tension specimen of, (fig.) 131

Post-yield strain gage, 109

Power, units of, 432

Precision:
  control of, 75
  definition of, 10

Precision ratio, 49

Prefixes (SI), 429, (table) 431

Preforming of wire rope, 325

Preparation of specimens, 73–74

Presentation of results (*see* Reports)

Presentations, oral, 66

Preservatives, 405

Pressure, units of, 432

Prestressing strands, 325

Principles of testing, 4

Probabilities of deviations, (fig.) 47

Probability curve, 42

Probability paper, 62

Probable error, 49–50

Procedure:
  for bending tests, 186–187
  for Brinell hardness test, 206
  for brittle coating analysis, 119–120
  for compression tests, 156–157
  for creep tests, 268–270
  for fatigue tests, 253–255
  for impact tests, 233–234
  Marshall, 384–386
  for Rockwell hardness test, 211, (fig.) 212
  for scleroscope hardness test, 216
  for tension tests, 135–137
  for torsion tests, 172–174

Process metallurgy, 307

Processing:
  of metal, 308
  of plastics, 416–419
  of steel, 316

Production:
  of asphalt, 353–354
  of concrete mixes, 338–339
  of metals, 307–311
    iron, 308
    steel (*see* Steel, production of)
  of plastics, 413–416
  of plywood, 407
  of portland cement, 331–332
  of wood, 391–392

Progressive fracture, 244

Proof stress, 27

Proof tests, 293

Properties:
  of asphalt, 356–370
  of cast iron, 311–312
  classification of, 5
  of concrete, 342–349
  of concrete mixes, 340–342
  of materials, control of, 3–16
  mechanical, 68–69
  of nonferrous alloys, 318–321
  of plastics, 421–425
  of plywood, 409–410
  of steel, 313–317
  of wood, 395–404

Proportional limit, 25–26
  in shear, 170

Proportioning concrete mix, 336–339

Propylene, 415

Protections, corrosion, 322–323

Proving ring, 93–94, (fig.) 96

Punching shear test, 167

Pure shear, 165

Pyrometallurgy, 307

Quarter-wave plates, 298–299

Quench-bend test, 192

Quenching, 309–311

Radiography, 275–279
  apparatus for, 276–277

Ramberg, W., 23

Range:
  of data, 45
  loading, 96
Range ratio, 245
Rate of growth of wood, 393, 395, 397
Rays (in wood), 392–393
Reactive aggregates, 336
Readout devices, 89–90, 93
Rebound hardness, 198, 202–203,
      215–218
Recoil absorption, 82
Recorder (readout device), 89, 93
  strip-chart, 93
Recorder (team member), 76–77
Rectification of signal, 115
Reduction of data, 37
Reference gages, 102
References:
  in reports, 63
  on SI, 430
  text, 448–453
Reflection polariscope, 300
Regression coefficient, 50
Regression line, 50–51
Reinforcements for plastics, 416, 419
Relative cumulative frequency, 41
Relative density, 17, 413
Relative error, 49
Relative frequency, 41
Repeated stresses [*see* Stress(es),
      repeated]
Reports, 56–67
  arrangement of, 56–57
    appendixes, 57
  figures, 60–63
  importance of, 78
  narration, 57–58
  mathematics, 58–59
  mechanics of preparation, 63–66
    headings, 63
    abstracts, 63
  narrations, 57–58
  oral presentations, 66
  tables, 59–60
Residual stresses, 28

Residuals, 49
Residuum, 353
Resilience, 32, 221
  modulus of, 32
Resin ducts, 392–393
Resins, 412
Resistance:
  abrasion (*see* Abrasion resistance)
  frictional, 166
  and strain, 107
Resistance wires, (table) 108
Rigidity, modulus of, 19
Ring-porous woods, 393
Rock asphalt, 352
Rockwell hardness test, 199, 210–215
  apparatus for test, 210, (fig.) 211
  observations in, 211–214
  specimens for, 215
  superficial-hardness, 199, 211, 214
  variables in, 215
Rolling thin film oven test, 362
Rope:
  efficiency of, 325–326
  lay of, 324–325
Rosette strain gage, 109–110
Roughness, surface (*see* Surface
      roughness)
Roundwood, 391
Rupture:
  of beams, 187–188
  under impact, 222, 241
  modulus of, 29, 174, 181, 188

Sample, 38
Sampling:
  principles of, 48–49
  of specimens, 72
Sand equivalent test, 376–377
Sapwood, 395
Sawmill, 391–392
Sawnwood, 392
Saybolt-Furol test, 362
Scale(s), 61
  mirror, 99, (fig.) 100
  Mohs', 198, 203, (table) 204

Scatter diagram, 50–51
Schenk-Peterson diagram, 257, (fig.) 258
Schmidt hammer, 198, 292
Scientific testing, 7
Scleroscope, shore, 199, 203, 215–218
Scleroscope hardness test, 215–218
  apparatus for, 215–216
  observations, 216–218
  procedure, 216
  specimens for, 215
Scratch-hardness tests, 198, 203, (table)
  204
Screen (sieve) sizes, 439
Screw-gear machines, 80, 86–88
  load measurement in, 86
Seal coatings, 353
Seasoning of wood, 392
Secant modulus, 19, 344
Selection:
  of asphalt mix, 387–389
  of materials, 6
  of specimens, 72–73
Sensitivity:
  of instrument, 75
  notch, 225, 237–238, (fig.) 240
  strain, 107–108
  velocity, 222
Separation, 31
Serviceability, 6
Set:
  of cement, 332
  permanent, 18
Settlement test, 362
Shape of aggregate particles, 375
Shear, 164–178
  direct, 164
  direct shear test (see Shear tests, direct)
  failure in, 174–175
  pure shear stress, 165
  scope and applicability of, 166–168
  shearing strain, 166
  single, 164, 170
  torsion, 164–165
    apparatus for, 171–172
    procedures in, 172–174

Shear (Cont.):
  specimens for, 170–171
Shear and moment diagrams, 180, 441
Shear tests:
  direct, 167–170
    for metals, 168–169
    nature of, 69, 164, 167
    for plywood, 409
    procedure for, 168–170
    punching, 167
    for wood, 169–170
  observations in, 174–176
  transverse, 167
  variables in, 176–177
Shear tool, Johnson, 168
Shearing strain, 166
Shewhart, W. A., 51
Shore scleroscope, 199, 203, 215–218
Shot-blasting, 256
Shrinkage:
  of concrete, 348–349
  of plywood, 409
  of wood, 396
SI (Système International d'Unités),
  429–435
  computations in, 434–435
  general features, 429–431
    derived units, 429, (table) 430
  style and usage, 433–434
  symbols, 59, 430–434
Sieve analysis, 335, 373–374
Sieve sizes, (table) 439
Significance of tests, 9–10
Silicon, 315, 420
Simple gages, 103
Single shear, 164, 170
Single-blow pendulum machines,
  226–227, (figs.) 228, 229
Skewness factor, 47–48
Slip or flow, 27, 31
Slump test, 340
S-N diagrams (fatigue), 246–247, 254
  for metals, (fig.) 254
Softening point of asphalts, 359–360
Softwood, 391

Solubility test, 354
Solutions (alloys), 305
Solvents for asphalt, 355
Sonic analyses, 288-291
  of stiffness, 291
Sonizon apparatus, 290-291
Sonntag machine, 253
Soundness:
  of aggregates, 376
  of cement, 332
Specific surface, 332
Specific gravity (relative density), 17, 413
Specific strength, 421, 425
Specification of materials, 10-12
  for plywood, 407-409
  standard, 11-12
  for steel, 316
Specimens, 72-74
  for bending tests, 183-184
    wood, 184
  for Brinell hardness test, 204-205
  compact tension (CTS), 250
  for compression tests, 152-154
    wood, 153
  concrete, 152-153
  for fatigue tests, 249-250
  for impact tests, 231, (figs.) 232,
    237-238
  metallic, 72-73, 129-130, 152-153,
    249-250
  plastic, 249
  for Rockwell hardness test, 210
  sampling of, 72
  for scleroscope hardness test, 215
  selection of, 72-73
  for tension tests, 127-131
    wood, (fig.) 130
  test, 72-74
  for torsion tests, 170-171
  tubular torsion, 170, 176
  wood, 73, (figs.) 130, 153
Speed of testing:
  in bending, 191
  in compression, 156-157, 162
  idling or no-load, 136n.

Speed of testing (*Cont.*):
  in tension, 136, 144-146
  in torsion, 173-174
Speed adjustment, 90
Sperry apparatus, 285-286
Spread of data, 45
SR-4 gage, 108
Stability:
  of asphalt mixtures, 378-379, 384-389
  of beams, 190
  of columns, 442
  of emulsified asphalts, 362
Stabilizers for plastics, 416
Stabilometer, 384-385
Standard deviation, 46
Standard error of the mean, 49
Standard specifications, 11-12
Standardizing agencies, 12-15
Standards, ASTM, 14, 444-448
Static bending (*see* Bending; Bending tests)
Static compressions (*see* Compression
    tests)
Static shear (*see* Shear; Shear tests)
Static tension (*see* Tension; Tension
    specimens; Tension tests)
Static testing machines, 85-90
Static tests, definition of, 70
Statistical methods, 38
  descriptive, 38
  inference, 38
Statistical summaries, 53
Steel, 306
  alloying of, 313-316
  aluminum in, 315
  carbon content of, 313
  copper in, 314
  corrosion of, 319
  creep resistance of, (fig.) 263, (fig.)
    270, 271-272
  density of, 316
  designations for, 316
  effect of heating rate on, 148-149
  effect of L/d ratio on, 142
  effects on Charpy values of, (tables)
    238, (tables) 240

Steel (*Cont.*):
  hardening of, 309–310
  hardness of, (table) 217, (table) 317
  impact tests on, (tables) 238, (figs.)
    240
  processing of, 316
  production of, 308–310
    furnaces for, 308
  properties of, 313–317
  specifications for, 316
  stiffness of, 313, 317
  strength of, 313–317
  stress-strain diagram of, (fig.) 314
  tensile failure of, 139–140, 146
  tension-test log sheet for, (fig.) 138
  thermal expansion of, 316
  (*See also* Metals)
Steel wire, 324
Steelyard, 83
Stiffness, 19, 68–69
  of concrete, 344–345
  of nonferrous alloys, 320–321
  of plywood, 409–410
  sonic determination of, 291
  of steels, 313, 317
  of wire rope, 326
  of wood, 398–400, (tables) 403–404
Strain:
  definition of, 18
  measurement of, 107–120, 295–300
  natural, 18, 22
  and resistance, 107
  shearing, 166
Strain-gage instrumentation, 111–115
Strain gages:
  electric resistance, 107–115, 268
  rosette, 109–110
  Tuckerman's optical, 106
Strain indicator, 113–114
Strain sensitivity, 107–108
Strain-hardening, 27, 262
Strainometers, 104–105
  collar extensometer (compressometer),
    104
  lateral, 104–105

Strength, 68
  compressive, 30
  of concrete, 342–344
  fatigue, 30
  hardness as measure of, 68
  of metals as function of temperature,
    147
  of nonferrous alloys, 320–321
  of plastics, 422–425
  of plywood, 409–410
  of portland cement, 333
  specific, 421, 425
  of steels, 313–317
  tensile (*see* Tensile strength)
  ultimate, 28–29
  of wire rope, 326–328
  of wood, 401–404
Stress(es), 17–18
  bending, 179–182
  repeated, 245–246
    classification of, 245, (table) 246
  torsional, 172, 176
  true, 18, 22
  units of, 432
  yield, 27
Stress analysis, experimental, 7, 275
Stress conditions:
  complex, 70, (fig.) 166
  formulas for, 440
Stress grading, 392
Stress-strain diagrams, 20–24
  of concrete, (fig.) 345
  for ductile materials, (fig.) 29, 139
  nondimensionalization of, 23–24
  for nonductile materials, (fig.) 29
  plotting of, 20–22
  slope of, 19
  of steels, (fig.) 314
  true, 22–23
Stress relaxation, 263, (fig.) 264
Stress variation, 245, (table) 246, 254
Strip-chart recorder, 93
Stripping of aggregates, 377–378
Structure:
  of asphalt mixtures, internal, 378

Structure (*Cont.*):
  of metals, 244
Sulfur, 315
Summary:
  of statistics, 53
  of test results, 37, 57
Supplementary units (SI), 429, (table) 430
Surface:
  neutral, 181
  specific, 332
Surface area, 375
Surface moisture, 334
Surface of pavements, 379
Surface roughness, measurement of:
  electronic, 286–287
  visual, 293
Surface texture, 375
Sustained yield method, 391
Swain, George Fillmore, 78
Symbols in equations, 58–59
Systematic errors, 75
Système International d'Unités (*see* SI)

Tables in reports, 59–60
Tag open cup (TOC) test, 357
Tangent modulus, 19, 344–345
Tar, 351–352
Team organization, 76
Technique of testing, 3
Teflon, 415
Temperature:
  in creep tests, 261–262, 271
  equicohesive, 262–263
  in impact tests, 234–235, 238–241
  strength of metals as function of, 147
  units of, 433
Temperature conditions, 70–71
Temperature susceptibility of asphalt, 358–359
Tempering, 309–310
Templin grips, (fig.) 133
Tenderness, 256
Tensile failure of steel, 139–140, 146

Tensile fractures of metals, 139–140
Tensile strength, 28–29
  of wood, 126
Tension:
  diagonal, 168
  fracture in, 136, 146
Tension specimens, 127–131
  of metals, (figs.) 129–130
    cast iron, 130
  of molded electrical insulating materials, (fig.) 131
Tension tests, 69, 125–149
  apparatus for, 131–135
  eccentricity in, 144
  failure in, 139–140
  low- and high-temperature, 147–149
  for metals, 126
  observations in, 137–140
  procedure for, 135–137
  specimens for, 127–131
  speed in, 136, 144–146
  for steel, log sheet for, (fig.) 138
  variables in, 140–146
Teredos, 405
Termites, 405
Test conditions, 70–71
Test hammer method for portland cement concrete, 198, 292
Test results, summary of, 37, 57
Test specimens, 72–74
  preparation of, 73–74
  selection of, 72–73
Testers, microhardness, 200–201
Testing procedures, 68–79
  apparatus, selection of, 74
  common sense, in, 77–78
  conducting tests, 76–78n.
  destructive vs. nondestructive, 8
  vs. inspection, 8-9
  measurements, 74–76
  mechanical (*see* Mechanical testing)
  mechanical nondestructive, 291–293
  objectives of, 6–7
  role of, 3–4
  scientific, 7

Testing procedures (*Cont.*):
specimens, 72–74
speed of (*see* Speed of testing)
technique of, 3
(*See also* Testing apparatus; Tests)
Testing machines, 80–98
accessories for, 82, 91–93
calibration of, 93–97
capacity of, 86
cyclic loading, 90–91
determination of load, 82–85
hydraulic devices, 83–84
mechanical devices, 84–85
dynamic, 90–91, 251
hydraulic (*see* Hydraulic machines)
mechanical, 80, 81, 86–88
static, 85–90
types of, 81–82
early, (fig.) 81
requirements for, 81–82
verification of, 96
Tests:
accelerated soundness, 376
bend, 183, 191–193
bending (*see* Bending tests)
Brinell hardness (*see* Brinell hardness test)
Charpy, 224, 231–233
CKE, 375
cement mixing (asphalt), 362
classification of, 6–9
cold-bend, 191–192
commercial, 7
compression (*see* Compression tests)
conduct of, 76–78
creep (*see* Creep tests)
demulsibility, 362
design of, 71–72
dynamic (*see* Dynamic tests)
dynamic-hardness, 198, 202–203
vs. experiments, 7
fatigue (*see* Fatigue tests)
field, 7–8
file, 203
flexure, 183, 191–193

Tests (*Cont.*):
float, 360
Hatt-Turner, 225, 231, 233–234
high-heating-rate, 148
high-temperature, 147–149, 265
hot-bend, 192
impact (*see* Impact tests)
indirect tensile, 386–387
Izod, 224, 231–233
long-time, 70
Los Angeles abrasion, 376
low-temperature, 147, 234
mechanical, 69–72
nick-bend, 192
nondestructive (*see* Nondestructive tests)
notched-bar, 222
Rockwell hardness (*see* Rockwell hardness test)
rolling thin film oven, 362
sand equivalent, 376–377
Saybolt-Furol, 362
scleroscope hardness, 215–216
scratch-hardness, 198, 203, (table) 204
settlement, 362
shear (*see* Shear tests)
significance of, 9–10
slump, 340
solubility, 354
static, definition of, 70
tag open cup (TOC), 357
tension (*see* Tension tests)
thin film oven, 362
(*See also* Testing)
Thermal expansion:
coefficient of, 311n, 433
of concrete, 349
of nonferrous alloys, 319
of plastics, 422
of steels, 316
of woods, 396
Thermometers, 234
Thermoplastics, 415, 419–420, (table) 423
Thermosetting plastics, 415, 420, (table) 424

Thickness, ultrasonic determination of, 290-291
Thin film oven test, 362
Thomson, William (Lord Kelvin), 107
Timber, 392
Time, units of, 432
TOC (tag open cup) test, 357
Tolerance in gages, 103
Toluene, 415
Torsion machines, load measurement in, 171-172
Torsion tests:
  apparatus for, 171-172
  description of, 167-168, 170-176
  failure in, 174, (fig.) 175
  fracture in, 174-175
  nature of, 70, 164-165, 167-168
  specimens for, 170-171
  speed in, 173-174
Torsional stresses, 172, 176
Toughness, 33-34, 222
  of aggregates, 376
  modulus of, 33
  units of, 433
Tracheids, 392, 393
Transition-temperature range, 239, (fig.) 240
Transverse shear test, 167
Trees, deciduous, 391
Trial mixes of asphalt, 381-387
Triaxial compression test, 164n.
Trichloroethylene, 357, 369-370
Triggering in oscilloscopes, 115-116
Triphenyl phosphate, 416
Troptometers, 172, (fig.) 173
True stress–natural strain relations, 22-23
Tubular torsion specimens, 170, 176
Tuckerman's optical strain gage, 106
Tukon hardness tester, 200
Tungsten, 315
Twist, angle of, 173
Two-point loading, 186

Ultimate strength, 28-29
Ultrasonic analyses, 288-291

Ultrasonic determination of thickness, 290-291
Ultrasonic testing, apparatus for, 288-291
Ungrouped data, 38
Units:
  for cross-sectional properties, 187, 433
  of density, 431
  of energy, 432
  of force, 432
  of fracture toughness, 433
  of hardness values, 432-433
  in hardness tests, 196
  load, 91
  of mass, 431
  meter, 102n.
  of moment, 432
  of power, 432
  SI, 429, (table) 430
  of stress, 432
  of temperature, 433
  of time, 432
  of toughness, 433
  of viscosity, 433
  of volume, 433
Universal testing machines (UTM), 80

Vacuum viscometer:
  Asphalt Institute, 360-361
  Cannon-Manning, 360-361
Vanadium, 315
Variability of materials, 4
Variables:
  in bending tests, 189-191
  in Brinell hardness test, 208-210
  classification of, 69
  in compression tests, 159-162
  in creep tests, 265, 271-272
  in fatigue tests, 255-259
  in impact tests, 237-241
  relations between, 38
  in Rockwell hardness test, 215
  in shear tests, 176
  in tension tests, 140-146
Variate, 40

Variation:
  coefficient of, 46
  permissible, 96
  of results, 10, 38–39
  stress, 245, (table) 246, 254
Velocity of impact testers, 229, 237
Velocity sensitivity, 222
Veneer, 391, 406
Verification of testing machines, 96
Vernier, 99–100
Vessels (in wood), 392
Vickers hardness, 200
Viscoelasticity, 145
Viscometers:
  capillary tube, 360
  vacuum: Asphalt Institute, 360–361
    Cannon-Manning, 360–361
  Zeitfuchs cross-arm, 361–362
Viscosity:
  of asphalt, 358–363
  types of, 360
    dynamic (absolute), 360–361, 433
    kinematic (relative), 360–362, 433
  units of, 433
Visual inspection, 293–295
Voids in aggregates, 371, 383–384
Volume, units of, 433
Volumetric relations in asphalt mixtures,
    383
Vose's interferometer, (fig.) 105

Warwick, C. L., 15
Water for concrete, 333
Weight vs. mass, 17, 196, 431
Weights, use of, 82–83
Wheatstone bridge, 112
Wire(s):
  isoelastic, 108
  resistance, (table) 108
  snubbing device for, (fig.) 133
  steel, 324
Wire rope, 323–328
  abrasion resistance of, 326
  durability of, 327
  modulus of, 326

Wire rope (*Cont.*):
  socket for, (fig.) 134
  stiffness of, 326
  strength of, 326–328
Wood, 391–411
  bending specimens of, 184
  compression specimens of, 153
  compressive failure of, 158, (fig.) 159
  creep of, 402
  direct shear test for, 169–170
  durability of, 395, 402, 405
  fire resistance of, 405
  flexural failure of, 188
  fracture of, 158, (fig.) 159
  hardness of, 202
  impact specimen of, (fig.) 232
  impact test for, 225, 231
  mechanical properties of, 397–402
    density, 397–398
    moisture content, 398
    stiffness, 398–400, (tables) 403–404
    strength, 401–404
  moisture determination in, 287–288
  physical properties of, 395–397
    density, 396
    hygroscopicity (moisture content),
      395–396
    shrinkage, 396
    thermal expansion, 396
  structure of, 392–395
    diffuse porous, 393
    grain in, 395
    growth rings in, 393
    roundwood, 393
    sapwood, 395
  types and production of, 391–392
    deciduous trees, 391
    plywood (*see* Plywood)
    sawnwood, 392
    seasoning of, 392
    softwood, 391
    sustained yield method, 391
  (*See also specific woods*)
Wood fibers, 292–293
Wood specimens, 73, (figs.) 130, 153

Xeroradiography, 277
X-rays, 275–279
  Coolidge X-ray tube, 276–277
X-Y recorder, 93
Xylene, 415

Yield points, 27
Yield strength, 26

Yield stress, 27
Yielding, 25–27
  in beams, 189
Young's modulus, 19

Zeitfuchs cross-arm viscometer, 361–362
Zinc, 307
Zinc chloride, 405
Zirconium, 307

MOMENT	POWER	STRESS

lbf·in	N·m	hp	kW	ksi	MPa

1 lbf·in = 0.113 N·m

8.85 lbf·in = 1 N·m

1 hp = 0.746 kW

1.34 hp = 1 kW

1 ksi = 6.89 MPa

145 psi = 1 MPa